Springer Proceedings in Mathematics & Statistics

Volume 97

More information about this series at http://www.springer.com/series/10533

Springer Proceedings in Mathematics & Statistics

This book series features volumes composed of select contributions from workshops and conferences in all areas of current research in mathematics and statistics, including OR and optimization. In addition to an overall evaluation of the interest, scientific quality, and timeliness of each proposal at the hands of the publisher, individual contributions are all refereed to the high quality standards of leading journals in the field. Thus, this series provides the research community with well-edited, authoritative reports on developments in the most exciting areas of mathematical and statistical research today.

Slawomir Koziel • Leifur Leifsson
Xin-She Yang
Editors

Solving Computationally Expensive Engineering Problems

Methods and Applications

Springer

Editors
Slawomir Koziel
School of Science and Engineering
Reykjavik University
Reykjavik, Iceland

Leifur Leifsson
School of Science and Engineering
Reykjavik University
Reykjavik, Iceland

Xin-She Yang
School of Science and Technology
Middlesex University
London, United Kingdom

ISSN 2194-1009 ISSN 2194-1017 (electronic)
ISBN 978-3-319-08984-3 ISBN 978-3-319-08985-0 (eBook)
DOI 10.1007/978-3-319-08985-0
Springer Cham Heidelberg New York Dordrecht London

Library of Congress Control Number: 2014950072

Mathematics Subject Classification (2010): 97M10, 65K10, 65D17, 90C26, 90C31, 49Q10, 76B75, 74P05, 74P10

Printed on acid-free paper

Springer is part of Springer Science+Business Media (www.springer.com)

Preface

The costs of extensive computational simulations used for engineering designs can be very expensive, and thus can be a serious bottleneck for the design process in many applications. Nowadays prototyping is heavily involved in design and verification using computer models, and such computational approaches can have many advantages such as the reduction of the overall design costs and design cycles as well as finding good solutions to 'what-if' scenarios. However, the computation costs incurred by extensive computational time can still be very high. Though the speed of the computer power has steadily increased over past the decades, computationally extensive tasks are still a challenging issue. One of the reasons is the ever-increasing demand of the high-accuracy, high-fidelity models for simulating complex systems. For many applications such as those in aerospace engineering, microwave engineering and biological applications, a single simulation task can take hours, even days or weeks on modern computers. While in other applications such as combinatorial optimization problems, the evaluations of every possible combination can be prohibitive because such numbers of combinations can be astronomical. For continuous problems such as computational fluid dynamics and electromagnetic wave simulation, some forms of efficient approximations such as surrogate-based models are needed, while for combinatorial problems, efficient algorithms should be used, though there are no efficient algorithms for genuinely NP-hard problems.

In addition, other challenges associated with such problems include numerical noise in the simulation data, multimodality with multiple local optimum designs due to high nonlinearity, as well as multiple (potentially conflicting) objectives. All these make computationally expensive design tasks even more challenging. Thus, it is timely to edit a book to address such problems with the focus on the latest developments.

From the computational point of view, three key issues should be emphasized: approximation models, optimization algorithms and multi-objectives. Approximation models often use the so-called surrogates that can reliably represent the expensive, simulation-based model of the system/device of interest. If such surrogates are designed properly, they can speed up the simulation significantly. However, such surrogates tend to work for the local, smooth design landscape, and for multimodal problems, good approximations are not easy to construct. This book will include some of the latest developments in this area when dealing with nonlinear problems with complex design objectives.

Even with efficient, computational models, efficient optimization algorithms are also crucial to ensure design optimization that can be carried out successfully in a practically acceptable time scale. Traditional algorithms such as the trust-region method, the interior-point method and gradient-based algorithms can work well for local search, but for multimodal global optimization, heuristic and metaheuristic algorithms start to demonstrate their efficiency. Swarm intelligence based approaches will be introduced and reviewed in this book.

In almost all engineering applications, there are multiple design objectives and these objectives can often be conflicting, resulting in very complex objective landscapes in the design space. In addition, complex constraints can often modify the search regions significantly and thus make it even more challenging for search algorithms. Furthermore, the computational costs for multi-objective optimization will increase multifold, compared to the counterpart of single objective optimization problems. For example, multi-objective optimization can be very challenging in image processing applications, and we will also briefly touch this area in this book.

This edited book provides a timely snapshot of some of the latest developments in surrogate-based models, optimization algorithms and multi-objective design applications. Topics include surrogate models in engineering design, surrogate-based and PDE-constrained models in climate applications, shape-preserving response predictions, simulation-driven design for antenna designs, space dimension reduction for multi-objective design, large-scale optimization via swarm intelligence, clustering of radar images, classification of laser point clouds, knowledge-based modelling by artificial neural networks and others. However, as the length of the book is limited, it is not our intention to cover everything. As a result, many topics that are very active in the field may not be covered at all. But we hope all the topics we have covered can form a basis with enough literature for further research in the relevant areas.

The editors hope that topics covered in this book will allow the readers to gain understanding of basic mechanisms of surrogate modeling process and surrogate-based optimization algorithms, to follow the trend of swarm intelligence and image processing, and to see the ways of dealing with multi-objective optimization. Ultimately, this may help to reduce the costs of the design process aided by computer simulations. Therefore, this book can serve as a timely reference to researchers, lecturers and engineers in engineering design, modelling and optimization as well as industry.

May 2014 — Slawomir Koziel
Reykjavik, Iceland — Leifur Leifsson
London, UK — Xin-She Yang

Contents

Surrogate-Based and One-Shot Optimization Methods for PDE-Constrained Problems with an Application in Climate Models

Thomas Slawig, Malte Prieß, and Claudia Kratzenstein

Abstract We discuss PDE-constrained optimization problems with iterative state solvers. As typical and challenging example, we present an application in climate research, namely a parameter optimization problem for a marine ecosystem model. Therein, a periodic state is obtained via a slowly convergent fixed-point type iteration. We recall the algorithm that results from a direct or black-box optimization of such kind of problems, and discuss ways to obtain derivative information to use in gradient-based methods. Then we describe two optimization approaches, the One-shot and the Surrogate-based Optimization method. Both methods aim to reduce the high computational effort caused by the slow state iteration. The idea of the One-shot approach is to construct a combined iteration for state, adjoint and parameters, thus avoiding expensive forward and reverse computations of a standard adjoint method. In the Surrogate-based Optimization method, the original model is replaced by a surrogate which is here based on a truncated iteration with fewer steps. We compare both approaches, provide implementation details for the presented application, and give some numerical results.

Keywords Optimization • Climate model • Marine ecosystem model • One-shot method • Surrogate-based optimization

1 Introduction

Climate simulations are a very challenging task in applied mathematics and scientific computing. The underlying mathematical systems have a high number of uncertainties with respect to initial values, model parameters, or the relevant processes to be included. Moreover, the state equation solvers often involve iterative algorithms to compute steady or periodic solutions. To identify model parameters and to assess the models, model-to-data misfit functions are minimized using

T. Slawig (✉) • M. Prieß • C. Kratzenstein
Department of Computer Science and KMS Centre for Interdisciplinary Marine Science, Christian-Albrechts-Universität zu Kiel, 24098 Kiel, Germany
e-mail: ts@informatik.uni-kiel.de

S. Koziel et al. (eds.), *Solving Computationally Expensive Engineering Problems*, Springer Proceedings in Mathematics & Statistics 97,
DOI 10.1007/978-3-319-08985-0_1

numerical methods. Since iterative processes are involved, it is crucial to derive highly efficient optimizers.

In this respect, such kind of models and the corresponding optimization or control problems are only one representative for PDE-constrained optimization problems with iterative state solvers. Similar problems arise in Computational Fluid Mechanics and many other application areas. Thus, we will consider here a more general class of PDE-constrained optimization problems.

2 PDE-Constrained Optimization Problems with Iterative State Equation Solvers

We study optimization problems governed by partial differential equations (PDEs), given in the following general form

$$\min_{(y,u)\in Y\times U_{ad}} \quad J(y,u) \tag{1}$$

$$\text{s.t.} \quad e(y,u) = 0 \text{ in } Z. \tag{2}$$

Here $J : Y \times U \to \mathbb{R}$ is the cost or objective function defined on the cartesian product of state and control (or parameter) spaces Y and U, respectively. The admissible set U_{ad} characterizes additional constraints on the controls (parameters) u.

- either as infinite-dimensional function spaces
- or, e.g., when studying the discretized problem, as finite-dimensional spaces, being isomorphic to $\mathbb{R}^{n_Y}, \mathbb{R}^{n_U}$ with n_Y, n_U the dimensions of Y, U, respectively.

It is also possible (and often the case) that Y is a function space and U is finite-dimensional, e.g., a space of real-valued parameters. In many cases, the admissible set is then given by simple bounds, i.e., as

$$U_{ad} := \{u \in U : u_{min} \leq u \leq u_{max}\}$$

with some fixed $u_{min}, u_{max} \in \mathbb{R}^{n_U}$, and the inequalities meant component-wise.

The state equation is defined by the often nonlinear mapping $e : Y \times U \to Z$. If it is given in an infinite-dimensional setting, e.g., as weak form of a PDE, the space Z is the dual of the test space. When considering an already discretized state equation, Z is isomorphic to $\mathbb{R}^{n_Z}$ with appropriate dimension n_Z.

We will also use the reduced cost functional $\hat{J} : U \to \mathbb{R}$ defined by

$$\hat{J}(u) := J(y(u), u) \quad \text{where } y = y(u) \Longleftrightarrow e(y,u) = 0. \tag{3}$$

It is characteristic for the two optimization methods we describe that the solution of the state equation (2) is computed by a fixed-point type iteration of the form

$$y_j = G(y_{j-1}, u), \quad j = 1, 2, \ldots, \tag{4}$$

with iteration function $G : Y \times U \to Y$. In climate models, this iteration is called *spin-up*. Each iteration step includes one or more time-integration steps with constant or periodic external forcing, thus leading to a steady solution, in climate models often a steady annual cycle. A similar procedure is also very common in fluid dynamics to compute stationary or periodic solutions with transient solvers. This fact also motivates the notion *pseudo time-stepping scheme* for (4).

We will briefly write $G^j(y, u)$ for the result of j subsequent iterations with the same control variable u, i.e.

$$y_j = G^j(y_0, u), \quad j = 1, 2, \ldots \tag{5}$$

The solution of the state equation (2) is now given as the limit

$$y^* = \lim_{j \to \infty} y_j, \tag{6}$$

We will assume here that this limit exists for all feasible controls $u \in U_{ad}$, guaranteed, for example, by some contraction or quasi-contraction property of G. Assuming that G is continuous w.r.t. its first argument, (6) implies

$$e(y, u) = 0 \text{ in } Z \iff y = G(y, u) \text{ in } Y. \tag{7}$$

In practice, the iteration (4) has to be terminated when a stopping criterion is satisfied. Thus, instead of using the limit y^* from (6) in the evaluation of J, an approximation $\hat{y} := y_{j_{max}}$ with an appropriate value of j_{max} is taken.

3 Exemplary Application: Parameter Optimization in a Marine Ecosystem Model

In this section we show a model problem from climate research that fits in the above general setting, and where both methods can be and have been applied. The model problem is an application in marine science. It deals with the identification of climate model parameters using experiment or model data. In climate models, the iterative computation of the state variables is rather common and usually very time-consuming. Three-dimensional climate model simulations may take several days or more of computer time. In this section, we introduce the underlying model, i.e. the state equation, its discretization, and the resulting iterative scheme (4).

3.1 Marine Ecosystem Models as Example for Climate Models

We here give the basic structure of marine ecosystem models which serve as one possible example for climate models. A typical optimization problem for this kind of models is parameter identification or optimization, i.e., poorly known or not measurable model parameters shall be adjusted such that the model output fits given observational (or other model) data. This task is also called model calibration. We give one example that we actually used in both optimization strategies.

Marine ecosystem models consist of two parts, namely the ocean circulation and the biogeochemistry. The former is basically given by the Navier-Stokes equations including temperature and salinity transport, whereas the latter describes the reaction of and interactions between nutrients and different species of ocean biota, e.g., photosynthesis, dying and growth of plankton species, etc. The marine ecosystem plays an import role within the global carbon cycle, but its complex organic and inorganic cycles are challenging when formulating a comprehensive biogeochemical model, see for example, [23].

The coupling between ocean circulation and the biogeochemical interactions is mostly regarded as one-way coupling. The influence of the circulation (including temperature and salinity distribution) on the biota is assumed to be much more important as vice versa. This is mainly motivated by the high complexity and the enormous computational effort that is necessary to solve the time-dependent and spatially three-dimensional coupled system of (1) an ocean circulation model (consisting of the Navier-Stokes equations with free ocean surface, energy and salinity transport equations) together with (2) the biogeochemical model (consisting of between two and about 50 transport equations for the different species, depending on the chosen model). Thus often an *off-line computation* is performed: Velocity, turbulent diffusion, temperature, and salinity fields are computed beforehand by the ocean circulation model and used as input or *forcing* data for the biogeochemical simulations. This significantly reduces the amount of computation. By using pre-computed circulation data, all tracers necessarily are regarded as *passive*, i.e., they do not have any influence on the circulation.

In our example, we use this off-line mode. The model equations then form a system of coupled transport or advection-diffusion-reaction equations, with reaction terms given by the biogeochemical processes. Our model equations then read

$$\frac{\partial y_i}{\partial t} = \mathrm{div}(\kappa \nabla y_i) - \mathrm{div}(v y_i) + q_i(y, u), \quad \text{in } \Omega \times [0, T], \quad i = 1, \ldots, n_{state} \quad (8)$$

together with given initial data $y_{init} = y(t = 0) \in Y$ and usually Neumann boundary conditions. Here, $\Omega \subset \mathbb{R}^3$ is the spatial domain, $[0, T]$ the considered time interval, $y = (y_i)_{i=1,\ldots,n_{state}}$ the vector of state variables (biogeochemical tracers), where $y_i(x, t)$ denotes a single tracer concentration at (x, t).

The time dependent turbulent mixing or diffusion coefficient κ and the velocity vector field v, together with temperature and salinity distributions (entering in the q_i, but omitted for brevity in the notation above) are precomputed data from an ocean model and thus *not* subject to identification here. The turbulent mixing dominates the molecular tracer diffusion in this application, and thus κ is the same for all tracers. Since velocity v and κ are given, the nonlinearity in the above system w.r.t. to the state variables $y = (y_i)_i$ comes from the nonlinear coupling terms q_i, whereas the transport part (diffusion and advection, the first two terms on the right) is linear.

The parameters to be identified are summarized in the vector u and appear in the nonlinear biogeochemical coupling terms q_i. These are non-autonomous extensions of predator–prey models, since, for example, the growth rate of phytoplankton (algae) depends on the sunlight and thus on space and time. Nearly all of these coupling terms q_i are spatially local, i.e., they describe processes happening at point x and not depending on neighborhood points. Some of them include sinking processes and thus become non-local. The sinking velocity of dead material, for example, is one parameter that is crucial to identify.

3.2 Example: The N-DOP Model

In this section we describe the biogeochemical model we used both in the One-shot and the Surrogate-based Optimization approach. Since there are many different biogeochemical models, the one used here is only an example. Modelers are interested not only in the right parameters for a single model, but moreover try to assess and compare different models, a task where parameter identification becomes important.

The N-DOP model (see [19]) consists only of the two tracers phosphate (nutrients, N) and dissolved organic phosphorus (DOP), denoted by $y = (y_1, y_2)$. Thus $n_{state} = 2$ in (8), which is at the lower limit of complexity for such kind of model. Typical for a biogeochemical model are different biogeochemical interactions in two horizontal layers in the ocean, namely the upper, *euphotic zone* Ω_1 (where light enables photosynthesis) and the lower, non-euphotic zone Ω_2. The model consists of the following coupling terms:

$$q_1(y,u) = \begin{cases} -g(y_1, I) + \lambda y_2 & \text{in } \Omega_1 \\ (1-\sigma)\dfrac{\partial \tilde{g}}{\partial x_3}(y_1, I) + \lambda y_2 & \text{in } \Omega_2 \end{cases}$$

$$q_2(y,u) = \begin{cases} \sigma g(y_1, I) - \lambda y_2 & \text{in } \Omega_1 \\ -\lambda y_2 & \text{in } \Omega_2. \end{cases}$$

The vertical spatial coordinate here is x_3. On the left-hand side, we have summarized here as above all parameters (see below) in the vector u. The *biological production*

$$g(y_1, I) = \alpha \frac{y_1}{y_1 + K_N} \frac{I}{I + K_I}$$

depends on nutrients y_1 and light I, and is limited by a maximum production rate parameter α. Light is computed from the short wave radiation (as a function of latitude and time, thus making g a non-autonomous function), the photosynthetically available radiation, the ice cover and the exponential attenuation of water K_{H2O}. A fraction σ of the biological production remains suspended in the water column as dissolved organic phosphorus, which remineralizes with rate λ. The remainder of the production sinks as particulate to the bottom where it is remineralized according to an empirical power law relationship:

$$\tilde{g}(y_1, I) = \left(\frac{x_3}{x_{depth}}\right)^{-b} \int_0^{x_{depth}} g(y_1(x,t), I(x,t))\, dx_3.$$

Here $x_{depth} = x_{depth}(x_1, x_2)$ is the depth of the upper layer Ω_1 which depends on the horizontal coordinate (x_1, x_2). The parameters to be optimized are given in Table 1.

Here, the parameters are assumed to be constant w.r.t. to space and time, i.e., the parameter space U equals $\mathbb{R}^{n_U}$, with U_{ad} defined by box constraints that are also given in Table 1. For further model details we refer to [19].

3.3 Discretization

In this section we describe a discretization scheme that is adapted to the mentioned one-way coupling and was used in our numerical tests. It is built upon a matrix representation of the linear part of system (8), namely the pure transport operators.

The *Transport Matrix Method (TMM)* introduced in [13] computes the effect of the ocean circulation on the tracer distributions. It avoids using ocean circulation data $\kappa, \mathbf{v}$ directly and discretizing the corresponding diffusion and advection operators in the tracer transport simulation. In contrast, the TMM builds up a set of pairs of explicit and implicit matrices (corresponding to the discretization in the ocean model which is based on an operator splitting) in every time step. The transport matrices are generated by several runs of one time step of the ocean model, each for a given initial tracer distribution (designed similar to a linear finite element ansatz function) in every grid point. Each resulting tracer distribution builds one column of the pair of transport matrices for one ocean model time step, and all evaluations together build up the whole matrix pair. Since the discretization of the transport in ocean models typically involves nonlinear schemes (like flux limiters etc.), this generation of the matrix pair can be seen as a way of linearization of the scheme. Since the external forcing depends on time, even with climatological (i.e., annually periodic) forcing data and due to the typical time step-size of 3 h, the amount of storage for all these matrices would be prohibitively large, even though the matrices are block-diagonal and sparse. Thus, the matrices are usually averaged in time. In our case, they are monthly averaged. More details on the temporal and spatial discretization and the evaluation of transport matrices, especially in combination with operator splitting schemes, can be found in [13].

Table 1 Parameters to be optimized in the N-DOP model with bounds $u_{i,min}$, $u_{i,max}$, $u_{i,guess}$ (used in the regularization) and initial value u_{i0} for the optimization methods

u_i	Symbol	Description	Unit	$u_{i,min}$	$u_{i,max}$	$u_{i,guess}$	u_{i0}
u_1	λ	Remineralization rate of DOP	d^{-1}	0.25	0.75	0.5	0.3
u_2	α	Maximum community production rate	d^{-1}	1.5	200	2.0	5.0
u_3	σ	Fraction of DOP	–	0.05	0.95	0.67	0.40
u_4	K_N	Half saturation constant of N	$mmolPm^{-3}$	0.25	1.5	0.5	0.8
u_5	K_I	Half saturation constant of light	Wm^{-2}	10.0	50.0	30.0	25.0
u_6	K_{H2O}	Attenuation of water	m^{-1}	0.01	0.05	0.02	0.04
u_7	b	Sinking velocity exponent	–	0.7	1.5	0.858	0.78

We now turn to the resulting discretized version of (8). Let Ω_h be the set of discrete spatial points $x \in \overline{\Omega}$, usually arranged on a rectangular grid, adapted to the bottom ocean topography and coastlines. This set Ω_h is determined by the spatial discretization of the used ocean model which was used to compute the transport matrices. In our case, the latitudinal and longitudinal resolution of the underlying ocean model grid is 2.8125°, with 15 vertical levels. This results in a dimension of the discretized state space Y of $n_Y = 105,498$.

Let $\mathbf{y}_l$ be the appropriately arranged vector of values $(y_i(x,t_l))_i$ of all n_{state} tracers on all spatial grid points $x \in \Omega_h$, and, in a similar way, $\mathbf{q}_l(\mathbf{y}_l, u)$ the vector of discretized coupling terms q_i for all $x \in \Omega_h$, both at fixed time step l. The time integration scheme for 1 year model time with a fixed step-size τ then reads

$$\mathbf{y}_{l+1} = \mathbf{A}_{imp,l}\,(\mathbf{A}_{exp,l}\,\mathbf{y}_l + \tau\,\mathbf{q}_l(\mathbf{y}_l, u)) =: \varphi_{l+1}(\mathbf{y}_l, u), \quad l = 0, 1, \ldots, l_{year} - 1. \quad (9)$$

Here $\mathbf{A}_{imp,l}, \mathbf{A}_{exp,l}$ are the implicit and explicit transport matrices at time step l, which are linearly interpolated between the pre-computed set of monthly matrices to the corresponding time t_l. In our case, the number of time steps in 1 year model ranges may vary from $l_{year} = 45$ (a very coarse temporal resolution, resulting in a step-size of $\tau = 192$ h) to $l_{year} = 2,880$ (which is the original one of the model, resulting in $\tau = 3$ h). Each step in the time-integration scheme (9) now consists of the evaluation of the coupling term $\mathbf{q}_l$ and two matrix–vector multiplications.

In climate model calibration or parameter optimization, as a first step a steady annual cycle, in our case a periodic solution of (8), is computed and used in the cost function evaluation. As a consequence, the iteration function G in (4) is given as

$$G := \varphi_{n_{year}} \circ \ldots \circ \varphi_1.$$

We thus regard one step in (4) as 1 year model time, and j there counts model years. The set of all discrete time instants used in a simulation is denoted by

$$[0,T]_\tau := \left\{((j-1)l_{year} + l)\tau : j = 1, \ldots, j_{max}, l = 1, \ldots, l_{year}\right\},$$

where j_{max} is the number of model years simulated. For the numerical computation, the iteration starts with a constant distribution $\mathbf{y}_0$. It can be observed that after $j_{max} \approx 3,000$ to $10,000$ iterations of G in (4), i.e. years model time, an acceptable approximately steady periodic solution is obtained, see also [16]. Since a typical step-size is 3 h, this means that about 10^6–10^8 discrete time steps of the spatially three-dimensional system of transport equations are necessary to attain a steady periodic solution. Even on parallel high performance hardware, the computation of a numerically converged steady periodic solution may take several minutes. This is the reason for the high demand for fast optimization methods, since usually one optimization may take hundreds of function evaluations, i.e., the mentioned computations of steady periodic solutions. More details and results can be found in [21].

3.4 Parameter Optimization Problem

In our numerical tests, the considered minimization problem was a least squares cost functional with regularization term given by

$$J(y,u) := \frac{1}{2}\|y - y_{data}\|_Y^2 + \frac{\alpha}{2}\|u - u_{guess}\|_U^2, \quad \alpha > 0. \tag{10}$$

Working in the finite-dimensional spaces for the discretized models, the used norms are Euclidean vector norms, optionally with weighting coefficients. Components of u are the parameters in Table 1. We follow [16] in the choice of the initial parameter guess u_{guess} and took the initial value u_0 (both given in Table 1) for the parameters in both optimization methods. For the choice of the desired state or target data y_{data} we use here model-generated test data, obtained with parameter vector $u = u_d$, in order to evaluate the two methods and the quality of their results compared to the known optimal parameter values.

4 Direct Optimization

In this section we describe a direct optimization algorithm, i.e., a method where the state equation iteration (4) is numerically converged before the parameters u are updated. Such kind of algorithm is used for example, when a black-box optimizer is applied on the original problem (1–2). In an iterative optimization algorithm, there will be two nested iterations then, and it can be conceptually written as follows.

Optimization Algorithm with Iterative State Equation Solver:

1. Choose an initial value for the control u_0.
2. For $k = 0, 1, \ldots, k_{max}$:

   a. Choose an initial value y_0 for the state corresponding to control u_k.
   b. Compute an approximation of the state for u_k:

   $$\hat{y}_k = G^{j_k}(y_0, u_k).$$

   c. Optionally: Compute the gradient of the cost J w.r.t. u, i.e.,

   $$\frac{d}{du}J(y(u), u)\bigg|_{(\hat{y}_k, u_k)}, \tag{11}$$

   using

   - either the gradient $\hat{y}'_k := \dfrac{d\hat{y}_k}{du}\bigg|_{u=u_k}$
   - or the adjoint state $\bar{y}_k$.

d. Perform an update $u_k \to u_{k+1}$ of the control.
e. If some criterion for u or J or its gradient is satisfied, stop.

The number j_k of inner iterations in step 2b might be varying with k or be fixed beforehand to some value constant j_{max}. To make things simpler in notation, we will here assume in this section that $j_k = j_{max}$ is constant.

A typical case which motivates the two methods compared here is that the evaluation of G is costly and/or that the convergence in (4) and thus in step 2b (and then presumably also in the gradient iterations usually needed in step 2c) is slow. Both strategies aim to reduce this high computational effort by using some kind of reduced accuracy or low-fidelity approximation in step 2b by taking low values j_{max} there to reduce the computationally effort or to make the whole algorithm feasible at all. As a result, the $\hat{y}_k$ used to evaluate the cost are different (and eventually far away) from the limits

$$y_k^* := \lim_{j\to\infty} G(y_j, u_k). \tag{12}$$

We call the above algorithm a *direct* or *direct fine model optimization* if the number j_{max} of inner state iterations equals a high number denoted by j^f, for example given by the original model or simulation code before used in an optimization.

4.1 Gradient Evaluation

The optional gradient computation in step 2c of the above algorithm can be performed either by a sensitivity or by an adjoint approach. Here, we assume that all derivatives of G, J and y used below exist as Fréchet derivatives and that Y and U are Hilbert spaces.

4.1.1 Sensitivity Equation Approach

Using a sensitivity equation, the iteration (4), with $u = u_k$ and up to step $j = j_k$, is differentiated w.r.t. u. This leads to the following iteration for the derivatives:

$$y'_j := \left.\frac{dy_j}{du}\right|_{u=u_k} = G_y(y_{j-1}, u_k)y'_{j-1} + G_u(y_{j-1}, u_k), \quad j = 1, \ldots, j_k. \tag{13}$$

Here subscripts y, u denote partial derivatives of G. This iteration is initialized by $y'_0 = \frac{dy_0}{du}$ which usually is zero, except in the case when the initial data are to be optimized as well (which is possible).

Now, the two iterations in steps 2b and 2c can be computed in different ways and orders. We use here the notions introduced in [7, 11].

- In the *two-phase approach*, the two iterations are performed after another, i.e., the two steps 2b and c remain separate. It can be seen from (13), where the iterates y_j of the state are used, that these have to be stored during step 2b in order to use them in step 2c.
- In the *piggy-back approach*, both iterations are combined to

$$\left.\begin{aligned} y'_j &= G_y(y_{j=1}, u_k)y'_{j-1} + G_u(y_{j-1}, u_k), \\ y_j &= G(y_{j=1}, u_k) \end{aligned}\right\} \qquad j = 1, \dots, j_k.$$

  This approach avoids the storing of the state iterates.
- The *Christianson approach* presented in [2] performs the sensitivity iteration (13) with the previously computed (numerically) converged state $\hat{y}_k$ instead of using its iterates y_j, $j = 0, \dots, j_k - 1$, thus also avoiding storage of the iterates.

Once $\hat{y}'_k$ is computed, the gradient of $\hat{J}$ with respect to the control can be computed by the chain rule as

$$\left.\frac{d}{du} J(y(u), u)\right|_{(\hat{y}_k, u_k)} = J_y(\hat{y}_k, u_k)\hat{y}'_k + J_u(\hat{y}_k, u_k). \tag{14}$$

4.1.2 Adjoint Approach

In the adjoint approach, the Lagrangian associated with problem (1–2) is used to compute the gradient (14) without knowing or evaluating $\hat{y}'_k$. We compute the directional derivative of the state equation in its fixed-point form, namely the right-hand side of (7), w.r.t. u in direction v and obtain

$$y'(u)v = \frac{d}{du} G(y(u), u)v = G_y(y(u), u)y'(u)v + G_u(y(u), u)v \quad \text{in } Y, \tag{15}$$

i.e.,

$$\left[G_y(y(u), u) - Id_Y\right] y'(u)v = -G_u(y(u), u)v \quad \text{in } Y. \tag{16}$$

Here Id_Y is the identity in Y, and the subscripts y, u denote partial derivatives of G.

To eliminate $y'(u)$ (or its approximation $\hat{y}'_k$) from the last equation, we introduce the Lagrange multiplier or adjoint state $\bar{y} \in Y'$ (the dual of Y) and the Lagrangian $L : Y \times Y' \times U \to \mathbb{R}$ given by

$$L(y, \bar{y}, u) = J(y, u) + \langle \bar{y}, G(y, u) - y \rangle_{Y', Y}.$$

Here $\langle \cdot, \cdot \rangle_{Y',Y}$ denotes the dual pairing in the function space setting. It can be replaced by an inner product in $\mathbb{R}^{n_Y}$ in finite dimensions. In a solution point $(y^*, \bar{y}^*, u^*) \in Y \times Y' \times U$ of problem (1–2), the Lagrangian is stationary w.r.t. to variations in all three variables. This leads to the three Karush-Kuhn-Tucker (KKT) conditions:

$$\left.\begin{aligned} 0 = L_y(y^*, \bar{y}_*, u_*) &= J_y(y^*, u_*) + \bar{y}_* \circ G_y(y^*, u_*) - \bar{y}_* && \text{in } Y', \\ 0 = L_{\bar{y}}(y^*, \bar{y}_*, u_*) &= G(y^*, u_*) - y^* && \text{in } Y'' \cong Y, \\ 0 = L_u(y^*, \bar{y}_*, u_*) &= J_u(y^*, u_*) + \bar{y}_* \circ G_u(y^*, u_*) && \text{in } U'. \end{aligned}\right\} \quad (17)$$

For arbitrary state y and control u, the first equation (called the adjoint equation) can be used to compute the adjoint variable or state $\bar{y}$ from

$$\langle \bar{y}, \left[G_y(y,u) - Id_Y\right] w \rangle_{Y',Y} = -J_y(y,u)w \quad \text{for all } w \in Y. \quad (18)$$

With the adjoint state $\bar{y}$ computed, we take $w = y'(u)v$ in this equation and get with (16) the representation

$$J_y(y,u)y'(u)v = -\langle \bar{y}, \left[G_y(y,u) - Id_Y\right] y'(u)v \rangle_{Y',Y} = \langle \bar{y}, G_u(y,u)v \rangle_{Y',Y}. \quad (19)$$

Thus the gradient representation (14) can be written as

$$\frac{d}{du} J(y(u),u)\bigg|_{(\hat{y}_k, u_k)} = \bar{y}_k \circ G_u(\hat{y}_k, u_k) + J_u(\hat{y}_k, u_k). \quad (20)$$

5 One-Shot Optimization Method

The approach described here was in this form developed by Hamdi and Griewank, and can be seen as an extension of the piggy-back strategy. Theoretical results were published in [8, 9], and summarized also in [5]. An engineering application was presented in [18] and results from an ocean model calibration in [15]. Two examples in infinite-dimensional spaces are studied in [12]. In the One-shot approach described here, the motivation is to update the control u already during the state iteration. In the above algorithm, this means that the number j_{max} of steps in the state equation iteration (step 2b) is set to 1 or (in the so-called multistep One-shot method to some low value).

The motivation for this method is again taken from the KKT system (17). The adjoint equation, i.e., the first equation in (17), can be formulated (omitting the stars) as a fixed-point equation for the adjoint state:

$$\bar{y} = \bar{G}(y, \bar{y}, u) := J_y(y,u) + \bar{y} \circ G_y(y,u) \quad \text{in } Y, \quad (21)$$

and a corresponding fixed-point iteration (only for the adjoint, with state and control fixed) can be defined by

$$\bar{y}_j = \bar{G}(y, \bar{y}_{j-1}, u), \quad j = 1, \ldots, j_k. \tag{22}$$

Similar to the iteration for the state in (5), we write for j subsequent iterations:

$$\bar{y}_j = \bar{G}^j(y, \bar{y}_0, u), \quad j = 1, 2, \ldots \tag{23}$$

We now formulate the following algorithm:

Multistep One-Shot Optimization Algorithm:

1. Choose initial values for state, adjoint state, and control $(y_0, \bar{y}_0, u_0)$.
2. For $k = 0, 1, \ldots, k_{max}$:

   a. Compute an approximation of the state for u_k:

   $$y_{k+1} = G^{j_k}(y_k, u_k).$$

   b. Update the adjoint state:

   $$\bar{y}_{k+1} = \bar{G}^{\bar{j}_k}(y_{k+1}, \bar{y}_k, u_k).$$

   c. Update the control using the formula

   $$u_{k+1} = u_k - B_k^{-1}\left[J_u(y_{k+1}, u_k) + \bar{y}_{k+1} \circ G_u(y_{k+1}, u_k)\right]$$

   d. If some criterion for $(y, \bar{y}, u)$ or J is satisfied, stop.

The term *multistep* comes from the usage of the $j_k, \bar{j}_k$ subsequent state and adjoint updates in steps 2a and b, respectively, before a control update is performed in step 2c. The operators $B_k : U \to U'$ can be seen as control preconditioners. They are chosen such that the whole coupled iteration defined in step 2 converges. In the finite-dimensional setting, the B_k are matrices.

If contractivity of the state iteration is given, i.e., there exists $\rho < 1$ satisfying

$$\|G(y, u) - G(\tilde{y}, u)\| \le \rho \|y - \tilde{y}\|, \qquad \forall y, \tilde{y} \in Y, \tag{24}$$

the first equation in the coupled iteration (step 2a) converges linearly for fixed u. Although the second equation exhibits a certain time-lag, it converges with the same asymptotic R-factor (see [10]). For the coupled iteration, the goal is to find B_k that ensure that the spectral radius of the coupled iteration stays below 1 and as close as possible to the one of the Jacobian of the original iteration function G.

5.1 Choice of Preconditioner B_k

We now briefly describe the choice of appropriate preconditioners B_k according to [8,9] that we used in our study. For the derivation of B_k, the authors of [8,9] use the doubly augmented Lagrangian L^a, defined as

$$L^a(y,\bar{y},u) := L(y,\bar{y},u) + \frac{\alpha_L}{2}\|G(y,u) - y\|_Y^2 + \frac{\beta_L}{2}\|\bar{G}(y,\bar{y},u) - \bar{y}\|_{Y'}^2,$$

which is the Lagrangian of the original problem augmented by the errors in the state and adjoint fixed-point equations, with $\alpha_L, \beta_L > 0$ being weighting coefficients. In [9] it is proved that under certain conditions on α_L and β_L (see below), stationary points of problem (1–2) are also stationary points of L^a and that L^a is an exact penalty function. This leads to the idea to choose B_k as an approximation to the Hessian of L^a, i.e. $B_k \approx \frac{d^2}{du^2}L^a(y_k,\bar{y}_k,u_k)$. In [9], it is also proved that—in the finite-dimensional setting—descent of L^a is provided for any preconditioner B_k fulfilling

$$B_k \succeq B_0 := \frac{1}{\sigma}(\alpha_L G_u^\top G_u + \beta_L L_{yu}^\top L_{yu}) \tag{25}$$

i.e., $B_k - B_0$ is positive semidefinite, with

$$\sigma := 1 - \rho - \frac{(1 + \frac{\|L_{yy}\|}{2}\beta_L)^2}{\alpha_L\beta_L(1-\rho)}.$$

In order to make the maximal eigenvalue of B_0 as small as possible (but still positive), under the assumptions $\sqrt{\alpha_L\beta_L}(1-\rho) > 1 + \frac{\beta_L}{2}\|L_{yy}\|$ and $\|L_{yy}\| \neq 0$ the choice

$$\alpha_L = \frac{\|L_{yu}\|^2\beta_L(1 + \frac{\|L_{yy}\|}{2}\beta_L)}{\|G_u\|^2(1 - \frac{\|L_{yy}\|}{2}\beta_L)}, \qquad \beta_L = \frac{3}{\sqrt{\|L_{yy}\|^2 + 3\frac{\|L_{yu}\|^2}{\|G_u\|^2}(1-\rho)^2} + \frac{\|L_{yy}\|}{2}}$$

was made in [9]. At a stationary point of L^a the Hessian of L^a w.r.t. u is

$$\frac{d^2}{du^2}L^a = \alpha_L G_u^\top G_u + \beta_L L_{yu}^\top L_{yu} + L_{uu}.$$

As L^a is an exact penalty function, $\frac{d^2}{du^2}L^a \succ 0$ in a neighborhood of the constrained optimization solution. Assuming that also $\frac{d^2}{du^2}L \succ 0$ implies that the preconditioner

$$B = \frac{1}{\sigma}(\alpha_L G_u^\top G_u + \beta_L L_{yu}^\top L_{yu} + L_{uu})$$

fulfills (25) and thus the update in u of the coupled iteration yields descent on L^a. In the more recent paper [8], the same authors perform a different approach in the choice of the weighting factors and obtain two alternative versions, namely:

$$\sigma = 1, \quad \alpha_L = \frac{2\|L_{yy}\|}{(1-\rho)^2}, \beta_L = \frac{2}{\|L_{yy}\|} \text{ or } \alpha_L = \frac{6\|L_{yy}\|}{(1-\rho)^2}, \beta_L = \frac{6}{\|L_{yy}\|}. \quad (26)$$

To simplify the computations even more, in [8] the choice $\|L_{yy}\| = 1$ is proposed.

5.2 Required Derivatives and Automatic Differentiation

In the One-shot iteration including the preconditioner B_k, first and second order derivative information is needed. The cost for its calculation is small compared to the one of the direct method, since here only one iteration step, i.e., G and not G^{j_k}, has to be differentiated. In the discretized setting, the iteration function $G : \mathbb{R}^{n_Y \times n_U} \to \mathbb{R}^{n_Y}$ consists of up to 2,880 intermediate time steps, compare Sect. 3.3. Thus, the computation of the needed derivatives G_y, G_u using, for example, forward finite differences would mean to perform those time steps n_Y, n_U times only for G_y, G_u, with high computational costs. To reduce the effort and moreover avoid the approximation error of finite differences, we used here the technology of *Automatic or Algorithmic Differentiation (AD)*, see [11]. We used the tool *Transformation of Algorithm in Fortran* (*TAF*, [6]) on the nonlinear biogeochemical model terms q_i, whereas the linear transport matrix part was differentiated analytically.

There are two modes of AD, namely the *forward* and the *reverse mode* (corresponding to an adjoint equation). The forward mode enhances an iteration with the corresponding derivative iteration and thus is the discrete analogue of the sensitivity equation approach from Sect. 4.1.1. Here, the cost for evaluating derivatives increases linearly with the number of unknowns, in our case n_Y or n_U, which is comparable to a finite difference approximation, but avoids the approximation errors. In our application, it is only recommended for derivatives w.r.t u, since in our application $n_U = 7 \ll n_Y \approx 100{,}000$. In contrast, the reverse mode stores all intermediate variables of the function evaluation and then, in a reverse sweep reverting the order of operations, computes *all partial derivatives* of the function with respect to intermediate variables *at once*. It is the discrete analogue of the adjoint equation approach described in Sect. 4.1.2. In particular, the gradient of a scalar valued function as J can be evaluated as a cost independent from the number of independent variables. Therefore, the reverse mode is appropriate for derivatives w.r.t. y. The concatenation of a reverse and a forward sweep yields second order derivatives. Due to the very small number of parameters $n_U = 7$ and the complexity of the code, we chose the reverse sweep followed by a finite differences approach to compute second order derivatives. Table 2 summarizes the applied strategies for the computation of the needed derivatives.

Table 2 Computation of derivatives using different approaches

Derivative	Mode of computation
J_y	Analytically
J_u	Analytically
$\bar{y}^\top G_y, \bar{y}^\top G_u$	One reverse sweep of AD combined with transport matrices by hand
G_u	Forward mode of AD combined with transport matrices by hand
J_{yu}	Analytically (= 0)
$\bar{y}^\top G_{yu}$	After computation of $\bar{y}^\top G_y$, application of finite differences

For the computation of the weights σ, α_L and β_L of the preconditioner B_k, see Sect. 5.1, we chose the first of the cheaply computable versions defined in (26) and fixed $\|L_{yy}\| = 1$. Furthermore, we set the unknown contraction factor ρ of the state iteration function G to $\rho = 0.9$. We observed for the N-DOP model that the contraction property (24) is violated for some steps in the state iteration. However, it converges to a steady solution and the average contraction factor is close to, but less than 1. Fixing ρ in such a way simplifies the code. Another option is to update y and $\bar{y}$ without an update of u (i.e., increasing the iteration numbers $j_k, \bar{j}_k$) until the contraction factor ρ is less than 1 again.

6 Surrogate-Based Optimization

The surrogate-based optimization strategy (SBO, see, e.g., [1,4,17,22]) is built upon a *coarse or low-fidelity model* that can be evaluated much faster than the original, in this context then called *fine or high-fidelity model* used in the direct optimization approach. Since a coarse model naturally does not include as much information or have the accuracy of the fine one, an (ideally computationally cheap) *alignment* or *correction* of the coarse model is performed. The aim of this alignment is to keep the output of the aligned coarse model, the so-called *surrogate*, close to the output of the fine model, also when the optimization parameters are changed to a certain limit. When this optimization of the surrogate is numerically converged, the fine model is evaluated again and the alignment is updated. Then the surrogate is optimized again, and the process is iterated. The benefit is that fine model optimization runs are completely avoided, which reduces the overall effort tremendously.

Surrogates can be created by approximating sampled fine model data (*functional* surrogates). Popular techniques include polynomial regression, kriging, artificial neural networks, and support vector regression [22, 24, 25]. Another possibility, exploited in this work, is to construct the surrogate model through appropriate correction/alignment of a low-fidelity or coarse model (*physics-based* surrogates, [26]). Physics-based surrogates inherit physical characteristics of the original fine model so that only a few fine model data is necessary to ensure their good alignment with the fine model. Moreover, generalization capability of the physics-based models is typically much better than for functional ones. As a result, SBO schemes working with this type of surrogates normally require small number of fine model evaluations to yield a satisfactory solution. On the other hand, their transfer to other applications is less straightforward since the underlying coarse model and chosen correction approach is rather problem specific. The specific correction technique exploited in this work is described below (see also [20]).

In applications that use iterative state equation solvers, a simple way to construct a coarse model is just to stop the iteration after fewer steps or with a relaxed stopping criterion. Then, the term *coarse* refers not to a coarser discretization in space and/or time, but to a model and state with reduced accuracy compared to the original one. This way of constructing a coarse model is much simpler than to use a coarser discretization scheme, which of course is also possible, but involves prolongation and restriction operations on the model output. We now give the structure of an SBO algorithm based on this coarse model construction.

Surrogate-Based Optimization Algorithm (With Iterative Solver):

1. Choose initial value for the control u_0.
2. For $k = 0, 1, \ldots, k_{max}$:

   a. Compute an approximation of the state for u_k with a fine model:

   $$y_k^f = G^{j_k^f}(y_k, u_k).$$

   b. Compute an approximation of the state for u_k with a coarse model:

   $$y_k^c = G^{j_k^c}(y_k, u_k).$$

   c. Compute the correction or alignment operator

   $$A_k = A_k(y_k^f, y_k^c) : Y \to Y \qquad \text{satisfying} \qquad A_k y_k^c = y_k^f$$
   $$\text{and optionally} \qquad \frac{d A_k y_k^c}{du} = \frac{d y_k^f}{du}$$

   and define the surrogate

   $$s_k : U \to Y, \qquad s_k(u) := A_k G^{j_{k,c}}(y_k, u)$$

   d. Compute

   $$u_{k+1} = \operatorname*{argmin}_{u \in U_{ad}} J(s_k(u), u),$$

   or approximate it by i_k steps of an iterative optimization method.

   e. If some criterion for y, u or J is satisfied, stop.

The iteration numbers chosen in steps 2b (j_k^c) and step 2d (i_k) can be either kept constant or adapted. The latter variant is called a *hybrid SBO strategy*. The two conditions imposed on the alignment operator A_k in step 2c are called *zeroth* and *first order consistency*. The second one might be relaxed to be valid only approximately. If the surrogate s_k satisfies both conditions at $u = u_k$, the SBO algorithm is provable convergent to at least a local optimum under conditions regarding the coarse and fine model smoothness, and provided that the algorithm is enhanced by the trust-region safeguard, i.e., in step 2d the minimum is just taken over the set

$$U_k := \{u \in U_{ad} : \|u - u_k\|_U \leq \delta_k\}$$

with δ_k being a trust-region radius updated according to the usual trust region rules. We refer the reader to, e.g., [3, 14] for more details.

One example for the alignment operator we used in our application is the pointwise multiplicative operator

$$A_k(y_k^f, y_k^c)y(x,t) := \frac{y_k^f(x,t)}{y_k^c(x,t)}\, y(x,t) \qquad \text{for all grid} - \text{points } (x,t) \in \Omega_h \times [0,T]_\tau,$$

which is very easy to compute. It just satisfied the zeroth order consistency condition.

For the inner optimization iteration in step 2d, any algorithm is possible. We used the MATLAB (registered trademark of The MathWorks, Inc.) function `fmincon`, exploiting the active-set algorithm and using the option setting {'TolCon', 1e-6, 'TolX', 1e-6, 'TolFun', 1e-6}. To ensure convergence of the SBO, we enhanced each surrogate optimization in step 2d by restricting the current step-size to a certain trusted region δ_k. The gradients used in this inner optimization were supplied externally as finite difference approximations, but took special care about the step-sizes for their computation.

We used the absolute difference (measured in the Euclidean norm) between two successive iterates u_k and u_{k-1} as well as a lower bound for the trust-region radius δ_k as stopping criterion for the outer iteration (over k). The inner optimization for each surrogate (i.e., for each k) is terminated after $i_k = 10$ iterations (for all k), except in the examples below using a hybrid strategy (where it varies between $i_k = 2, 3$).

7 Optimization Results

In this section we present a brief summary of results of the two optimization methods for a parameter optimization problem for the above presented ecosystem model. For both methods, results for twin-data experiments are available. In such kind of experiments, model-generated data (with parameters u_d) are used to evaluate the applicability and computational efficiency of the methods.

7.1 Surrogate-Based Optimization

Results for the SBO method for this application were in detail presented in [21]. Therein, the coarse model uses a constant number $j^c = j_k^c = 25$ (for all k) of iteration steps in the state equation solver, compared to $j^f = j_k^f = 3,000$ (also constant for all k) ones in the original fine model. On these results, we thus here give only a brief summary. Additionally, we present some recent results obtained using a hybrid SBO strategy that uses different values of j_k^c in the different optimization steps.

The number $j^c = 25$ of coarse model iterations used in [21] leads to a poor identification of two of the seven parameters, see also Fig. 1. The usage of higher values of j^c or a hybrid strategy solved this problem. Concerning performance, Figs. 1 and 2 show that the gain compared to the direct optimization can be significantly enlarged when using a hybrid strategy. On the other hand, finding adequate sequences of inner optimization steps $(i_k)_k$ and number of coarse model iteration steps $(j_k^c)_k$ requires considerable testing or experience.

7.2 One-Shot Optimization

The results for the One-shot optimization available so far are preliminary and not that detailed as the one for the SBO approach. A difference by design of the method is that, in its current version, the One-shot method does not treat parameter bounds explicitly, whereas the SBO method can take them into account in the inner optimization loop. The considered model problem is the same as for the SBO method, concerning (model-generated) data y_d, the least-squares cost function (10), and the underlying simulation model. Here we show results for a version of the

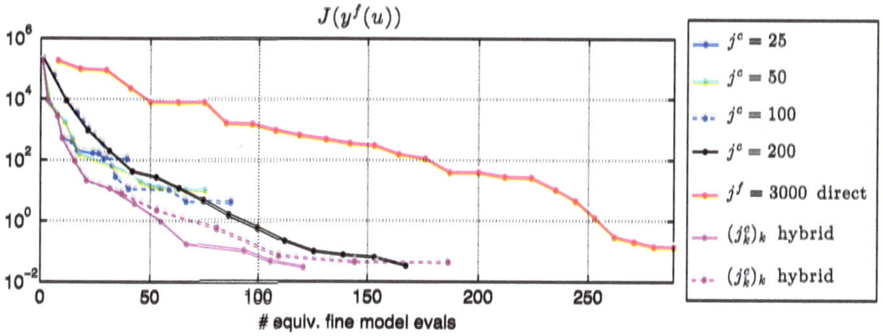

Fig. 1 Cost function value J during SBO runs using (1) different constant coarse models' iteration numbers j^c, (2) two hybrid strategies both with $(j_k^c)_k = (25, 100, 200, 200, \ldots)$, but once $i_k = 3$ for all k (*top*) and the other time $(i_k)_k = (3, 2, 2, \ldots)$, (3) a direct fine model optimization. The computational cost of the optimization is decreased by about 75–90 %

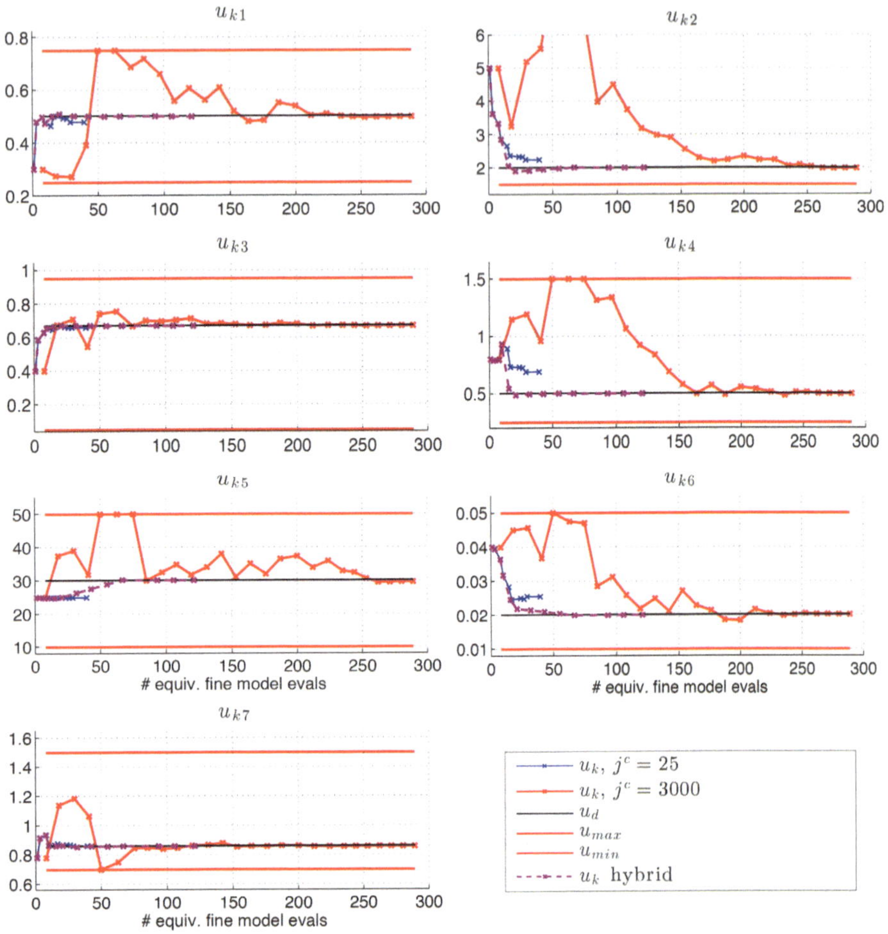

Fig. 2 Convergence of the model parameters $(u_{ki})_{i=1,\dots,7}$ during the optimization (step counter k), here only for one SBO run, one hybrid run and for the direct fine model optimization. Also shown are target parameter vector u_d and the constant bounds u_{min}, u_{max}

one-step variant ($j_k = \bar{j}_k = 1$), i.e. after one iteration of the state equation solver one for the adjoint and another for the parameter is performed. The only exception is the first step, where a higher number j_0 of state iterations are performed before the first adjoint step.

In this one-step variant, one of the seven parameters provides some problems. The parameters and therewith the tracer concentrations oscillate and the cost function is not reduced anymore after a certain time even though the weighting factor α was chosen very large ($\alpha = 100$) to force parameters towards u_{guess}, see the plot on the left of Fig. 3. The problematic parameter turned out to be $u_7 = b$,

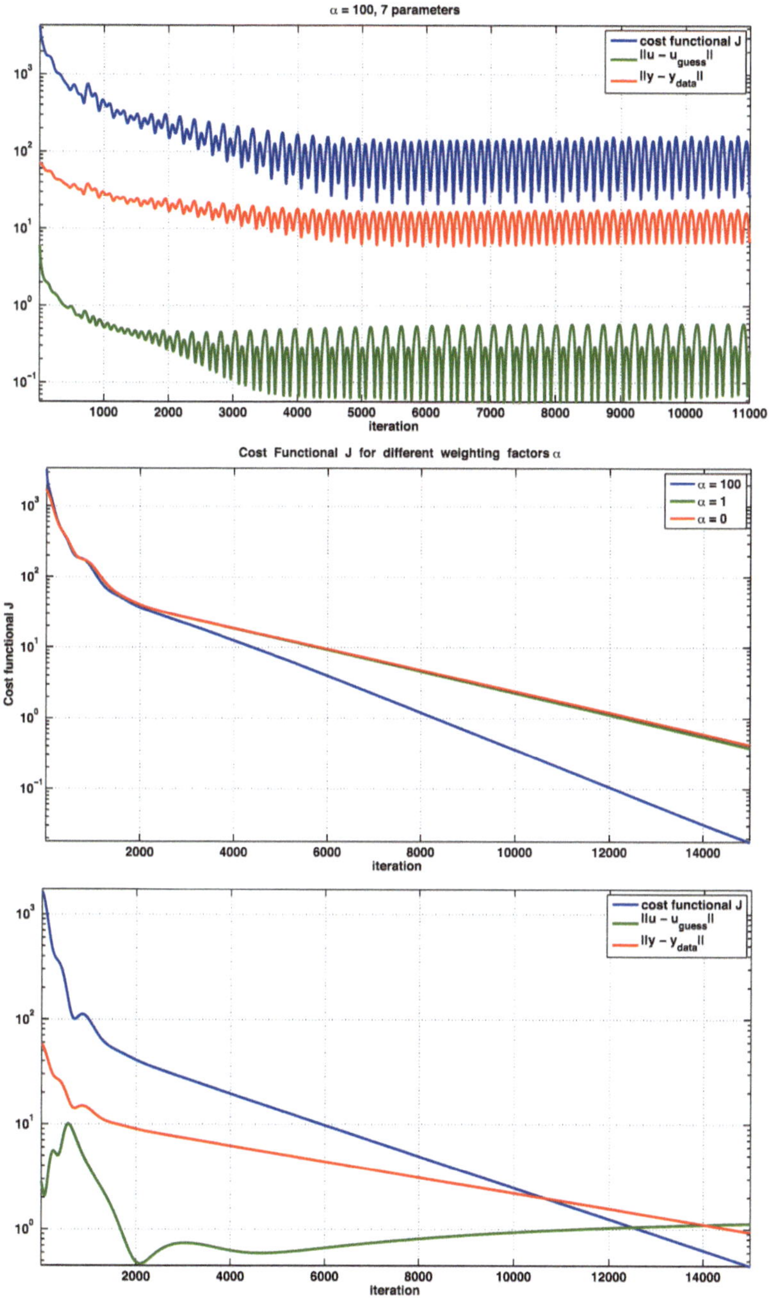

Fig. 3 Results of the One-shot method for seven parameters showing oscillations (*top*), for six parameters, $u_{guess} = u_d$ and different α (*middle*), and for $u_{guess} \neq u_d$ and $\alpha = 0.01$ (*bottom*)

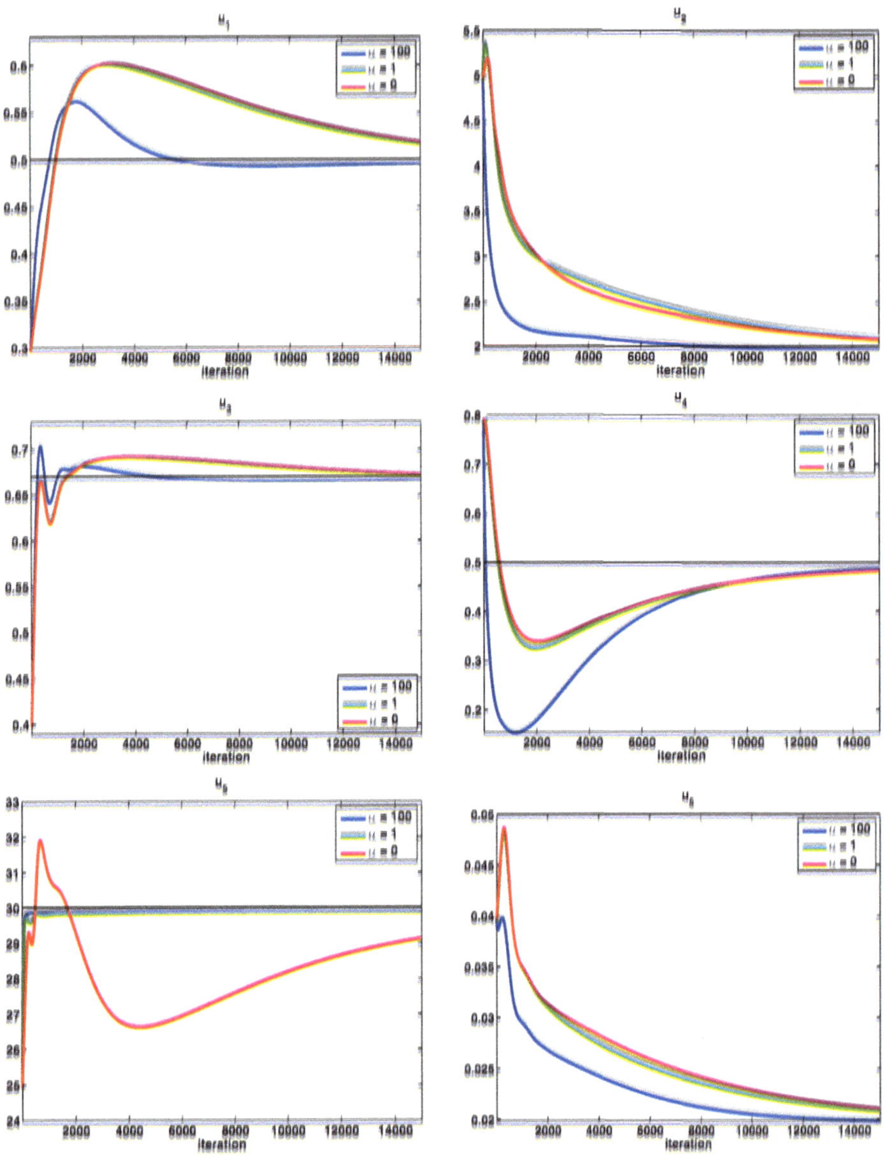

Fig. 4 Parameter values during the optimization for $u_{guess} = u_d$ and different α

the sinking exponent. Its influence demands further analysis. Numerical tests with the other parameters, fixing $u_7 = u_{d7}$ performed very well, both for $u_{guess} = u_d$ (the easier case), but also for $u_{guess} \neq u_d$ and even without any regularization ($\alpha = 0$). Figure 4 shows the convergence of the parameter values during the optimization process for different regularization parameters α. Not surprisingly, the larger α the better the data is fit and the cost function is reduced if $u_{guess} = u_d$. During the first One-shot iteration steps, parameter values may go astray optimal values, because at that stage of the optimization y and $\bar{y}$ are far away from optimality such that the preconditioner B_k will correct both via a big change in u. Concerning performance, the method needs about 15,000 iterations to give an acceptable solution. It has to be noticed that the computational effort in the adjoint step and the evaluation of the derivatives used in the algorithm parameters α_L, β_L has to be further quantified in order to give sound performance comparisons with the SBO method.

Acknowledgements The work was supported by DFG in the Cluster "Future Ocean" and the priority program 1253 "Optimization with Partial Differential Equations", and by the EU in the FP7 project "CarboChange".

References

1. Bandler, J.W., Cheng, Q.S., Dakroury, S.A., Mohamed, A.S., Bakr, M.H., Madsen, K., Søndergaard, J.: Space mapping: the state of the art. IEEE Trans. Microwave Theory Tech. **52**(1), 337–361 (2004)
2. Christianson, B.: Reverse accumulation and implicit functions. Optim. Methods Softw. **9**(4), 307–322 (1998)
3. Conn, A.R., Gould, N.I.M., Toint, P.L.: Trust-region methods. In: MPS-SIAM Series on Optimization. Society for Industrial and Applied Mathematics, Philadelphia (2000)
4. Forrester, A.I.J., Keane, A.J.: Recent advances in surrogate-based optimization. Prog. Aerosp. Sci. **45**(1–3), 50–79 (2009)
5. Gauger, N., Griewank, A., Hamdi, A., Kratzenstein, C., Özkaya, E., Slawig, T.: Automated extension of fixed point PDE solvers for optimal design with bounded retardation. In: Leugering, G., Engel, S., Griewank, A., Hinze, M., Rannacher, R., Schulz, V., Ulbrich, M., Ulbrich, S. (eds.) Constrained Optimization and Optimal Control for Partial Differential Equations. International Series of Numerical Mathematics, vol. 160. Birkhäuser, Basel (2011)
6. Giering, R., Kaminski, T.: Applying TAF to generate efficient derivative code of Fortran 77–95 programs. Proc. Appl. Math. Mech. **2**(1), 54–57 (2003)
7. Griewank, A.: Evaluating Derivatives Principles and Techniques of Algorithmic Differentiation. Frontiers in Applied Mathematics, vol. 19. SIAM, Philadelphia (2000)
8. Griewank, A., Hamdi, A.: Properties of an augmented Lagrangian for design optimization. Optim. Methods Softw. **25**(4), 645–664 (2010)
9. Griewank, A., Hamdi, A.: Reduced quasi-Newton method for simultaneous design and optimization. Comput. Optim. Appl. Online **49**, 521–548 (2011)
10. Griewank, A., Kressner, D.: Time-lag in derivative convergence for fixed point iterations. ARIMA Numéro spécial CARI'04, pp. 87–102 (2005)
11. Griewank, A., Walther, A.: Evaluating derivatives: principles and techniques of algorithmic differentiation. Society for Industrial and Applied Mathematics (SIAM), Philadelphia (2008)
12. Kaland, L., De Los Reyes, J.C., Gauger, N.: One shot methods in function space for pde-constrained optimal control problems. Optim. Methods Softw. **29**(2), 376–405 (2013)
13. Khatiwala, S., Visbeck, M., Cane, M.A.: Accelerated simulation of passive tracers in ocean circulation models. Ocean Model. **9**(1), 51–69 (2005)
14. Koziel, S., Bandler, J.W., Cheng, Q.S.: Robust trust-region space-mapping algorithms for microwave design optimization. IEEE Trans. Microw. Theory Tech. **58**(8), 2166–2174 (2010)

15. Kratzenstein, C., Slawig, T.: Simultaneous model spin-up and parameter identification with the one-shot method in a climate model example. Int. J. Optim. Control Theory Appl. **3**(2), 99–110 (2013)
16. Kriest, I., Khatiwala, S., Oschlies, A.: Towards an assessment of simple global marine biogeochemical models of different complexity. Prog. Oceanogr. **86**(3–4), 337–360 (2010)
17. Leifsson, L., Koziel, S.: Multi-fidelity design optimization of transonic airfoils using physics-based surrogate modeling and shape-preserving response prediction. J. Comput. Sci. **1**(2), 98–106, 6 (2010)
18. Özkaya, E., Gauger, N.: Single-step one-shot aerodynamic shape optimization. Technical report (2008)
19. Parekh, P., Follows, M.J., Boyle, E.A.: Decoupling of iron and phosphate in the global ocean. Glob. Biogeochem. Cycles **19**(2), GB2020 (2005)
20. Prieß, M., Koziel, S., Slawig, T.: Surrogate-based optimization of climate model parameters using response correction. J. Comput. Sci. **2**(4), 335–344, 12 (2011)
21. Prieß, M., Piwonski, J., Koziel, S., Oschlies, A., Slawig, T.: Accelerated parameter identification in a 3D marine biogeochemical model using surrogate-based optimization. Ocean Model. **68**, 22–36 (2013)
22. Queipo, N.V., Haftka, R.T., Shyy, W., Goel, T., Vaidyanathan, R., Tucker, P.K.: Surrogate-based analysis and optimization. Prog. Aerosp. Sci. **41**(1), 1–28 (2005)
23. Sarmiento, J.L., Gruber, N.: Ocean Biogeochemical Dynamics. Princeton University Press, Princeton (2006)
24. Simpson, T.W., Poplinski, J.D., Koch, P.N., Allen, J.K.: Metamodels for computer-based engineering design: Survey and recommendations. Eng. Comput. **17**, 129–150 (2001)
25. Smola, A.J., Schölkopf, B.: A tutorial on support vector regression. Stat. Comput. **14**, 199–222 (2004)
26. Søndergaard, J.: Optimization using surrogate models - by the space mapping technique. PhD thesis, Informatics and Mathematical Modelling, Technical University of Denmark, DTU (2003)

Shape-Preserving Response Prediction for Surrogate Modeling and Engineering Design Optimization

Slawomir Koziel and Leifur Leifsson

Abstract Computer simulation models are fundamental tools of contemporary engineering design. The components, structures, and systems considered in most engineering disciplines are far too complex to be accurately described using simple theoretical models. Therefore, numerical simulation is often the only way to evaluate the performance of the design with sufficient reliability. However, accurate, high-fidelity simulations are computationally expensive. Consequently, their use for design automation, especially when exploiting conventional optimization algorithms is often prohibitive. Availability of faster computers and more efficient simulation software does not always translate into computational speedup due to growing demand for improved accuracy and the need to evaluate larger and larger systems. Surrogate-based optimization (SBO) techniques belong to the most promising approaches capable of alleviating these difficulties. SBO allows for reducing the number of expensive objective function evaluations in a simulation-driven design process. This is obtained by replacing the direct optimization of the expensive model by iterative updating and re-optimization of its cheap surrogate model. Among proven SBO techniques, the methods exploiting physics-based low-fidelity models are probably the most efficient. This is because the knowledge about the system of interest embedded in the low-fidelity model allows constructing the surrogate model that has good generalization capability at a cost of just a few evaluations of the original model. This chapter reviews one of the most recent SBO techniques, the so-called shape-preserving response prediction (SPRP). We discuss the formulation of SPRP, its limitations, and generalizations, and, most importantly, demonstrate its applications to solve design problems in various engineering areas, including microwave engineering, antenna design, and aerodynamic shape optimization. We also discuss the use of SPRP for creating fast surrogate models with illustrations from the microwave engineering area.

S. Koziel (✉) • L. Leifsson
Engineering Optimization & Modeling Center, School of Science and Engineering, Reykjavik University, Menntavegur 1, 101, Reykjavik, Iceland
e-mail: koziel@ru.is; leifurth@ru.is

S. Koziel et al. (eds.), *Solving Computationally Expensive Engineering Problems*, Springer Proceedings in Mathematics & Statistics 97,
DOI 10.1007/978-3-319-08985-0_2

Keywords Microwave engineering • Aerodynamic optimization • Surrogate modeling • Surrogate-based optimization • Shape-preserving response prediction

1 Introduction

Computer simulations are one of the most important tools in contemporary science and engineering. From miniature electronic components and circuits, through complete systems such as aircraft, to large-scale physical phenomena (e.g., climate models), simulations are used to describe the behavior, evaluate the performance, and validate designs. Nowadays, commercial simulation packages have matured and the computing resources are cheaper and in abundance. In spite of this, in many cases, accurate, high-fidelity simulations are computationally expensive, to the extent that their use in the design process, e.g., by employing simulations directly in an automated design optimization loop, may be impractical. The primary reason is that conventional optimization algorithms, both gradient-based [1] and derivative-free [2] typically require a large number of objective function evaluations. In some cases, the use of adjoint sensitivity [3] can alleviate this problem; however, this technique is not always available through commercial simulation packages. Conversely, design automation is key in situations where simple theoretical models are no longer capable to adequately account for complex interactions between the system components and, therefore, only yield an initial approximation of the optimum design which consequently has to be tuned further in order to meet the given performance requirements. In practice, design "tuning" is often based on parametric studies guided by engineering experience. This combination is often sufficient to obtain satisfactory designs in a reasonable time; however, it is far from being an automated process.

Surrogate-based optimization (SBO) [4, 5] is one of the most promising approaches to alleviate the difficulties discussed in the previous paragraph. In SBO, direct optimization of an expensive high-fidelity simulation model is replaced by iterative updating and re-optimization of its computationally cheap representation, a surrogate. The high-fidelity model is referenced occasionally to verify the prediction produced by the surrogate and to improve the latter. The overall design cost can be greatly reduced, because the optimization burden is shifted to the surrogate.

SBO methods differ mostly in the way the surrogate is created. A large group of function approximation modeling techniques exist. Here the surrogate is created by approximating sampled high-fidelity model data and the most popular methods include polynomial approximation [5], radial basis function interpolation [6], kriging [7], support vector regression [8], and neural networks [9]. Approximation models are very fast, however, a large number of training samples—and a high CPU cost of gathering the simulation data—are necessary to ensure reasonable accuracy. Furthermore, the number of required samples grows exponentially with the dimensionality of the design space (the *curse of dimensionality*). Depending on the model purpose, this initial computational overhead may or may not be justified. This depends for example whether the models are for a multiple-use library or a one-time optimization.

Correcting an auxiliary low-fidelity (or coarse) model is another approach to SBO. A low-fidelity model is a reduced-accuracy but faster representation of the system of interest. Low-fidelity models can be developed in various ways, such as by using simplified-physics, leaving out certain second-order effects, or by describing the system on a different physical level (e.g., equivalent circuit versus full-wave electromagnetic simulation in case of microwave components). Engineers have been using simplified models for decades: before the computer era simplified models and physical experiments were the only tools available to perform the design process. Because of the fact that a low-fidelity model contains certain knowledge about the system of interest, physics-based surrogates offer good generalization capabilities and can be set up using a limited number of training points. These are their biggest advantages over purely approximation models.

Several techniques have been proposed to exploit physics-based surrogate models in the SBO process, such as the approximation model management optimization (AMMO) framework [10], space mapping (SM) [11], manifold mapping [12], and simulation-based tuning [13]. Several of these methods are based on correcting the low-fidelity model output (response). The SBO process is provably convergent to the high-fidelity model optimum [13] when embedded in the trust-region framework [14] and the correction is realized by ensuring both zero- and first-order consistency [10] between the surrogate and the high-fidelity model. In some cases (with a notable example of SM), the correction can be done by introducing a mapping between the parameter spaces of the low- and high-fidelity models.

The shape-preserving response prediction (SPRP) technique [15] is a recently developed approach which exploits physics-based low-fidelity models. The method was originally developed in the microwave engineering area [15], but has also been applied to problems in antenna design [16] and aerodynamic design [17]. SPRP is a parameter-less method where the surrogate model response is constructed by tracking the changes of the low-fidelity model response when moving from a certain reference design to another one, and applying those changes (represented by translation vectors) to a reference response of the high-fidelity model. The SPRP surrogate exploits the knowledge embedded in the low-fidelity model to a greater extent than other physic-based surrogate modeling approaches, e.g., SM. Therefore, the generalization capability of SPRP is usually better than that of SM [15]. In this chapter, we review the SPRP technique, its basic and generalized formulations, and attempt to give an intuitive explanation of its efficiency. We also illustrate its operation and performance using several design examples from various engineering disciplines.

The chapter is organized as follows. In Sect. 2, we formulate the engineering optimization problem, briefly recall the basics of SBO, and introduce the concept of the SPRP methodology. Section 3 demonstrates the use of SPRP for optimization of microwave filters. Application of SPRP for antenna design is discussed in Sect. 4.

Section 5 describes formulation and the use of SPRP for the design of transonic airfoils. Section 6 discusses the use of SPRP for surrogate modeling. Section 7 concludes the chapter.

2 Surrogate-Based Optimization and Shape-Preserving Response Prediction

In this section, we formulate the engineering design optimization problem, recall the concept of SBO, and discuss the SPRP methodology [15]. Examples illustrating application of SPRP in various engineering fields are provided in Sects. 3–6.

2.1 *Engineering Design Optimization. Problem Formulation*

The engineering design optimization problem can be defined as

$$\boldsymbol{x}_f^* = \arg\min_{\boldsymbol{x}} U(f(\boldsymbol{x})) \tag{1}$$

where f: $X_f \to R^m$, $X_f \subseteq R^n$, denotes the response vector of a high-fidelity (or fine) model of the device or system of interest; U: $R^m \to R$ is a given objective function, e.g., minimax [18]. In microwave engineering, the response vector may contain, for example, the values of transmission coefficient $|S_{21}|$ evaluated over certain frequency band.

2.2 *Surrogate-Based Optimization*

Because of the high computational cost of evaluating f, its direct optimization is replaced by an iterative procedure [5]

$$\boldsymbol{x}^{(i+1)} = \arg\min_{\boldsymbol{x}} U\left(s^{(i)}(\boldsymbol{x})\right) \tag{2}$$

that generates a sequence of points (designs) $\boldsymbol{x}^{(i)} \in X_f$, $i = 0, 1, \ldots$. Each $\boldsymbol{x}^{(i+1)}$ is the optimal design of the surrogate model $s^{(i)}$: $X_s^{(i)} \to R^m$, $X_s^{(i)} \subseteq R^n$, $i = 0, 1, \ldots$. $s^{(i)}$ is assumed to be a computationally cheap and sufficiently reliable representation of the fine model f, particularly in the neighborhood of the current design $\boldsymbol{x}^{(i)}$. Under these assumptions, the algorithm (2) is likely to produce a sequence of designs that quickly approach $\boldsymbol{x}_f^*$. Because f is evaluated rarely (usually once per iteration), the surrogate model is supposedly fast, and the number of iterations for a well-performing algorithm is substantially smaller than for most direct optimization methods, and the process (2) may lead to substantial reduction of the computational cost of solving (1). If the surrogate model satisfies zero- and first-order consistency conditions with the fine model, i.e., $s^{(i)}(\boldsymbol{x}^{(i)}) = f(\boldsymbol{x}^{(i)})$ and $(\partial s^{(i)}/\partial \boldsymbol{x})(\boldsymbol{x}^{(i)}) = (\partial f/\partial \boldsymbol{x})(\boldsymbol{x}^{(i)})$ (verification of the latter requires f sensitivity data), and the algorithm (2) is enhanced by the trust-region method [19], then it is provably convergent to a local optimum of the fine model [10]. Convergence can also be guaranteed if the algorithm (2) is enhanced by properly selected local search methods [20].

2.3 *Shape-Preserving Response Prediction: Concept [15]*

SPRP [15] has been initially introduced in microwave engineering to reduce the cost of optimizing electromagnetic (EM)-simulated structures such as filters [15]. In SPRP, the surrogate model is constructed assuming that the change of the fine model response due to the adjustment of the design variables from can be predicted using the actual response changes of the auxiliary low-fidelity (or coarse) model c: $X_c \rightarrow R^m$, $X_c \subseteq R^n$, that describes the same object as the high-fidelity model; c is less accurate but much faster to evaluate than f.

The choice of the coarse model very much depends on the engineering discipline. In microwave engineering, the coarse model might be an equivalent circuit of the considered microwave structure, that describes the structure using circuit theory methods rather than through solution of the Maxwell equations. It is critically important for SPRP that the coarse model is physically based, which ensures that the effect of the design parameter variations on the model response is similar for both the fine and coarse models. The change of the coarse model response is described by the translation vectors corresponding to certain (finite) number of characteristic points of the model's response. These translation vectors are subsequently used to predict the change of the fine model response with the actual response of f at the current iteration point, $f(\boldsymbol{x}^{(i)})$, treated as a reference.

Here, we explain the concept of SPRP using the specific case of a microwave filter. Figure 1a shows the example of the coarse model response, $|S_{21}|$ in the frequency range 8–18 GHz, at the design $\boldsymbol{x}^{(i)}$, as well as the coarse model response at some other design $\boldsymbol{x}$. The responses come from the double folded stub bandstop filter [15]. Circles denote five characteristic points of $c(\boldsymbol{x}^{(i)})$, here, selected to represent $|S_{21}| = -3$ dB, $|S_{21}| = -20$ dB, and the local $|S_{21}|$ maximum (at about 13 GHz). Squares denote corresponding characteristic points for $c(\boldsymbol{x})$, while small line segments represent the translation vectors that determine the "shift" of the characteristic points of c when changing the design variables from $\boldsymbol{x}^{(i)}$ to $\boldsymbol{x}$. Because the coarse model is physics-based, the fine model response at the given design, here, $\boldsymbol{x}$, can be predicted using the same translation vectors applied to the corresponding characteristic points of the fine model response at $\boldsymbol{x}^{(i)}$, $f(\boldsymbol{x}^{(i)})$. This is illustrated in Fig. 1b. Figure 2 shows the predicted fine model response at $\boldsymbol{x}$ as well as the actual response, $f(\boldsymbol{x})$, with a good agreement between both curves.

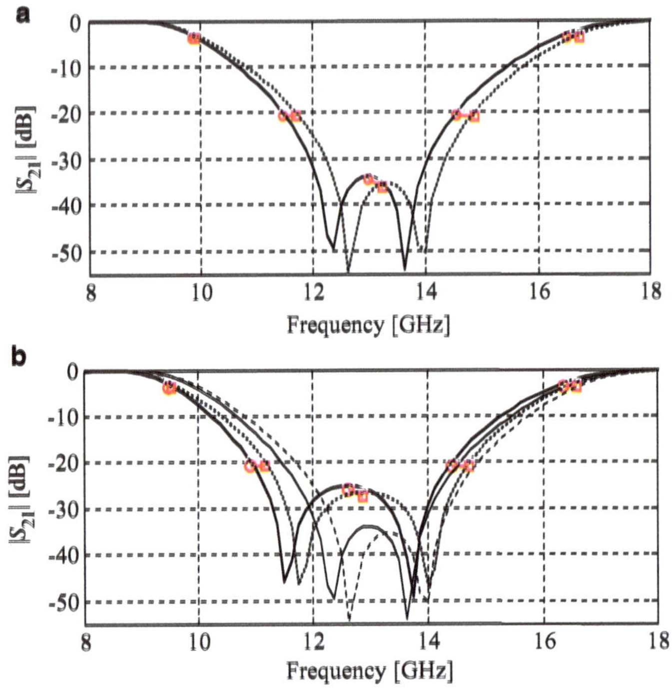

Fig. 1 The SPRP concept [15]: (**a**) Example coarse model response at the design $\boldsymbol{x}^{(i)}$, $c(\boldsymbol{x}^{(i)})$ (*solid line*), the coarse model response at $\boldsymbol{x}$, $c(\boldsymbol{x})$ (*dotted line*), characteristic points of $c(\boldsymbol{x}^{(i)})$ (o) and $c(\boldsymbol{x})$ (*square*), and the translation vectors (*short lines*); (**b**) Fine model response at $\boldsymbol{x}^{(i)}$, $f(\boldsymbol{x}^{(i)})$ (*solid line*) and the predicted fine model response at $\boldsymbol{x}$ (*dotted line*) obtained using SPRP based on characteristic points of this figure; characteristic points of $f(\boldsymbol{x}^{(i)})$ (o) and the translation vectors (*short lines*) were used to find the characteristic points (*square*) of the predicted fine model response; coarse model responses $c(\boldsymbol{x}^{(i)})$ and $c(\boldsymbol{x})$ are plotted using *thin solid* and *dotted line*, respectively [9]

2.4 *Shape-Preserving Response Prediction: Formulation [15]*

SPRP can be rigorously formulated as follows. Let $f(\boldsymbol{x}) = [f(\boldsymbol{x},\omega_1)\ \dots\ f(\boldsymbol{x},\omega_m)]^T$ and $c(\boldsymbol{x}) = [c(\boldsymbol{x},\omega_1)\ \dots\ c(\boldsymbol{x},\omega_m)]^T$, where ω_j, $j = 1, \dots, m$, is the frequency sweep (it can be assumed without loss of generality that the model responses are parameterized by frequency). Let $p_j^f = [\omega_j^f\ r_j^f]^T$, $p_j^{c0} = [\omega_j^{c0}\ r_j^{c0}]^T$, and $p_j^c = [\omega_j^c\ r_j^c]^T$, $j = 1, \dots, K$, denote the sets of characteristic points of $f(\boldsymbol{x}^{(i)})$, $c(\boldsymbol{x}^{(i)})$, and $c(\boldsymbol{x})$, respectively. Here, ω and r denote the frequency and magnitude components of the respective point. The translation vectors of the coarse model response are defined as $t_j = [\omega_j^t\ r_j^t]^T$, $j = 1, \dots, K$, where $\omega_j^t = \omega_j^c - \omega_j^{c0}$ and $r_j^t = r_j^c - r_j^{c0}$. The SPRP surrogate model is defined as follows

$$s^{(i)}(\boldsymbol{x}) = \left[s^{(i)}(\boldsymbol{x},\omega_1)\quad \dots\quad s^{(i)}(\boldsymbol{x},\omega_m)\right]^T \tag{3}$$

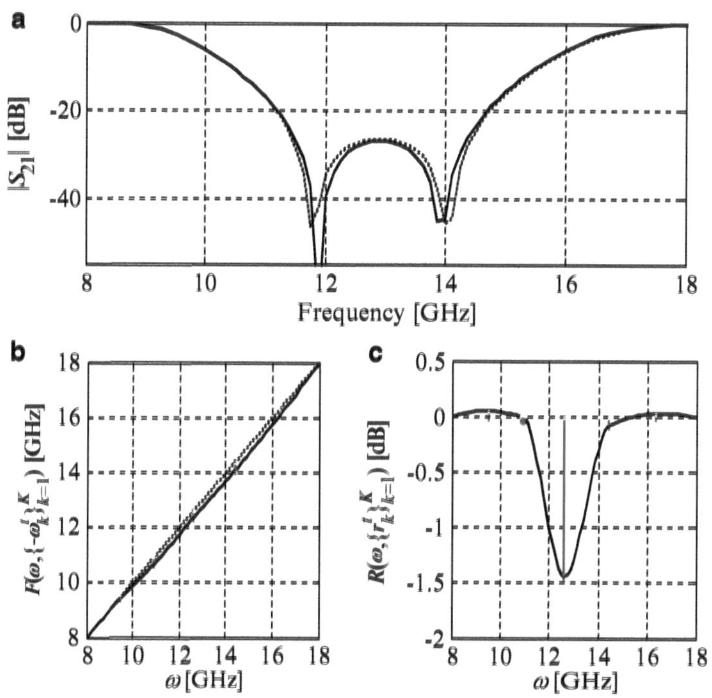

Fig. 2 (**a**) Fine model response at $\boldsymbol{x}$, $f(\boldsymbol{x})$ (*solid line*), and the fine model response at $\boldsymbol{x}$ obtained using the shape-preserving prediction (*dotted line*). Good agreement between both curves is observed, particularly in the areas corresponding to the characteristic points of the response; (**b**) Interpolating function F (*solid line*) corresponding to the fine/coarse model plots in Fig. 1; the identity function is denoted using the *dotted line*, the frequency components of the translation vectors are denoted as *short solid lines*; (**c**) Interpolating function R (*solid line*); the magnitude components of the translation vectors are denoted using *short solid lines*

where

$$s^{(i)}\left(\boldsymbol{x},\omega_j\right)=\overline{f}\left(\boldsymbol{x}^{(i)},F\left(\omega_j,\{-\omega_k^t\}_{k=1}^K\right)\right)+R\left(\omega_j,\{r_k^t\}_{k=1}^K\right) \tag{4}$$

for $j=1, \ldots, m$. $\overline{f}(\boldsymbol{x},\omega)$ is an interpolation of $\{f(\boldsymbol{x},\omega_1), \ldots, f(\boldsymbol{x},\omega_m)\}$ onto the frequency interval $[\omega_1,\omega_m]$. The scaling function F interpolates the data pairs $\{\omega_1,\omega_1\}$, $\{\omega_1{}^f,\omega_1{}^f-\omega_1{}^t\}$, $\ldots$, $\{\omega_K{}^f,\omega_K{}^f-\omega_K{}^t\}$, $\{\omega_m,\omega_m\}$, onto the frequency interval $[\omega_1,\omega_m]$. The function R does a similar interpolation for data pairs $\{\omega_1,r_1\}$, $\{\omega_1{}^f,r_1{}^f-r_1{}^t\}$, $\ldots$, $\{\omega_K{}^f,r_K{}^f-r_K{}^t\}$, $\{\omega_m,\omega r_m\}$; here $r_1=R_c(\boldsymbol{x},\omega_1)-R_c(\boldsymbol{x}^r,\omega_1)$ and $r_m=c(\boldsymbol{x},\omega_m)-c(\boldsymbol{x}^r,\omega_m)$. In other words, the function F translates the frequency components of the characteristic points of $f(\boldsymbol{x}^{(i)})$ to the frequencies at which they should be located according to the translation vectors t_j, while the function R adds the necessary magnitude component. The interpolation onto $[\omega_1,\omega_m]$ is necessary because the original frequency sweep is a discrete set.

Formally, both the translation vectors t_j and their components should have an additional index (i) indicating that they are determined at iteration i of the optimization algorithm (2), however, this was omitted for the sake of simplicity.

Figure 2 shows the plots of the functions F and R corresponding to the fine/coarse model response plots of Fig. 1. The interpolation of $\{f(\boldsymbol{x},\omega_1), \ldots, f(\boldsymbol{x},\omega_m)\}$, F, and R is implemented using cubic splines.

As follows from its formulation, SPRP is developed assuming that the frequency components of the translation vectors are zero at the edges of the frequency spectrum (i.e., at ω_1 and ω_m). This limitation can be easily overcome either by extending the frequency range of the covarse model and applying extrapolation (cf. [15]). Also, it is assumed that the overall shape of both the fine and coarse model response is similar. This means, in particular, that the characteristic points of responses of both the coarse model c and the fine model f are in one-to-one correspondence. If this assumption is not satisfied, the surrogate model (3), (4) cannot be evaluated because the translation vectors t_i are not well defined. Generalizations of SPRP that allow alleviating this difficulty in some cases can be found in [15].

3 SPRP for Microwave Design Optimization

In this section, we demonstrate the use of SPRP for the design optimization of microwave components. Consider the dual-band bandpass filter [21] (Fig. 3a). The design parameters are $\boldsymbol{x} = [L_1\ L_2\ S_1\ S_2\ S_3\ d\ g\ W]^T$ mm. The fine model is simulated in Sonnet ***em*** [22]. The design specifications are $|S_{21}| \geq -3$ dB for $0.85\ \text{GHz} \leq \omega \leq 0.95$ GHz and $1.75\ \text{GHz} \leq \omega \leq 1.85$ GHz, and $|S_{21}| \leq -20$ dB for $0.5\ \text{GHz} \leq \omega \leq 0.7$ GHz, $1.1\ \text{GHz} \leq \omega \leq 1.6$ GHz, and $2.0\ \text{GHz} \leq \omega \leq 2.2$ GHz. The coarse model is implemented in Agilent ADS [23] (Fig. 3b). The initial design is $\boldsymbol{x}^{(0)} = [16.14\ 17.28\ 1.16\ 0.38\ 1.18\ 0.98\ 0.98\ 0.20]^T$ mm (the optimal solution of c). The following characteristic points are selected to set up functions F and R: four points for which $|S_{21}| = -20$ dB, four points with $|S_{21}| = -5$ dB, as well as six additional points located between -5 dB points. For the purpose of optimization, the coarse model was enhanced by tuning the dielectric constants and the substrate heights of the microstrip models corresponding to the design variables L_1, L_2, d, and g (original values of ε_r and H were 10.2 and 0.635 mm, respectively) [15]. The filter was optimized using two versions of SPRP, a regular one and SPRP enhanced by input SM (cf. Table 1). Figure 4 shows the initial fine model response as well as the fine model response at the design obtained using the SPRP method.

As the second example, consider the third-order Chebyshev bandpass filter [29] shown in Fig. 5. The design parameters are $\boldsymbol{x} = [L_1\ L_2\ S_1\ S_2]^T$ mm; $W_1 = W_2 = 0.4$ mm. The fine model is simulated in Sonnet ***em*** [22]. The design specifications are $|S_{21}| \geq -3$ dB for $1.8\ \text{GHz} \leq \omega \leq 2.2$ GHz, and $|S_{21}| \leq -20$ dB

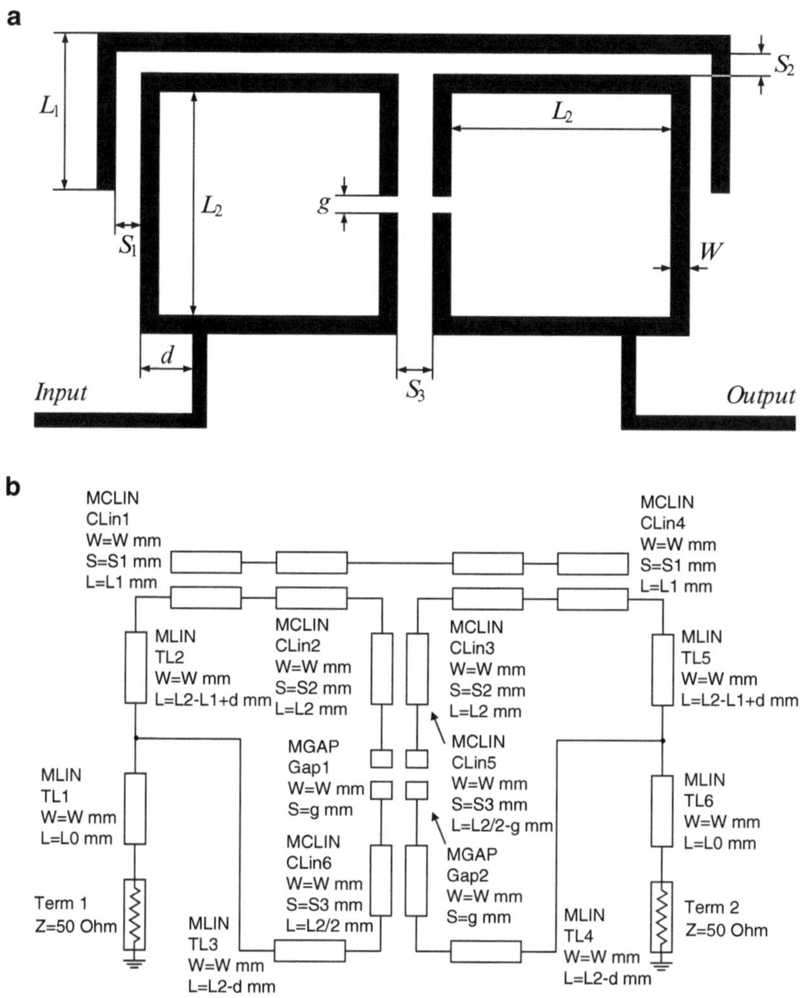

Fig. 3 Dual-band bandpass filter: (**a**) geometry [21], (**b**) coarse model (Agilent ADS)

for 1.0 GHz $\leq \omega \leq$ 1.6 GHz and 2.4 GHz $\leq \omega \leq$ 3.0 GHz. The coarse model is implemented in Agilent ADS [23] (Fig. 6). The initial design is $\boldsymbol{x}^{(0)} = [14.6\ 15.3\ 0.56\ 0.53]^T$ mm (the optimal solution of the coarse model c). The following characteristic points are selected to set up functions F and R: two points for which $|S_{21}| = -30$ dB, two points with $|S_{21}| = -20$ dB, two points with $|S_{21}| = -6$ dB, as well as ten additional points located between -6 dB points. Figure 7 shows the initial fine model response as well as the fine model response at the design obtained using SPRP. The numerical results including the design cost are presented in Table 2.

Table 1 Optimization results for dual-band bandpass filter

Algorithm	Final specification error (dB)	Number of fine model evaluations[a]
Shape-preserving response prediction	−2.0[b]	3
Shape-preserving response prediction + ISM[c]	−1.9[d]	2

[a]Excludes the fine model evaluation at the starting point
[b]Design specifications satisfied after the first iteration (spec. error −1.2 dB)
[c]The surrogate model is of the form $s^{(i)}(\boldsymbol{x}) = c(\boldsymbol{x} + \boldsymbol{c}^{(i)})$; $\boldsymbol{c}^{(i)}$ is found using parameter extraction [9]
[d]Design specifications satisfied after the first iteration (spec. error −1.0 dB)

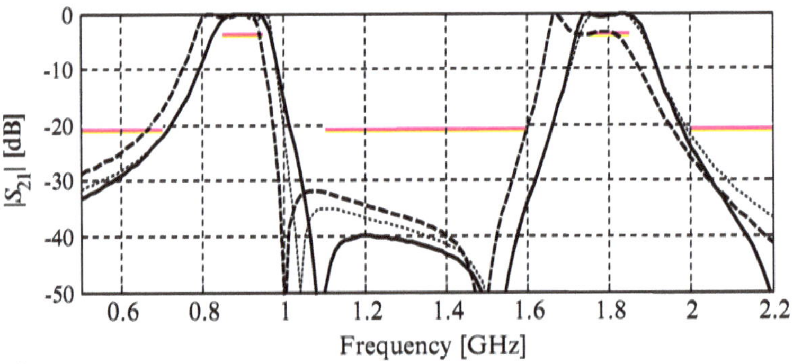

Fig. 4 Dual-band bandpass filter: fine model (*dashed line*) and coarse model (*thin dashed line*) response at $\boldsymbol{x}^{(0)}$, and the optimized fine model response (*solid line*) at the design obtained using shape-preserving response prediction

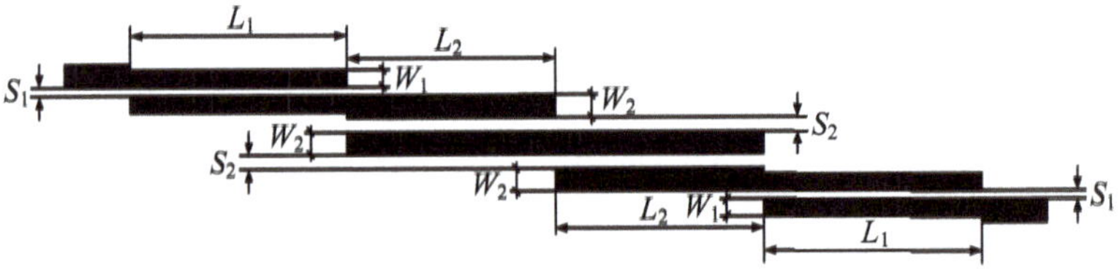

Fig. 5 Third-order Chebyshev bandpass filter: geometry [29]

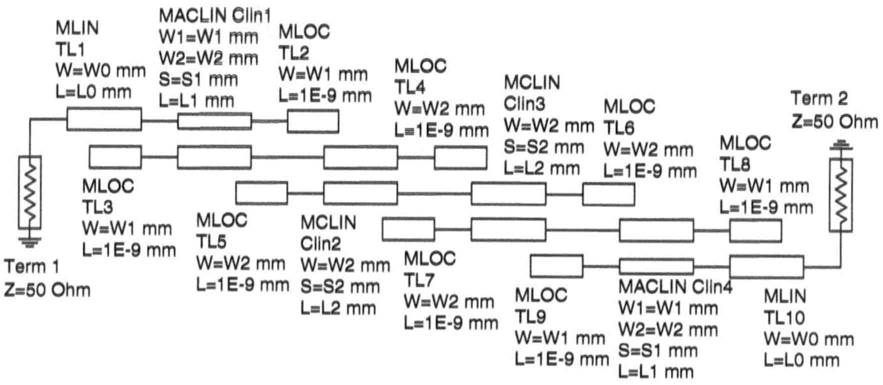

Fig. 6 Third-order Chebyshev filter: coarse model (Agilent ADS)

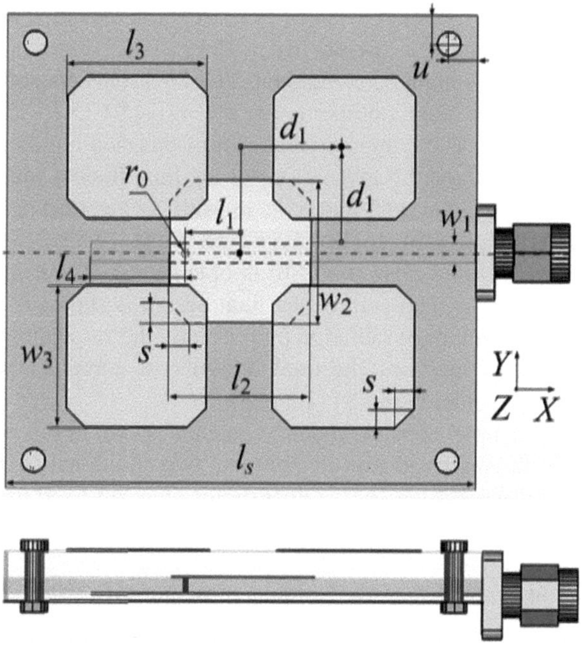

Fig. 7 Wideband microstrip antenna [24]: *top* and *side* views. The *dash-dot line* in the *top* view shows the magnetic symmetry wall (XOY)

Table 2 Optimization results for third-order Chebyshev filter

Algorithm	Final specification error (dB)	Number of fine model evaluations[a]
Shape-preserving response prediction	−1.8	2

[a]Excludes the fine model evaluation at the starting point

4 SPRP for Antenna Design

In this section, we illustrate the use of SPRP for the design of antenna structures. As an example, consider an antenna shown in Fig. 7 [24], where $\boldsymbol{x} = [l_1\ l_2\ l_3\ l_4\ w_2\ w_3\ d_1\ s]^T$ are the design variables. Multilayer substrate is $l_s \times l_s$ ($l_s = 30$ mm). The antenna stack (bottom-to-top) comprises: metal ground, 0.813 mm thick RO4003, microstrip trace ($w_1 = 1.1$ mm), 1.905 mm thick RO3006 and a trace-to-patch via ($r_0 = 0.25$ mm), driven patch, 3.048 mm thick RO4003, and four patches at the top. The antenna stack is fixed with four M1.6 bolts at the corners ($u = 3$ mm). Metallization is with thick 50 μm copper. Feeding is through an edge mount 50Ω SMA connector with the 10 × 10 × 2 mm flange.

The design objective is $|S_{11}| \le -10$ dB for 3.1–4.8 GHz. Realized gain not less than 5 dB for the zero zenith angle is an optimization constrain over the frequency band. The initial design is $\boldsymbol{x}^{init} = [-4\ 15\ 15\ 2\ 15\ 15\ 20\ 2]^T$ mm.

Both the high-fidelity model f (2,334,312 mesh cells at the initial design, 160 min of the evaluation time) and the low-fidelity model c (122,713 mesh cells, 3 min of the evaluation time) are simulated using the CST MWS transient solver [25]. Here, the first step is to find the rough optimum of c, $\boldsymbol{x}^{(0)} = [-4.91\ 15.15\ 15.07\ 2.56\ 14.21\ 14.23\ 21.07\ 2.67]^T$ mm. The computational cost of this step is 82 evaluations of c (which corresponds to about 1.5 evaluations of the high-fidelity model). Figure 8a shows the responses of f at $\boldsymbol{x}^{init}$ and $\boldsymbol{x}^{(0)}$, as well as the response of c at $\boldsymbol{x}^{(0)}$. The final design $\boldsymbol{x}^{(4)} = [-5.21\ 15.38\ 15.57\ 2.58\ 14.41\ 13.73\ 21.07\ 2.067]^T$ mm ($|S_{11}| \le -11$ dB for 3.1–4.8 GHz, Fig. 8b) is obtained after four iterations of the SPRP-based optimization. The gain of the final design is shown in Fig. 8c which illustrates that the maximum of radiation points along the zero zenith angle closely over the bandwidth of interest. The total design cost corresponds to about ten evaluations of the high-fidelity model (Table 3).

As the second example, consider a planar antenna shown in Fig. 9. It consists of a planar dipole as the main radiator element and two additional strips. The design variables are $\boldsymbol{x} = [l_0\ w_0\ a_0\ l_p\ w_p\ s_0]^T$. Other dimensions are fixed to: $a_1 = 0.5$ mm, $w_1 = 0.5$ mm, $l_s = 50$ mm, $w_s = 40$ mm, and $h = 1.58$ mm. Substrate material is Rogers RT5880 [30].

The high-fidelity model f of the antenna structure (10,250,412 mesh cells at the initial design, evaluation time of 44 min) is simulated using the CST MWS transient solver. The design objective is to obtain $|S_{11}| \le -12$ dB for 3.1–10.6 GHz. The initial design is $\boldsymbol{x}^{init} = [20\ 10\ 1\ 10\ 8\ 2]^T$ mm. The low-fidelity model c is also evaluated in CST but with coarser discretization (108,732 cells at $\boldsymbol{x}^{init}$, evaluated in 43 s). For this example, the approximate optimum of c, $\boldsymbol{x}^{(0)} = [18.66\ 12.98\ 0.526\ 13.717\ 8.00\ 1.094]^T$ mm, is found as the first design step. The computational cost is 127 evaluations of c, and it corresponds to about two evaluations of f. Figure 10a shows the reflection responses of $\boldsymbol{R}_f$ at both $\boldsymbol{x}^{init}$ and $\boldsymbol{x}^{(0)}$, as well as the response of c at $\boldsymbol{x}^{(0)}$. The final design $\boldsymbol{x}^{(2)} = [19.06\ 12.98\ 0.426\ 13.52\ 6.80$

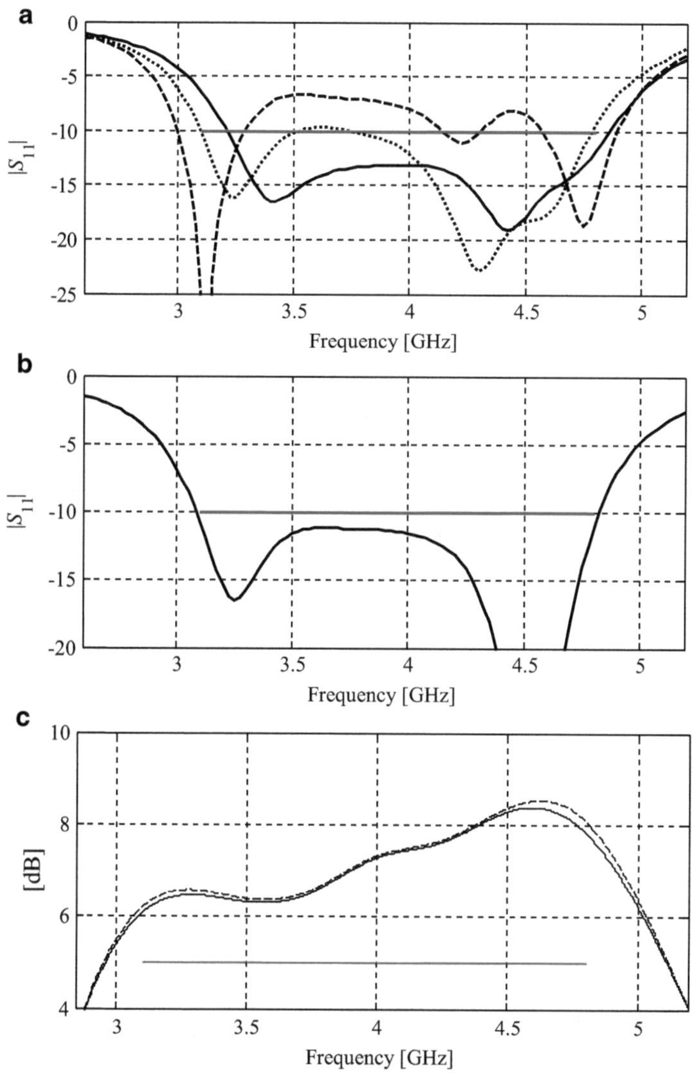

Fig. 8 Wideband microstrip antenna: (**a**) high-fidelity model response (*dashed line*) at the initial design x^{init}, and high- (*solid line*) and low-fidelity (*dotted line*) model responses at the approximate low-fidelity model optimum $x^{(0)}$; (**b**) high-fidelity model $|S_{11}|$ at the final design; (**c**) realized gain at the final design for the zero zenith angle (*solid line*, XOZ co-pol.) and realized peak gain (*dash line*). Design constrain is shown with the *horizontal line* at the 5 dB level

$1.094]^T$ mm ($|S_{11}| \leq -13.5$ dB for 3.1–10.6 GHz) is obtained after two iterations of the SPRP-based optimization with the total cost corresponding to about seven evaluations of the high-fidelity model (see Table 4). Figure 10b shows the reflection response and Fig. 11 shows the gain response of the final design $x^{(2)}$.

Table 3 Wideband microstrip antenna: optimization cost

		Evaluation time	
Algorithm component	Number of model evaluations	Absolute (h)	Relative to $\boldsymbol{R}_f$
Evaluation of $\boldsymbol{R}_{cd}$[a]	$289 \times \boldsymbol{R}_{cd}$	14.4	5.4
Evaluation of $\boldsymbol{R}_f$[b]	$5 \times \boldsymbol{R}_f$	13.3	5.0
Total optimization time	N/A	27.7	**10.4**

[a]Includes initial optimization of $\boldsymbol{R}_{cd}$ and optimization of SPRP surrogate
[b]Excludes evaluation of $\boldsymbol{R}_f$ at the initial design

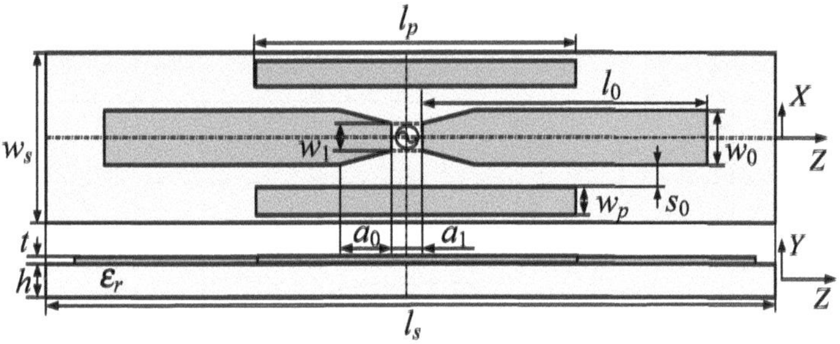

Fig. 9 UWB dipole antenna geometry: *top* and *side* views. The *dash-dot lines* show the electric (YOZ) and the magnetic (XOY) symmetry walls. The 50 Ω source impedance is not shown at the figure

5 SPRP for Aerodynamic Shape Optimization

The SPRP technique is illustrated here on aerodynamic design of airfoil sections at transonic flow conditions [17]. The airfoil shapes are parameterized with three parameters of the NACA four-digit method: m (the maximum ordinate of the mean camberline as a fraction of chord), p (the chordwise position of the maximum ordinate), and t/c (the thickness-to-chord ratio) [26]. The design variable vector is $\boldsymbol{x} = [m\ p\ t/c]^T$.

The airfoil performance is obtained through computational fluid dynamic (CFD) models which are implemented using the ICEM CFD [27] grid generator and the FLUENT [28] flow solver. The high-fidelity CFD model f is a two-dimensional steady-state Euler analysis with roughly 400,000 mesh cells and an overall simulation time around 67 min. The low-fidelity CFD model c is the same as the high-fidelity one, but with a coarser mesh (roughly 30,000 cells) and relaxed convergence criteria (100 flow solver iterations). The low-fidelity model is roughly 80 times faster than the high-fidelity one.

In aerodynamic shape optimization, the SPRP technique is applied to the pressure distribution ($C_p(\boldsymbol{x})$) on the airfoil surface [17]. Figure 12a shows the pressure distributions of two different designs obtained by the low-fidelity model. Shown are the characteristic points (red circles) and the translation vectors (blue lines) at important areas of the distributions. The application of the translation vectors to the high-fidelity model distributions is shown in Fig. 13b.

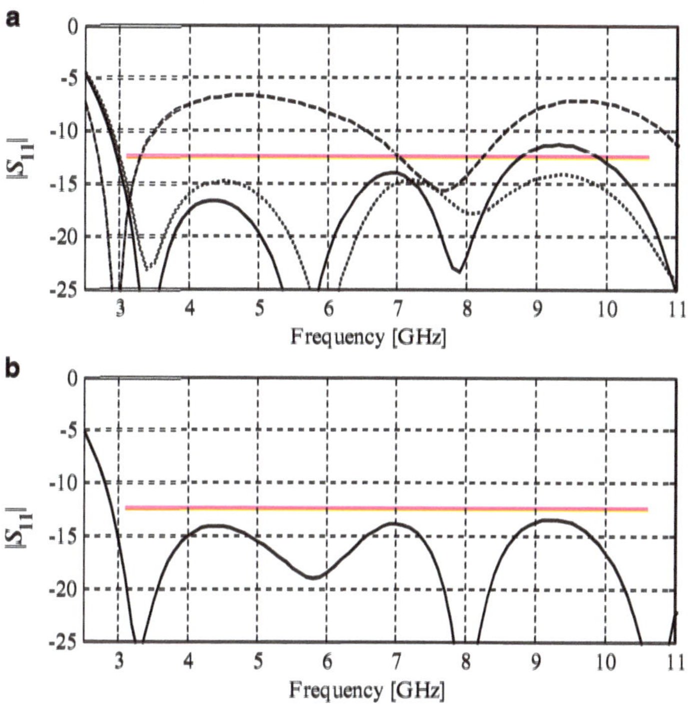

Fig. 10 UWB dipole antenna reflection response: (**a**) high-fidelity model response (*dashed line*) at the initial design $\boldsymbol{x}^{init}$, and high- (*solid line*) and low-fidelity (*dotted line*) model responses at the approximate low-fidelity model optimum $\boldsymbol{x}^{(0)}$; (**b**) high-fidelity model $|S_{11}|$ at the final design

Table 4 UWB dipole antenna: optimization cost

		Evaluation time	
Algorithm component	Number of model evaluations	Absolute (min)	Relative to $\boldsymbol{R}_f$
Evaluation of $\boldsymbol{R}_{cd}$[a]	$233 \times \boldsymbol{R}_{cd}$	167	3.8
Evaluation of $\boldsymbol{R}_f$[b]	$3 \times \boldsymbol{R}_f$	132	3.0
Total optimization time	N/A	299	**6.8**

[a]Includes initial optimization of $\boldsymbol{R}_{cd}$ and optimization of SPRP surrogate
[b]Excludes evaluation of $\boldsymbol{R}_f$ at the initial design

The design objective is to maximize the section lift coefficient ($C_l(\boldsymbol{x})$) subject to constraints on the section drag coefficient ($C_{dw}(\boldsymbol{x})$) and the non-dimensional cross-sectional area ($A(\boldsymbol{x})$). The problem is formulated as minimization of the high-fidelity model $f(\boldsymbol{x}) = -C_l(\boldsymbol{x})$ subject to $g_1(\boldsymbol{x}) = C_{dw}(\boldsymbol{x}) - C_{dw.max} \leq 0$, and $g_2(\boldsymbol{x}) = A_{min} - A(\boldsymbol{x}) \leq 0$, where $C_{dw.max} = 0.0041$ is the maximum drag and $A_{min} = 0.065$ the minimum cross-section. The free-stream Mach number is set $M_\infty = 0.75$ and the angle of attack $\alpha = 1°$. The design variable bounds are $0 \leq m \leq 0.1$, $0.2 \leq p \leq 0.8$, and $0.05 \leq t \leq 0.20$. The initial design is $\boldsymbol{x}^{init} = [0.03\ 0.2\ 0.1]^T$.

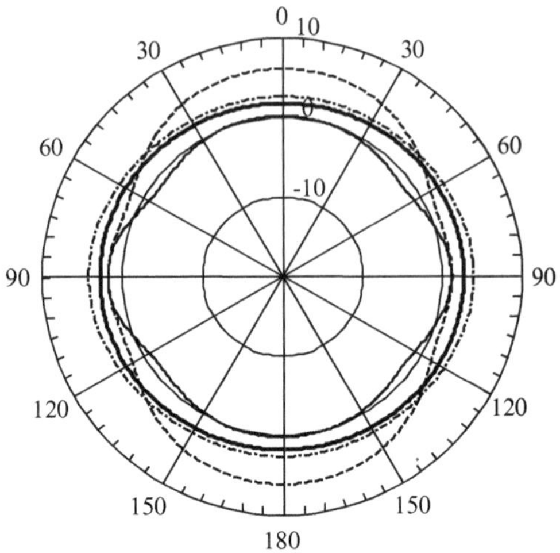

Fig. 11 UWB dipole antenna at the final design: IEEE gain pattern (x-pol.) in the XOY plane at 4 GHz (*thick solid*), 6 GHz (*dash-dot line*), 8 GHz (*dash line*), and 10 GHz (*solid line*)

Due to unavoidable misalignment between the pressure distributions of the high-fidelity model and its SPRP surrogate, it is not convenient to handle the drag constraint directly, because the design that is feasible for the surrogate model may not be feasible for the high-fidelity model. This problem is alleviated by implementing the drag constraint through a penalty function. More specifically, the objective function is defined as

$$H\left(C_p\left(\boldsymbol{x}\right)\right) = -C_{l.s}\left(C_p\left(\boldsymbol{x}\right)\right) + \beta\left[\Delta C_{dw.s}\left(C_p\left(\boldsymbol{x}\right)\right)\right]^2 \tag{5}$$

where $\Delta C_{dw.s} = 0$ if $C_{dw.s} \leq C_{dw.s.max}$ and $\Delta C_{dw.s} = C_{dw.s} - C_{dw.s.max}$ otherwise. The cross-sectional area constraint is handled directly. We use $\beta = 1{,}000$ in the numerical study. Here, the pressure distribution for the surrogate model is $C_p = C_{p.s}$, and for the high-fidelity model $C_p = C_{p.f}$. Also, $C_{l.s}$ and $C_{dw.s}$ denote the lift and drag coefficients for the surrogate.

The optimization problem is solved by the direct optimization of the high-fidelity model using the pattern-search algorithm, as well as by the SPRP algorithm. The results are presented in Table 5. It can be seen that both approaches are able to meet the design goals and produce similar optimized airfoil shapes. The direct approach requires 120 high-fidelity model evaluations (N_f). The SPRP algorithm requires 330 low-fidelity model evaluations (N_c) and 11 high-fidelity ones, yielding a total cost of less than 18 equivalent high-fidelity model evaluations.

To meet the design goals, the optimizer does three fundamental shape changes: (1) the maximum ordinate of the mean camber line (m) is reduced, (2) the location of the maximum ordinate of the mean camber line (p) is moved aft, thus increasing the trailing-edge camber, and (3) the thickness-to-chord ratio (t/c) is reduced. Shape

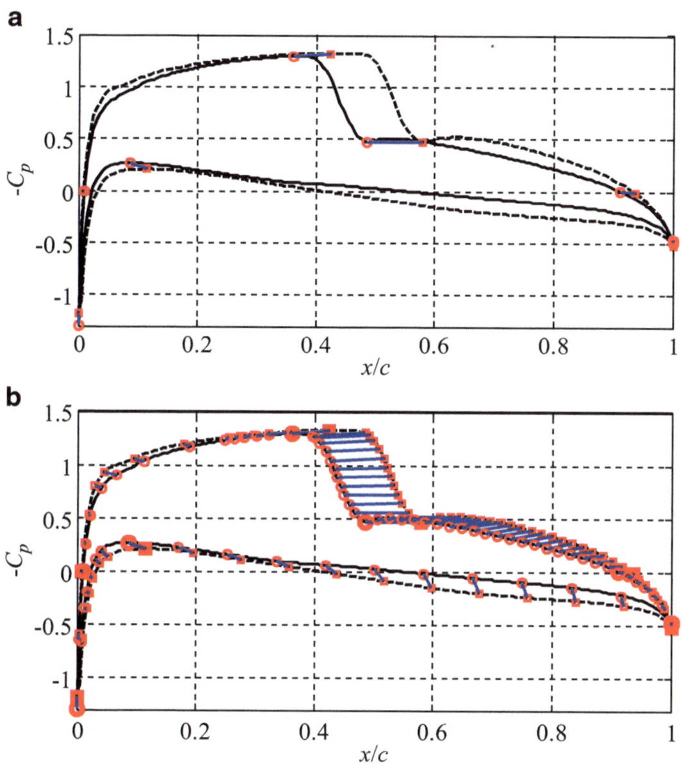

Fig. 12 An illustration of the SPRP technique applied to the pressure distributions obtained by the low-fidelity CFD models of two designs, (**a**) initial characteristic points and translation vectors, (**b**) additional points

changes (1) and (3) reduce the shock strength and, thus, reduce the drag coefficient. The associated change in the pressure distribution reduces the lift coefficient. However, shape change (2) improves (or recovers a part of) the lift by opening up the pressure distribution behind the shock. These effects can be seen in the pressure distribution plot in Fig. 14, and the Mach contour plots in Figs. 15 and 16.

6 Fast Surrogate Modeling Using SPRP

In this section, we illustrate the use of SPRP for modeling of microwave components. We consider two versions of SPRP surrogates: the basic one and the modified implementation that exploits multiple training points. Further discussion on the recent developments of SPRP models can be found in [31].

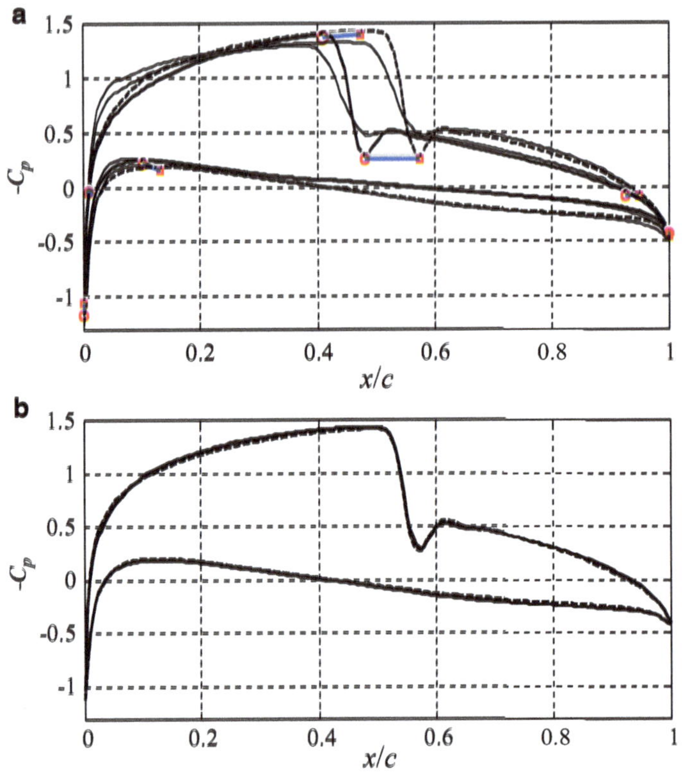

Fig. 13 Application of SPRP to the high-fidelity CFD model responses (*thick lines*) with (**a**) initial characteristic points and translation vectors (coarse model distributions are shown with *thin lines*), and (**b**) comparison of the actual and the predicted (*dash*) high-fidelity response

Table 5 Numerical results for the airfoil design optimization

Variable	Initial	Direct	SPRP
m	0.0300	0.0080	0.0090
p	0.2000	0.6859	0.6732
t/c	0.1000	0.1044	0.1010
C_l	0.8035	0.4641	0.4872
C_{dw}	0.0410	0.0041	0.0040
A	0.0675	0.0703	0.0680
N_c	N/A	0	330
N_f	N/A	120	11
Total cost	N/A	120	<18

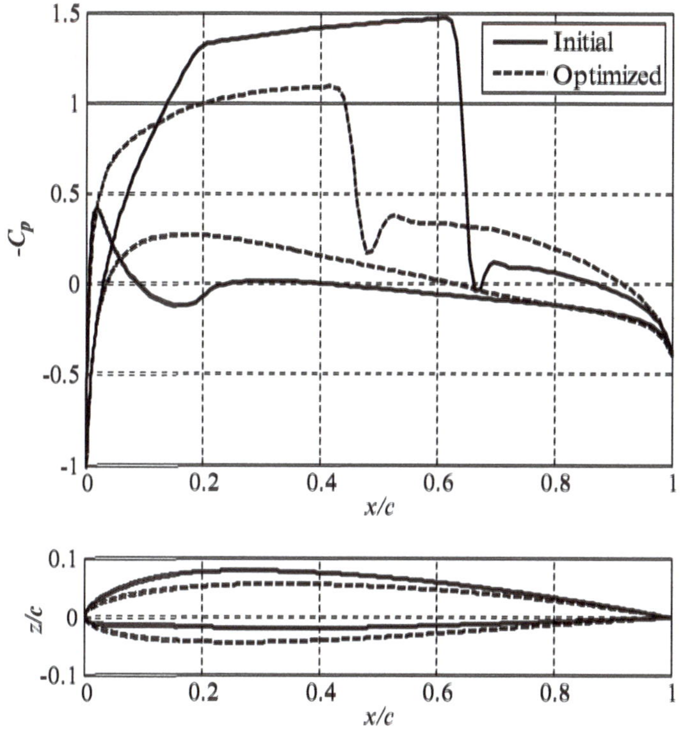

Fig. 14 Airfoil optimization results: initial and optimized airfoils pressure distributions and shapes

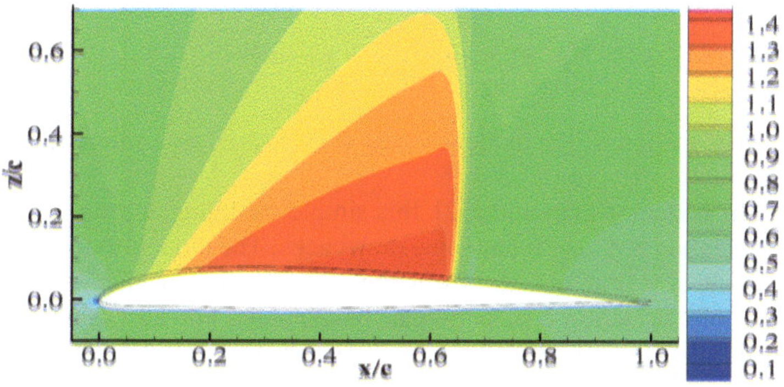

Fig. 15 Airfoil optimization results: Mach contours at the initial design

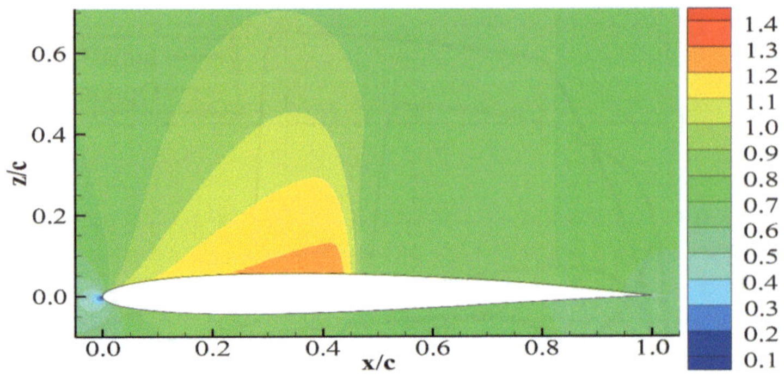

Fig. 16 Airfoil optimization results: Mach contours at the optimized design

6.1 SPRP Modeling: Basic Version [32]

Let $X_R \subseteq X$ be the region of interest where we want the surrogate model to be valid. Typically, X_R is an n-dimensional interval in R^n with center at reference point $\boldsymbol{x}^0 = [x_{0.1} \ \ldots \ x_{0.n}]^T \in R^n$ and size $\boldsymbol{\delta} = [\delta_1 \ \ldots \ \delta_n]^T$. Let $X_B = \{\boldsymbol{x}^1, \boldsymbol{x}^2, \ \ldots, \boldsymbol{x}^N\} \subset X_R$ be the base set, such that the fine model response is known at all points $\boldsymbol{x}^j$, $j = 1$, 2, ..., N. Here, the base points are allocated using so-called star-distribution [33], which is a design of experiments traditionally used by space mapping.

The SPRP surrogate model is defined as follows:

$$s(\boldsymbol{x}) = S(\boldsymbol{x}, \boldsymbol{x}^r) \tag{6}$$

where $\boldsymbol{x}^r$ is the base point that is the closest to $\boldsymbol{x}$, i.e.,

$$\boldsymbol{x}^r = \arg \min_{\boldsymbol{y} \in X_B} ||\, \boldsymbol{x} - \boldsymbol{y} \,|| \tag{7}$$

whereas $S(\boldsymbol{x},\boldsymbol{x}^r)$ is the SPRP model created with $\boldsymbol{x}^r$ used as a reference design (cf. Sect. 2.3).

Although, as demonstrated in [32], this simple modeling approach proves to be more accurate than SM, and it has some drawbacks. The model (6), (7) utilizes only one base point at a time. As shown in Fig. 17a, the region of interest is divided into regions of "attractions" of particular base points. For all evaluation points $\boldsymbol{x}$ located in a given region of "attraction," the surrogate model (6) is determined using the same single base point as a reference design. Due to this, the surrogate does not utilize all available f-model data at a time. Also, the surrogate model is discontinuous at the borders of the areas of "attraction" because the solution to (6) is not unique at these points. This may cause some problems while using the surrogate for design optimization.

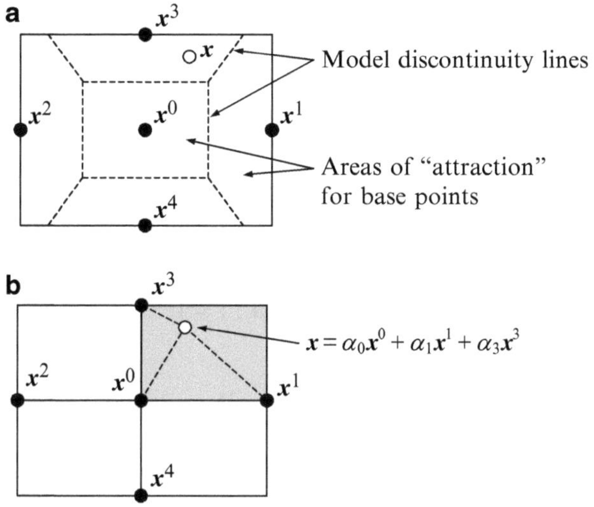

Fig. 17 SPRP modeling ($n = 2$): (**a**) Original: Star-distributed base points are denoted using *black circles*. The region of interest is divided into areas of "attraction" of particular base points, determined by the Euclidean distance. An example evaluation design $\boldsymbol{x}$ is close to the base design $\boldsymbol{x}^3$, and this point becomes a reference design for SPRP model; (**b**) Modified: Base points are denoted using *black circles*. A shaded area denotes a hypercube defined by a subset X_S of base points being the closest to an example evaluation design $\boldsymbol{x}$. The surrogate at $\boldsymbol{x}$ is defined as a linear combination of SPRP models using all base points from X_S as reference designs. Coefficients of this linear combination are calculated by representing $\boldsymbol{x}$ through all points from X_S

6.2 *Modified SPRP Modeling [34]*

Here, a modified SPRP modeling technique is proposed that utilizes multiple reference designs and solves the discontinuity problem described in the previous section. Again, the base set is assumed to be allocated using star-distribution [33]; however, the model can also be formulated for more general setups.

The concept of the SPRP model exploiting multiple reference designs is explained in Fig. 17b. For an evaluation point $\boldsymbol{x}$, we find a subset X_S of the base set X_B that defines a rectangular area (hypercube) of the region of interest containing $\boldsymbol{x}$. The surrogate model is set up using all points from X_S. The star-distribution base set contains $N = 2n + 1$ points as illustrated in Fig. 17a for $n = 2$. Without loss of generality, we can assume that $X_S = \{\boldsymbol{x}^0, \boldsymbol{x}^1, \ldots, \boldsymbol{x}^n\}$. We have

$$\boldsymbol{x} = \boldsymbol{x}^0 + \beta_1 \boldsymbol{v}_1 + \beta_2 \boldsymbol{v}_2 + \cdots + \beta_n \boldsymbol{v}_n \tag{8}$$

where $\beta_1, \ldots, \beta_n$ determines a unique representation of $\boldsymbol{x} - \boldsymbol{x}^0$ using vectors $\boldsymbol{v}_i = \boldsymbol{x}^i - \boldsymbol{x}^0$, $i = 1, \ldots, n$. Coefficients β_i can be found as

$$\begin{bmatrix} \beta_1 \\ \beta_2 \\ \\ \beta_n \end{bmatrix} = [v_1 \ \ v_2 \ \ \ldots \ \ v_n]^{-1} \cdot \left(x - x^0\right) \tag{9}$$

The vector x can be the unique represented as

$$x = \alpha_0 x^0 + \alpha_1 x^1 + \alpha_2 x^2 + \cdots + \alpha_n x^n \tag{10}$$

where $\alpha_0 = 1 - (\alpha_1 + \ldots + \alpha_n)$, and $\alpha_i = \beta_i$, $i = 1, \ldots, n$. The modified SPRP surrogate model is then defined as

$$\hat{s}(x) = \alpha_0 S\left(x, x^0\right) + \alpha_1 S\left(x, x^1\right) + \cdots + \alpha_n S\left(x, x^n\right) \tag{11}$$

with $S(x, x^i)$, $i = 0, 1, \ldots, n$, being the SPRP models (1) determined using respective reference designs.

It can be verified that the model (11) is continuous with respect to x provided that both f and c are continuous functions of x. Also, it is expected to be more accurate than the model (6), (7) because it exploits the available fine model data in a more comprehensive way.

6.3 *Verification: Fourth-Order Ring Resonator Bandpass Filter [35]*

In this section we illustrate the use of SPRP for modeling of a microwave filter. We also compare both basic and modified SPRP with surrogate modeling using standard space mapping [33]. The standard SM model is quite involved because it is using input and output SM of the form $A \cdot c(B \cdot x + c)$, enhanced by the implicit and frequency space mapping [33]. All surrogate models are set up using the same base set consisting of $N = 2n + 1$ points allocated according to the star-distribution [33]. The quality of the models is assessed using a relative error measure $||f(x) - s(x)||/||f(x)||$ expressed in percent.

Consider the fourth-order ring resonator bandpass filter [35] (Fig. 18a). The design parameters are $x = [L_1\ L_2\ L_3\ S_1\ S_2]^T$ mm. The fine model f is simulated in FEKO [36]. The coarse model, Fig. 18b, is implemented in Agilent ADS [23]. The region of interest is defined by the reference point $x^0 = [24.0\ 21.0\ 26.0\ 0.2\ 0.1]^T$ mm, and the region size $\delta = [2.0\ 2.0\ 2.0\ 0.1\ 0.05]^T$ mm.

The modeling accuracy has been verified using 50 random test points. The results shown in Table 6 and in Fig. 19 indicate that the modified SPRP model ensures better accuracy than both the standard SM model and the original version of SPRP [32].

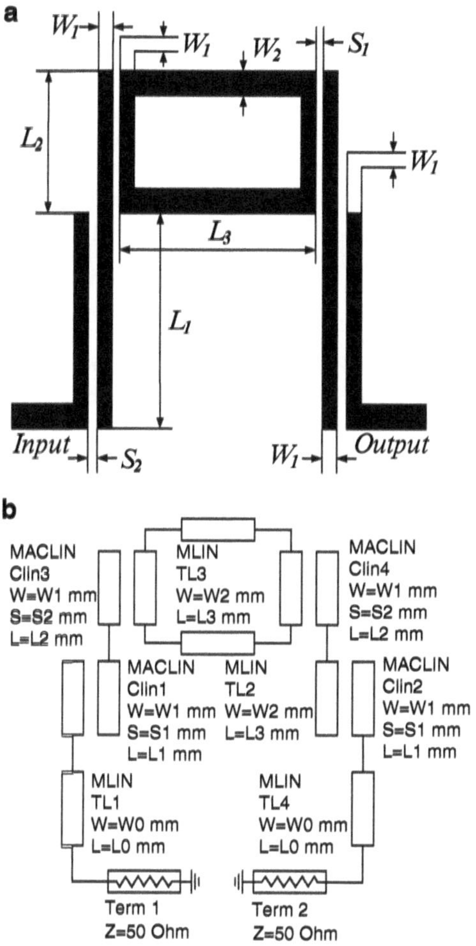

Fig. 18 Fourth-order ring resonator bandpass filter: (**a**) geometry [35], (**b**) coarse model (Agilent ADS)

Table 6 Fourth-order ring resonator filter: modeling results

Model	Average error (%)	Maximum error (%)
SM	1.8	4.5
SPRP (Basic version [32])	1.1	2.7
SPRP (Modified version)	0.3	0.6

As an application example, the modified SPRP surrogate was utilized to optimize the filter with respect to the following design specifications: $|S_{21}| \geq -1$ dB for 1.75 GHz $\leq \omega \leq$ 2.25 GHz, and $|S_{21}| \leq -20$ dB for 1.0 GHz $\leq \omega \leq$ 1.5 GHz and 2.5 GHz $\leq \omega \leq$ 3.0 GHz. The initial design was $\boldsymbol{x}^0 = [24.0\ 21.0\ 26.0\ 0.2\ 0.1]^T$ mm.

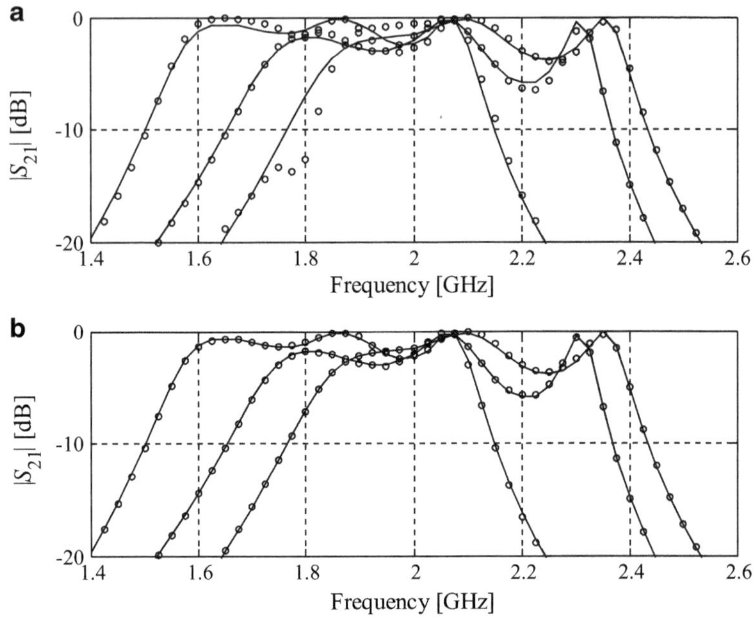

Fig. 19 Fourth-order ring resonator bandpass filter: fine model (*solid line*) and surrogate model (*circles*) responses at three selected test points for: (**a**) standard SM model, (**b**) modified SPRP surrogate model

Figure 20a shows the fine model response of the filter at the initial design and at the design $\boldsymbol{x}^* = [22.61\ 20.11\ 26.626\ 0.156\ 0.040]^T$ mm obtained by optimizing the surrogate. The specification error at the optimized design is −0.45 dB.

The SPRP model was also used to estimate yield at the optimized design, assuming 0.2 mm deviation for length parameters (L_1 to L_2) and 0.02 mm for spacing parameters (S_1 and S_2). The yield estimation based on 200 random samples is 68 % (Fig. 9b). This value is very close to the yield estimated directly using the fine model (70 %). The estimation performed with the SM model is less accurate (50 %). Note that the total computational cost of building the surrogate model, design closure, and statistical analysis is only 11 full-wave simulations of the filter structure!

7 Conclusion

A review of SPRP and its applications to solving simulation-driven design problems in various engineering disciplined has been presented. SPRP exploits the knowledge embedded in the low-fidelity model of the structure under consideration in order to predict the response of the expensive high-fidelity model. As a result, SPRP is capable of yielding a satisfactory design at a low computational cost as demonstrated

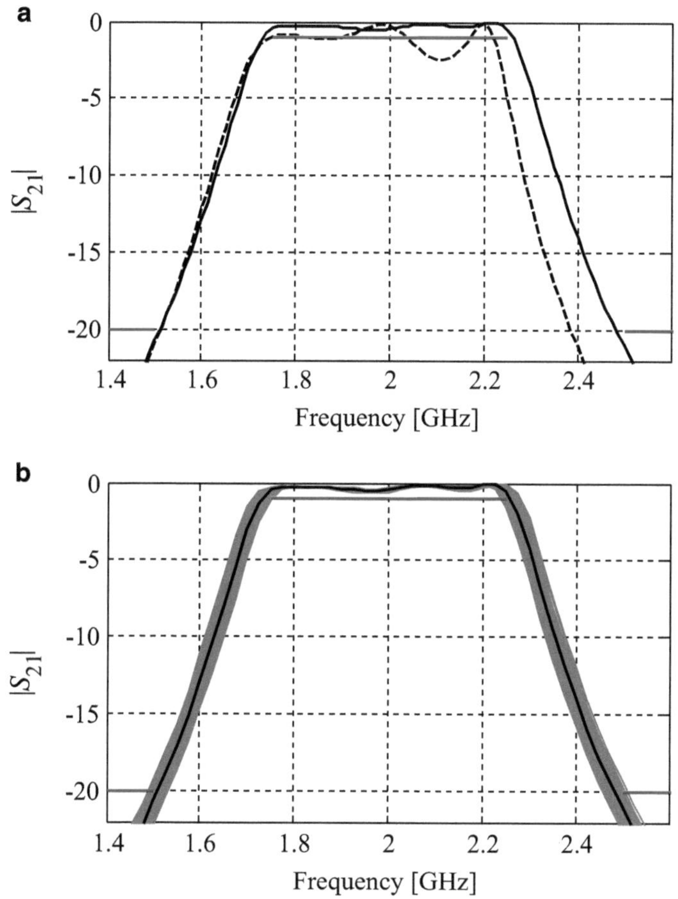

Fig. 20 Fourth-order ring resonator bandpass filter: (**a**) fine model responses at the reference point $\boldsymbol{x}^0$ (*dashed line*) and at the optimal solution $\boldsymbol{x}^*$ of the modified shape-preserving response prediction surrogate model (*solid line*); (**b**) statistical analysis at $\boldsymbol{x}^*$ using the modified shape-preserving response prediction model. Estimated yield is 68 %. *Thick black solid line* denotes the fine model response at optimal design $\boldsymbol{x}^*$

using several examples involving design problems in electrical and mechanical engineering. As indicated in Sect. 5, SPRP can also be used to construct accurate global or quasi-global surrogate models. SPRP is a relatively novel technique that is still under development. Recent papers provide various enhancement of the technique in the context of both optimization (e.g., [37]) and modeling (e.g., [31]). It should also be mentioned that a potential limitation of SPRP is the fact that one-to-one correspondence of all the model (both low- and high-fidelity ones) responses involved in the process of creating the surrogate model is an important prerequisite for the technique to work. Various ways of ensuring such a correspondence can be found in the literature (e.g., [15]).

References

1. Nocedal, J., Wright, S.: Numerical Optimization, 2nd edn. Springer, New York (2006)
2. Conn, A.R., Scheinberg, K., Vicente, L.N.: Introduction to Derivative-Free Optimization. MPS-SIAM Series on Optimization, MPS-SIAM (2009)
3. El Sabbagh, M.A., Bakr, M.H., Nikolova, N.K.: Sensitivity analysis of the scattering parameters of microwave filters using the adjoint network method. Int. J. RF Microw. Comput. Aided Eng. **16**, 596–606 (2006)
4. Koziel, S., Echeverría-Ciaurri, D., Leifsson, L.: Surrogate-based methods. In: Koziel, S., Yang, X.S. (eds.) Computational Optimization, Methods and Algorithms, Series: Studies in Computational Intelligence, pp. 33–60. Springer, New York (2011)
5. Forrester, A.I.J., Keane, A.J.: Recent advances in surrogate-based optimization. Prog. Aerosp. Sci. **45**, 50–79 (2009)
6. Simpson, T.W., Peplinski, J., Koch, P.N., Allen, J.K.: Metamodels for computer-based engineering design: survey and recommendations. Eng. Comput. **17**, 129–150 (2001)
7. Queipo, N.V., Haftka, R.T., Shyy, W., Goel, T., Vaidynathan, R., Tucker, P.K.: Surrogate-based analysis and optimization. Prog. Aerosp. Sci. **41**, 1–28 (2005)
8. Smola, A.J., Schölkopf, B.: A tutorial on support vector regression. Stat. Comput. **14**, 199–222 (2004)
9. Rayas-Sánchez, J.E.: EM-based optimization of microwave circuits using artificial neural networks: the state-of-the-art. IEEE Trans. Microw. Theory Tech. **52**, 420–435 (2004)
10. Alexandrov, N.M., Dennis, J.E., Lewis, R.M., Torczon, V.: A trust region framework for managing use of approximation models in optimization. Struct. Multidiscip. Optim. **15**(1), 16–23 (1998)
11. Cheng, Q.S., Bandler, J.W., Koziel, S., Bakr, M.H., Ogurtsov, S.: The state of the art of microwave CAD: EM-based optimization and modeling. Int. J. RF Microw. Comput. Aided Eng. **20**, 475–491 (2010)
12. Echeverria, D., Hemker, P.W.: Space mapping and defect correction. CMAM Int. Math. J. Comput. Methods Appl. Math. **5**(2), 107–136 (2005)
13. Rautio, J.C.: Perfectly calibrated internal ports in EM analysis of planar circuits. In: IEEE MTT-S Int. Microwave Symp. Dig., Atlanta, pp. 1373–1376 (2008)
14. Conn, A.R., Gould, N.I.M., Toint, P.L.: Trust Region Methods, MPS-SIAM Series on Optimization (2000)
15. Koziel, S.: Shape-preserving response prediction for microwave design optimization. IEEE Trans. Microw. Theory Tech. **58**, 2829–2837 (2010)
16. Koziel, S., Ogurtsov, S., Szczepanski, S.: Rapid antenna design optimization using shape-preserving response prediction. Bull. Pol. Acad. Sci. Technical Sci. **60**, 143–149 (2012)
17. Koziel, S., Leifsson, L.: Transonic airfoil shape optimization using variable-resolution models and pressure distribution alignment. In: AIAA Applied Aerodynamic Conference, Honolulu, 27–30 June 2011, AIAA-2011-3177
18. Bandler, J.W., Cheng, Q.S., Dakroury, S.A., Mohamed, A.S., Bakr, M.H., Madsen, K., Sondergaard, J.: Space mapping: the state of the art. IEEE Trans. Microw. Theory Tech. **52**, 337–361 (2004)
19. Koziel, S., Bandler, J.W., Cheng, Q.S.: Robust trust-region space-mapping algorithms for microwave design optimization. IEEE Trans. Microw. Theory Tech. **58**, 2166–2174 (2010)
20. Booker, A.J., Dennis Jr., J.E., Frank, P.D., Serafini, D.B., Torczon, V., Trosset, M.W.: A rigorous framework for optimization of expensive functions by surrogates. Struct. Optim. **17**, 1–13 (1999)
21. Guan, X., Ma, Z., Cai, P., Anada, T., Hagiwara, G.: A microstrip dual-band bandpass filter with reduced size and improved stopband characteristics. Microw. Opt. Tech. Lett. **50**, 618–620 (2008)
22. ***em***™ Version 12.54, Sonnet Software, Inc., 100 Elwood Davis Road, North Syracuse, NY 13212, USA, 2010

23. Agilent ADS, Version 2011, Agilent Technologies, 1400 Fountaingrove Parkway, Santa Rosa, CA 95403–1799 (2011)
24. Chen, Z.N.: Wideband microstrip antennas with sandwich substrate. IET Microw. Ant. Prop. **2**, 538–546 (2008)
25. CST Microwave Studio, CST AG, Bad Nauheimer Str. 19, D-64289 Darmstadt, Germany (2011)
26. Abbott, I.H., Von Doenhoff, A.E.: Theory of Wing Sections. Dover Publications, New York (1959)
27. ICEM CFD, ver. 14, ANSYS Inc., Southpointe, 275 Technology Drive, Canonsburg, PA 15317 (2011)
28. FLUENT, ver. 14, ANSYS Inc., Southpointe, 275 Technology Drive, Canonsburg, PA 15317 (2011)
29. Kuo, J.T., Chen, S.P., Jiang, M.: Parallel-coupled microstrip filters with over-coupled end stages for suppression of spurious responses. IEEE Microw. Wirel. Comput. Lett. **13**, 440–442 (2003)
30. RT/duroid® 5870/5880 High Frequency Laminates, Data Sheet, Rogers Corporation, Publication #92-101 (2010)
31. Koziel, S., Leifsson, L.: Generalized shape-preserving response prediction for accurate modeling of microwave structures. IET Microw. Ant. Prop. **6**, 1332–1339 (2012)
32. Koziel, S.: Shape-preserving response prediction for microwave circuit modeling. In: IEEE MTT-S Int. Microw. Symp. Dig, Anaheim, pp. 1660–1663 (2010)
33. Bandler, J.W., Cheng, Q.S., Koziel, S.: Simplified space mapping approach to enhancement of microwave device models. Int. J. RF Microw. Comput. Aided Eng. **16**, 518–535 (2006)
34. Koziel, S., Szczepanski, S.: Accurate modeling of microwave structures using shape-preserving response prediction. IET Microw. Antennas Propag. **5**, 1116–1122 (2011)
35. Salleh, M.K.M., Pringent, G., Pigaglio, O., Crampagne, R.: Quarter-wavelength side-coupled ring resonator for bandpass filters. IEEE Trans. Microw. Theory Tech. **56**, 156–162 (2008)
36. FEKO® *User's Manual*, Suite 5.4, 2008, EM Software & Systems-S.A. (Pty) Ltd., 32 Techno Lane, Technopark, Stellenbosch, 7600, South Africa
37. Koziel, S., Ogurtsov, S., Cheng, Q.S., Bandler, J.W.: Rapid electromagnetic-based microwave design optimisation exploiting shape-preserving response prediction and adjoint sensititivies. IET Microwaves Antennas Prop. **8**, 775–781 (2014)

Nested Space Mapping Technique for Design and Optimization of Complex Microwave Structures with Enhanced Functionality

Slawomir Koziel, Adrian Bekasiewicz, and Piotr Kurgan

Abstract In this work, we discuss a robust simulation-driven methodology for rapid and reliable design of complex microwave/RF circuits with enhanced functionality. Our approach exploits nested space mapping (NSM) technology, which is dedicated to expedite simulation-driven design optimization of computationally demanding microwave structures with complex topologies. The enhanced functionality of the developed circuits is achieved by means of slow-wave resonant structures (SWRSs), used as replacement components for conventional transmission lines. The NSM is a hierarchical, bottom-up methodology, in which the inner space mapping layer is applied to improve generalization capabilities of the equivalent circuit constructed on the SWRS level, whereas the outer layer is used to enhance the surrogate model of the entire structure of interest. We demonstrate that the NSM significantly improves the performance of traditional surrogate-based optimization routines applied to the design problem of computationally expensive microwave/RF structures with modular topology. The proposed technique is used to design three exemplary microwave/RF circuits with enhanced functionality: two abbreviated microstrip matching transformers and a miniaturized rat-race coupler with harmonic suppression. We also provide a comprehensive comparison with other surrogate-assisted methods, as well as supply the reader with basic design guidelines for the state-of-the-art SWRS-based microwave/RF circuits.

Keywords Microwave/RF circuit • Surrogate model • Electromagnetic (EM) simulation • Nested space mapping • Surrogate-based optimization • Slow-wave resonant structure (SWRS) • Uniform transmission line • Two-level modeling • Microstrip circuit

S. Koziel (✉)
Engineering Optimization & Modeling Center, School of Science and Engineering, Reykjavik University, Menntavegur 1, 101 Reykjavik, Iceland
e-mail: koziel@ru.is

A. Bekasiewicz (✉) • P. Kurgan
Faculty of Electronics, Telecommunications and Informatics, Gdansk University of Technology, Narutowicza 11/12, 80-233 Gdansk, Poland
e-mail: adrian.bekasiewicz@pg.gda.pl

S. Koziel et al. (eds.), *Solving Computationally Expensive Engineering Problems*, Springer Proceedings in Mathematics & Statistics 97,
DOI 10.1007/978-3-319-08985-0_3

1 Introduction

Modern wireless communication systems impose stringent requirements upon microwave and radio-frequency (RF) blocks, placing particular emphasis on passive components. These commonly used circuits are required to satisfy strictly defined system specifications, e.g., multiband [1–3], or wideband operation [4–6], attenuation of harmonic frequencies [7–9], high isolation [10–12], etc. Moreover, physical dimensions of passive components are also regulated by available, often limited estate area [13–15]. In general, traditional theory-based design routines are incapable of providing reliable design solutions when circuit size and its performance are simultaneously taken into consideration [16, 17]. Among variety of techniques dedicated to enhance the functionality of conventional passive components [18–22], the modification of circuit's geometry by means of intentional perturbations, defects, or discontinuities—implemented in either metallization plane—has gained increased attention as the most promising method to perform a cost-efficient microwave/RF circuit refinement [23–25].

Although the implementation of various perturbations and discontinuities may be extremely beneficial from the perspective of the functionality of microwave/RF structures—both geometrical- and performance-wise—it simultaneously hinders the design process due to the increased number of designable parameters that have to be simultaneously adjusted to yield a proper operation of the circuit [26, 27]. A typical experience-based design approach using repetitive parameter sweeps is suitable for tuning only one parameter at a time and, therefore, its utilization is limited for multi-dimensional design spaces of microwave/RF circuits with complex topologies. Consequently, the design of microwave/RF structures with enhanced functionality is considered to be a multifaceted problem that may be addressed only by means of numerical optimization.

Reliable design optimization of highly miniaturized microwave/RF components is an extremely challenging issue of contemporary wireless communication engineering. The main reason for it is the lack of computationally cheap and accurate theoretical models representing the behavior of such unconventional structures. Unfortunately, a reliable performance evaluation of complex microwave/RF components, and—consequently—their design, can only be achieved through CPU-intensive electromagnetic (EM) simulations. As opposed to conventional microwave/RF circuits, EM models of sophisticated structures with enhanced functionality are, in general, computationally expensive, which is another crucial factor hindering the design process. Additionally, a large number of independent designable parameters involved in structure optimization significantly increases numerical complexity of the process, as well as the number of EM evaluations necessary to complete the optimization task. Hence, direct EM-based optimization using conventional gradient [28] or derivative-free [29] algorithms is normally prohibitive. On the other hand, techniques such as adjoint sensitivity [30, 31] allow for low-cost derivative evaluation, which may lead to substantial cost reduction of gradient-based search procedures [32, 33]. However, this technology is not yet widely available in commercial computer-aided design (CAD) software. Another important issue related to conventional optimization techniques is their local convergence properties, i.e., the ability to find only a local optimum (usually the one closest to the initial design). The aforementioned difficulty may be partially addressed by global optimization methods. In practice, this means resorting to population-based metaheuristics, which are even more expensive—computational-wise—than local-search algorithms [34, 35]. For that reason, the utilization of direct optimization techniques in the design and optimization of complex circuits is usually impractical.

High-computational cost related to the design of compact microwave/RF structures may be partially alleviated by means of surrogate-based optimization (SBO) techniques [36], including, among others, manifold mapping [37, 38], shape preserving response prediction [39, 40], or space mapping [41, 42]. The attractiveness of the SBO lies in its ability to iteratively correct/enhance a low-fidelity model using a limited amount of data acquired from simulations of a high-fidelity model [43]. SBO methods gained a considerable attention in diverse engineering fields [44–46] and proved to be very efficient design methodologies, capable of yielding desired solutions at the cost of only a few simulations of respective high-fidelity models. Space mapping—originated in the field of microwave technology—is particularly interesting in the context of numerically complex circuit design, especially due to its simple implementation [47, 48], high efficiency [43, 49], and successful validation on a variety of microwave structures [27, 43, 49]. On the other hand, SBO techniques, especially space mapping algorithms, are mostly used for the expedited design and optimization of conventional microwave/RF circuits [47, 49]. Although several methods regarding this concept have been proposed for optimization of complex structures [27, 50] they require inconvenient manual setup of multiple optimization problems and are problematic when large number of parameters is involved [51].

In this chapter, we provide general guidelines for the development of unconventional microwave/RF circuits with enhanced functionality. This is achieved through the decomposition of a conventional circuit into its elementary building blocks, more particularly uniform transmission lines (TLs) and their subsequent replacement with unconventional (e.g., shortened, dual-band, etc.) slow-wave structures. Moreover, challenges and benefits regarding the design of SWRS-based circuit are presented. Next, we introduce a nested space mapping (NSM) technology aimed at fast and accurate design of computationally expensive planar microwave/RF components. NSM constructs a two-stage low-fidelity model, with the inner space mapping layer applied at the level of the decomposed TL, and the outer space mapping layer applied for the entire circuit. The proposed technique mitigates the problem of surrogate model inaccuracy resulting from complex layouts of unconventional circuits and enables its rapid optimization in a single run of the algorithm.

The chapter is organized as follows. In Sect. 2, we discuss the concept of circuit functionality enhancements based on its decomposition and a subsequent refinement of its elementary building blocks using SWRSs. We also explain techniques for the construction of SWRSs and their influence on the behavior of the entire microwave/RF circuit. Sect. 3 is devoted to the problem of complex microwave circuits design using surrogate-based optimization. We briefly formulate a SBO design task and introduce the concept of NSM, numerical methods used to construct an accurate surrogate model, as well as generalization capabilities of the developed surrogate models. Verification of the introduced methodology on the basis of several illustrative examples is given is Sect. 4. Two shortened matching transformers and a miniaturized rat-race coupler with harmonic suppression are considered for the design and optimization using the NSM. Section 5 concludes the chapter with a discussion and recommendations for the future research related to fast design and optimization of complex structures with enhanced functionality.

2 Design of Complex Microwave Circuits: Methodology

Design of complex microwave/RF circuits with enhanced functionality is troublesome, especially due to the lack of universal strategies for determination of their topology. In general, three approaches for the design of unconventional circuits are available, including: (1) manual, experience-based construction [52, 53] (2) structure decomposition and substitution of its sections with unconventional structures [54, 55], and (3) automated design by means of metaheuristic algorithms [56, 57].

While the first technique benefits from many degrees of freedom that allow for the construction of novel structures with unusual properties (both geometry- and performance-wise), possible results strongly depend on engineering experience. In such a setup, the design process is conducted using cut and trial technique, often in conjunction with repetitive parameter sweeps. Therefore, the method is laborious and prone to failure, which restricts its applicability to designs with relatively simple topologies with up to several independent parameters [58, 59].

Improved properties of the circuit may also be obtained through decomposition of a conventional structure into a set of TL sections. Each TL may be subsequently replaced with its discontinuous counterpart and modified sections may be utilized for the construction of an unconventional structure. Despite the manual setup of the aforementioned steps, the risk of design failure in such a scheme is alleviated by using discontinuous TL components with simplified geometry that mimic the behavior of their conventional TLs. Furthermore, a lot of perforations from the literature may be directly utilized to substitute typical TL sections [60–63], which makes the technique useful, even for less experienced engineers. One should note that the method is restricted only to certain microwave/RF structures that may be decomposed, however a variety of passive circuits fall into this category, e.g., matching transformers, hybrid couplers, Butler matrices, phase shifters, or planar filters.

Although methods based on the manual design of circuits with unconventional properties are most commonly used, some automated approaches for a construction of such structures are also available [56, 64–66]. Automated design is especially attractive for inexperienced engineers as it reduces the interference into the design process only to formulating the desired performance specifications. Design of a microwave/RF structure in such a setup may be conducted using either EM model generated from binary matrix [56, 64] or interconnected TL sections [65, 66]. While the former technique allows for a construction of very compact circuits, it requires a number of computationally expensive EM simulations for a metaheuristic algorithm to complete. The latter method is considerably faster because it exploits circuit simulator instead of EM one. On the other hand, it suffers from lack of support for the reduction of the overall circuit size.

In this section, we give general guidelines for the construction of microwave/RF circuits with enhanced functionality. Moreover, we instruct how to identify TL sections of a conventional circuit that may be extracted from the design. Two main approaches for the determination of sufficient perforations are discussed.

2.1 Construction of a Circuit with Enhanced Functionality

Modifications of a microwave/RF structure by a substitution of its TL sections with their corresponding perforations may be utilized to obtain some unconventional circuit behavior (e.g., attenuation of harmonic frequencies [50], high circuit selectivity [67], and/or broad operational band [68]) or advantageous physical dimensions [69]. One should bear in mind that techniques mentioned in this section require the preparation of a conventional circuit for its further modifications, however theory-based design of microstrip structures is well described in the literature (e.g., in [17, 70, 71]) and, for the sake of brevity, we omit details of their formulation.

The general flow (see Fig. 1 for a detailed block diagram with conceptual explanation of each step) of an unconventional structure design may be summarized as follows:

1. Define design specification of a circuit with enhanced functionality;
2. Synthesize conventional circuit using theory-based approach;
3. Decompose circuit into k sections that may be substituted with respective perforations;
4. Construct n abbreviated sections suitable for UTL replacement;
5. Perform surrogate-based design and optimization of the circuit.

In general, a conventional structure is composed of various building blocks, including: conventional lines (TL and/or coupled line sections) as well as their interconnections (e.g., bends, tees, crosses, etc.) [27]. This modular design allows to perform a so-called circuit decomposition step, which is a procedure—guided by engineering experience—aimed at identification of sections that may be important for functionality enhancements of the structure. While interconnections between consecutive sections are considered irrelevant, a total of k (where $k = 1, \ldots, K$) coupled lines and/or TLs may be distinguished and isolated from the circuit.

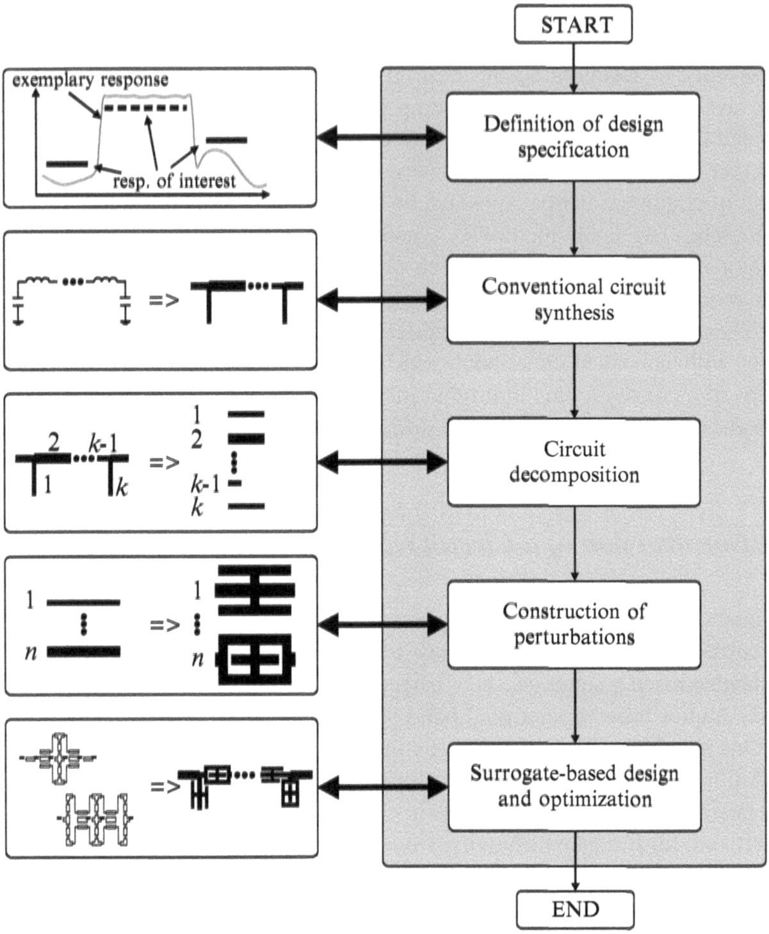

Fig. 1 Construction of a circuit with enhanced functionality by substituting conventional TL sections with perturbations—a design flow. In the first step, design specification is defined. Subsequently, general circuit line theory is utilized for a synthesis of a reference microstrip circuit. Next, the structure is decomposed into *k* conventional lines. In the fourth step, a set of *n* perturbations is designed in order to substitute their conventional counterparts. Finally, surrogate-based design driven by algorithm described in Sect. 3 is performed

Identification of respective sections is a crucial step for the determination of n (where $n = 1, \ldots, N$) different perturbations that are necessary to substitute their conventional counterparts. Exemplary microwave/RF circuits realized in microstrip technology with highlight of decomposition-ready sections are shown in Fig. 2.

A number of perturbations necessary to achieve desired functionality depend on such global factors such as desired bandwidth and/or geometry of the structure [69, 72], as well as local properties regarding characteristic impedance Z_C, electrical length θ (for TLs), and the coupling coefficient (for coupled lines). For that reason,

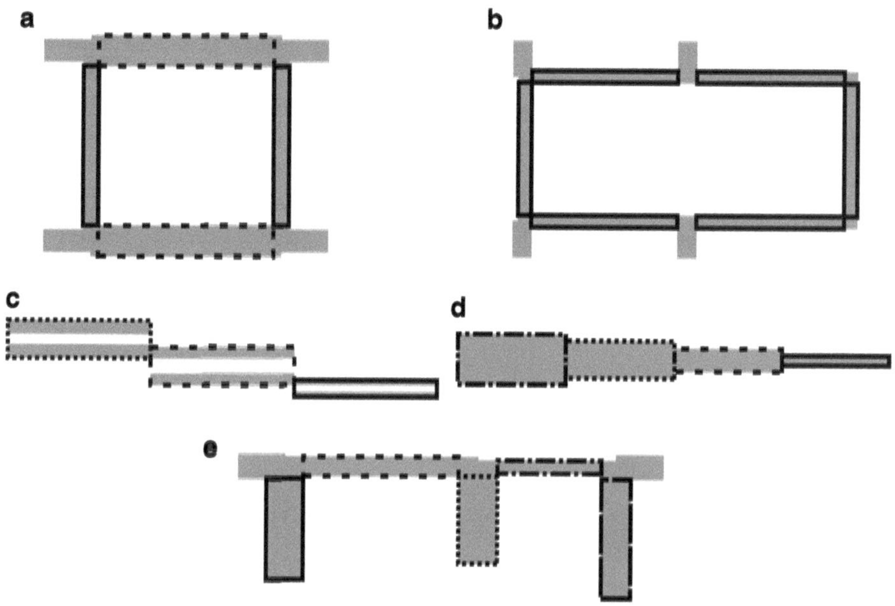

Fig. 2 Exemplary conventional microstrip circuits with highlighted relevant section: (**a**) brachline coupler —two TL sections with different characteristic impedance: (—) and (– – –); (**b**) rat-race coupler —one TL section: (—); (**c**) band-pass filter —three coupled line sections: (—), (– – –) and (•••); (**d**) impedance matching transformer —four TL section: (—), (—), (•••), (• – • –); (**e**) open-stub filter —five TL sections: (—), (– – –), (•••) (• – • –), (— —)

a number of necessary perturbations as well as their geometry should be carefully chosen for realization of specified design requirements. Although substantial research effort has been devoted towards the determination of perturbations with novel topologies over the years [60–63], only a few works attempted to address—in a systematic manner—the issue of their applicability for various unconventional designs [27, 50, 73]. However, in general, this problem may be solved only by means of engineering expertise, aided by some guidelines pointed out below:

- Substitution of conventional section with a cascade connection of perturbations allows for the bandwidth enhancement [24].
- Single perturbation is capable of realizing a range of Z_C and θ parameters and, therefore, it may substitute variety of corresponding TL sections [27].
- Miniaturization of asymmetric conventional circuit enforces preparation of a set of perturbations to mimic TL with equal electrical properties [69].

Several classes of perturbations, including: defected ground structures (DGS) [24], fractal space filing curves [67], or slow-wave resonant structures (SWRS) [50] may be exploited to mimic the behavior of conventional microwave/RF circuit components in a restricted frequency range. On the other hand, implementation of perforations not only increases the complexity of the structure, but also introduces multiple independent design parameters that highly influence its performance. Design and optimization of such circuits is considered to be a computationally expensive problem that may be partially addressed using SBO algorithms [27]. Perturbations in the form of SWRS are particularly attractive for SBO setup because—in contrary to fractal curves and DGSs—they may be designed using a circuit simulator. Moreover, SWRS introduces slow-wave phenomenon that allows for decreasing the phase velocity, and consequently such perforations are shorter than conventional transmission lines. Nonetheless, a large number of variables associated with such circuits introduce serious problems with convergence of conventional SBO algorithms [51]. Techniques for the design of SWRS are described in Sects. 2.2 and 2.3, whereas SBO-based design and optimization of unconventional circuits with multiple perturbations are addressed in Sect. 3.

2.2 *Design of Slow-Wave Resonant Structures: Database Approach*

The choice of proper SWRS for a construction of a microwave/RF circuit with enhanced functionality is troublesome. The design of SWRS that may be considered to be a sufficient replacement of its corresponding UTL (or coupled line) is mostly conducted using cut and trial technique guided by engineering experience, which is a time consuming process involving numerous EM simulations. On the other hand, determination of appropriate SWRS may be conducted by gathering EM models of various predefined components. Such a database comprises an extensive description of each SWRS, including a number and range of design variables, as well as electric properties. Moreover, it provides tools for the identification of structures being most appropriate for realization of desired circuit behavior [73]. Several approaches that exploit database for the construction of unconventional circuits are available in literature [27, 50, 73]. Figure 3 depicts a set of exemplary SWRSs that may be utilized for the construction of a database.

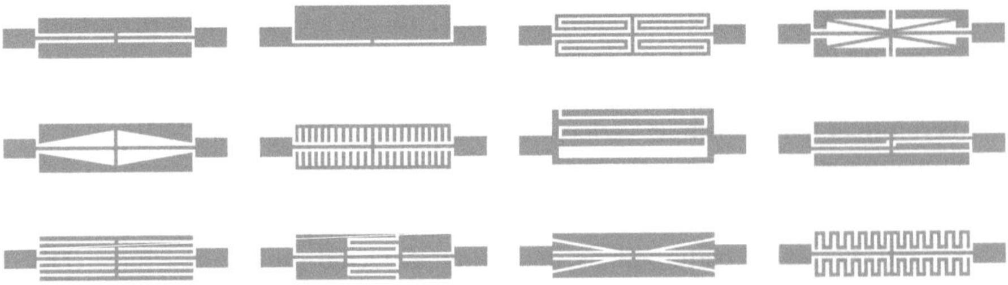

Fig. 3 An exemplary database containing 12 EM models of SWRSs [73]

The most appropriate SWRS is selected from the database by means of cell assessment with respect to defined efficiency coefficients, which may refer to local properties (e.g., range of Z_C and θ realizations, or transverse dimension of the SWRS) as well as global properties (e.g., bandwidth, or overall size) of the circuit (cf. Sect. 2.1). Each SWRS is evaluated with respect to all of its coefficients by computation of their weighted mean and a component with the best value is considered as sufficient to substitute respective conventional section. This technique may be also utilized for the selection of the most versatile SWRS to work as a substitute component of conventional sections with varied electrical parameters (c.f. Sect. 2.1). More detailed explanation of SWRS determination technique by means of a database utilization is presented in [73]. One should emphasize that a construction of a database comprising a number of EM models of SWRS requires considerable computational effort, however, once prepared, it may be reused multiple times with no extra computational cost.

2.3 *Design of Slow-Wave Resonant Structures: Knowledge-Based Approach*

Although design of unconventional circuit constituted by SWRS obtained from a predefined database is considerably easier in comparison to knowledge-based approach, it suffers from a smaller number of degrees of freedom in the process of forming a component into the desired shape. SWRSs gathered in database exhibit similarities that prevent their utilization for the design of structures with very unusual properties (e.g., very compact circuit [69], or wide range of electrical properties [27]). Therefore, SWRS designed exclusively for a specific circuit may provide the best results regarding desired specification. Additionally, a knowledge-based approach allows for a construction of complementary SWRSs (i.e., cells that geometrically supplement each other [69]), which is especially useful in the design of very compact circuits (e.g., [69, 73, 74]).

Despite the manual nature of the described SWRS design technique, some mathematical models aimed at the determination of its initial dimensions may be provided. In the lossless case, the response $\boldsymbol{R}_U$ of TL section may be described in the form of ABCD matrix:

$$\boldsymbol{R}_U = \begin{bmatrix} cos\,(\theta) & jZ_C sin\,(\theta) \\ \frac{j}{Z_C} sin\,(\theta) & cos\,(\theta) \end{bmatrix} \tag{1}$$

The performance of a distributed TL section may be mirrored at the given operating frequency by its corresponding SWRS section in the form of T-type distributed-element circuit, which is composed of interconnected high-impedance Z_H and stepped-impedance sections with low-impedance Z_L stub. Realization of SWRS in such a configuration is particularly attractive due to its considerable slow-wave

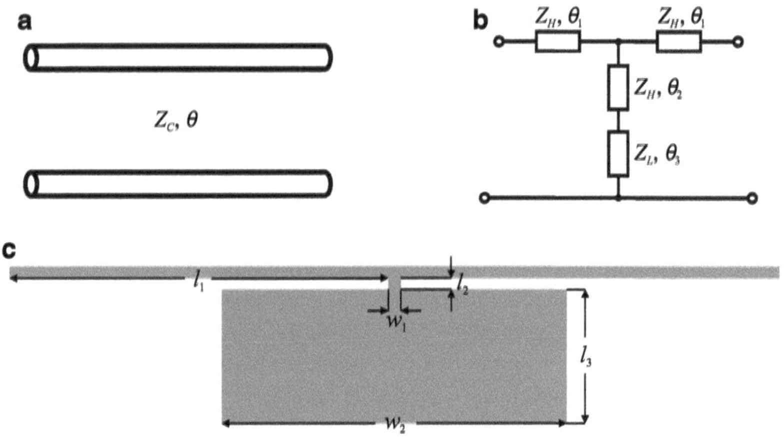

Fig. 4 Various models of a component: (**a**) distributed-element model of TL; (**b**) distributed model of T-type SWRS structure; (**c**) microstrip model of T-type SWRS—parameters w_1, w_2, and l_2 are set based on technology limitations

properties, as well as great usefulness for circuit miniaturization [69, 74, 75]. A conceptual illustration of a TL section and its interchangeable SWRS (both the composite and microstrip realizations) is shown in Fig. 4.

The response of T-type SWRS may be described by the following set of equations:

$$\boldsymbol{R}_T = \begin{bmatrix} cos(\theta_1) & jZ_H sin(\theta_1) \\ \frac{j}{Z_H} sin(\theta_1) & cos(\theta_1) \end{bmatrix} \begin{bmatrix} 1 & 0 \\ \frac{j}{Z_1} & 1 \end{bmatrix} \begin{bmatrix} cos(\theta_1) & jZ_H sin(\theta_1) \\ \frac{j}{Z_H} sin(\theta_1) & cos(\theta_1) \end{bmatrix} \quad (2)$$

$$Z_1 = Z_H \frac{Z_2 + jZ_H tan(\theta_2)}{Z_H + jZ_2 tan(\theta_2)} \quad (3)$$

$$Z_2 = Z_H \frac{Z_L}{jtan(\theta_3)} \quad (4)$$

where θ_1 stands for the electrical length of a high-impedance section, θ_2 and θ_3 denote electrical length of a high-impedance interconnection between the Z_H line and the low-impedance stub, respectively. One should note that parameters Z_H, Z_L and θ_2 may be determined a priori based on technology limitations of microstrip line circuits (minimal/maximal values of line length/width allowed for fabrication) [69]. Parameters θ_1 and θ_3 may be found numerically by solving $\boldsymbol{R}_U = \boldsymbol{R}_T$ for the given operational frequency (parameters Z_C and θ of UTL section are known). Subsequently, geometrical dimensions of SWRS are calculated using general microstrip equations [17].

Although the designed SWRS may be directly utilized for a construction of an unconventional circuit, its shape may not be optimal for some applications. Therefore, Z_H and Z_L sections of the cell may be manually formed—assuming preservation of their electrical properties—to fulfill the specified requirements regarding circuit functionality (e.g., compact size of hybrid couplers [74] or Butler matrices [76]). A more detailed explanation of the knowledge-based construction of SWRS can be found in [72].

3 Surrogate-Based Design and Optimization of Complex Circuits

Design and optimization of unconventional circuits with enhanced functionality is clearly a complex process that involves not only engineering knowledge, but also considerable computational resources. The cost-related issues may be partially alleviated by means of surrogate-based optimization. Here, we discuss a NSM methodology, which is suitable for the design of complex microstrip circuits with SWRS perturbations. We also demonstrate the performance of the NSM technique as well as its advantages over conventional space mapping modeling and optimization.

3.1 Surrogate-Based Optimization

The design of microwave/RF circuit driven by surrogate-based optimization algorithm requires two representations of the same microwave structure at different levels of fidelity. Let $\boldsymbol{R}_f(x)$ be a response vector of a high-fidelity EM model of a complex microwave/RF structure with enhanced functionality, whereas the vector $\boldsymbol{x}$ denotes independent design parameters of the respective circuit. Unfortunately, $\boldsymbol{R}_f$ model is computationally too expensive to be directly used in the numerical optimization process [77]. Instead, a physics-based low-fidelity model $\boldsymbol{R}_s$ in the form of equivalent circuit representation of the respective structure may be—upon suitable correction—utilized to reduce the computational cost of structure optimization. For the sake of brevity, we omit details regarding construction of surrogate model using circuit representation. A comprehensive explanation of this process is available in literature (e.g., [48, 78, 79]).

The design process of microwave/RF circuits may be formulated as a nonlinear minimization problem of the following form:

$$\mathbf{x}^* = \arg\min_{\mathbf{x}} U\left(\mathbf{R}_f(\mathbf{x})\right) \tag{5}$$

where U is a scalar merit function (e.g., a minimax function with upper and lower specifications) that implements given design specifications. Vector $\boldsymbol{x}^*$ is the optimal design to be determined. A high-computational cost of a single EM simulation makes the utilization of conventional optimization to handle (5) impractical, because both gradient-based (e.g., Quasi-Newton [28]) and derivative-free (pattern search [80], genetic algorithms [81]) methods usually require a substantial number of objective function (and thus, high-fidelity model) evaluations. In order to reduce the CPU expense, a direct optimization of a computationally expensive model may be replaced by the following iterative procedure [24, 82]:

$$\mathbf{x}^{(i+1)} = \arg\min_{\mathbf{x}} U\left(\mathbf{R}_s^{(i)}(\mathbf{x})\right) \tag{6}$$

that generates a sequence of approximate solutions $\boldsymbol{x}^{(i)}$ ($i = 0, 1, \ldots$) to the original design problem of (5). The surrogate model at iteration i, $\boldsymbol{R}_s^{(i)}$, is constructed from the low-fidelity model so that the misalignment between $\boldsymbol{R}_s^{(i)}$ and the fine model is reduced using so-called parameter extraction process. The latter is a nonlinear minimization problem by itself [36]. A conceptual flow of SBO is shown in Fig. 5.

For a well working SBO algorithm, only a few iterations of (6) are necessary to find a satisfactory solution. Also, the fine model is typically evaluated only

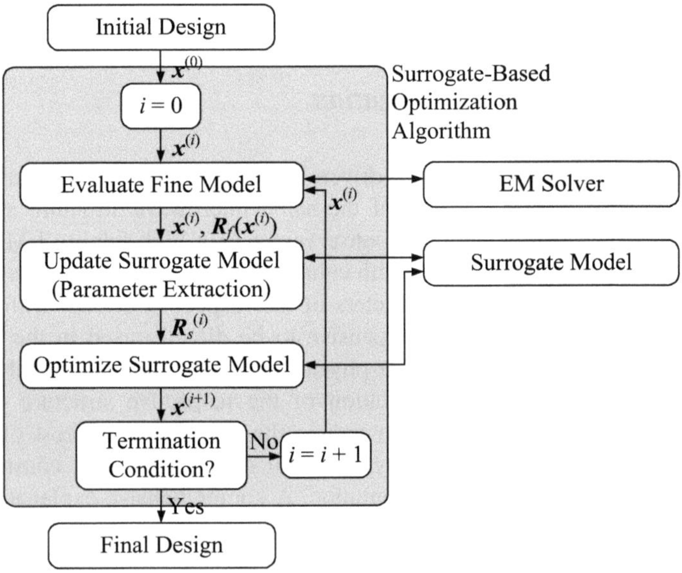

Fig. 5 A conceptual flow of surrogate-based optimization: the optimization burden is shifted to the computationally cheap surrogate model which is updated and re-optimized at each iteration of the main optimization loop. High-fidelity EM simulation is only performed once per iteration to verify the design produced by the surrogate model and to update the surrogate itself. The number of iterations for a well-performing SBO algorithm is substantially smaller than for conventional techniques

once per iteration [83]. However, conventional SBO algorithms—particularly space mapping—suffer from convergence problems or relatively large number of EM model evaluations necessary to conclude the process, if complex microwave/RF designs are considered. These difficulties are alleviated here using a NSM approach.

3.2 NSM Modeling

The NSM technique [51] is a two-level modeling methodology, with the first (inner) space mapping layer applied at the level of component (a so-called local model), and the second (outer) layer applied at the level of the entire structure (so-called global model). The purpose of NSM is to improve the generalization capability of the surrogate model and facilitate the parameter extraction process. Consequently the cost of the design optimization process using NSM can be greatly reduced compared to conventional space mapping applied only at the level of the entire structure [51].

Let $\boldsymbol{R}_{f.cell}(\boldsymbol{y})$ and $\boldsymbol{R}_{c.cell}(\boldsymbol{y})$ be responses (here, S-parameters) of the high-fidelity (i.e., EM-simulated) and low-fidelity—circuit-simulated—models of the local component (here, SWRS cell). The vector $\boldsymbol{y}$ represents geometry parameters of the cell, whereas $\boldsymbol{R}_f(\boldsymbol{x})$ and $\boldsymbol{R}_s(\boldsymbol{x})$ denote high- and low-fidelity response of the entire microwave/RF structure with $\boldsymbol{x}$ being a corresponding vector of geometrical parameters. In NSM approach the surrogate model of the entire circuit is constructed starting from the component level, i.e., each SWRS surrogate being relevant for circuit functionality enhancements. A surrogate of inner generic component denoted as $\boldsymbol{R}_{s.g.cell}(\boldsymbol{y},\boldsymbol{p}^*)$ stands for a composition of $\boldsymbol{R}_{c.cell}$ and suitable space mapping transformations; the vector $\boldsymbol{p}^*$ denotes the set of extractable space mapping parameters of the model. The space mapping surrogate $\boldsymbol{R}_{s.cell}$ of a SWRS component is obtained as

$$\mathbf{R}_{s.cell}(\mathbf{y}) = \mathbf{R}_{s.g.cell}(\mathbf{y}, \mathbf{p}^*) \tag{7}$$

The parameter vector $\boldsymbol{p}^*$ is obtained by solving the following nonlinear parameter extraction process

$$\mathbf{p}^* = \arg\min_{\mathbf{p}} \sum_{k=1}^{N_{cell}} \left|\left|\mathbf{R}_{s.g.cell}\left(\mathbf{y}^{(k)}, \mathbf{p}\right) - \mathbf{R}_f\left(\mathbf{y}^{(k)}\right)\right|\right| \tag{8}$$

where vectors $\boldsymbol{y}^{(k)}$, $(k = 1, \ldots, N_{cell})$ denote the training (or base) solutions obtained using a so-called star-distribution scheme [36]. A base obtained using star-distribution method is composed of $N_{cell} = 2d + 1$, where d is design space dimensionality (i.e., a number of independent design variables). Although local generic surrogate of each SWRS depends on much smaller number of parameters than the structure optimized using conventional space mapping technique (usually up to a few independent variables rather than a few dozen [51]), it exploits a combination of various space mapping methods (c.f. Sect. 3.3) [41]. Therefore, a local surrogate model $\boldsymbol{R}_{s.cell}$ implemented within a global model of the whole microwave/RF circuit should be valid for the entire range of parameters $\boldsymbol{y}$.

A generic space mapping surrogate of the entire unconventional circuit denoted by $\boldsymbol{R}_{s.g}(\boldsymbol{x},\boldsymbol{P})$ is composed of the local models of individual SWRS cells:

$$R_{s.g}(x, P) = R_{s.g}([y_1; \ldots; y_p], P) = F(R_{s.g.cell}(y_1, p^*), \ldots, R_{s.g.cell}(y_p, p^*), P) \tag{9}$$

where F realizes a sufficient connection between respective sections of a structure with enhanced functionality. Vector $\boldsymbol{x}$ represents a concatenation of component parameter vectors $\boldsymbol{y}_k$, while vector $\boldsymbol{P}$ stands for space mapping parameters at the outer level. One should note that $\boldsymbol{P}$ is usually defined as perturbations (with respect to $\boldsymbol{p}^*$) of selected space mapping parameters of individual components.

The outer space mapping level is applied to the global model $\boldsymbol{R}_{s.g}(\boldsymbol{x},\boldsymbol{P})$, so that the final surrogate $\boldsymbol{R}_s^{(i)}$ utilized in the ith iteration of the SBO scheme (6) is defined as follows

$$\mathbf{R}_s^{(i)}(\mathbf{x}) = \mathbf{R}_{s.g}\left(\mathbf{x}^{(i)}, \mathbf{P}^{(i)}\right) \tag{10}$$

where

$$\mathbf{P}^{(i)} = \arg\min_{\mathbf{P}} \| \mathbf{R}_{s.g}\left(\mathbf{x}^{(i)}, \mathbf{P}\right) - \mathbf{R}_f\left(\mathbf{x}^{(i)}\right) \| \tag{11}$$

It should be emphasized that the number of parameters in $\boldsymbol{P}$ is much smaller than the combined number of space mapping parameters of SWRS components (i.e., multiple copies of a vector $\boldsymbol{p}^*$). This is normally sufficient because the inner space mapping layer already provides good alignment between the $\boldsymbol{R}_{s.cell}$ and $\boldsymbol{R}_{f.cell}$ so that the role of (10), (11) is mainly to account for possible interactions between SWRS components that are considered by the composing function F. The algorithm may be summarized as follows (see Fig. 6 for conceptual illustration):

1. Construct circuit model $\boldsymbol{R}_{c.cell}$ of the nth SWRS component;
2. Obtain inner space mapping surrogate $\boldsymbol{R}_{s.cell}$ of nth SWRS by performing multipoint extraction of $\boldsymbol{p}^*$ parameters;
3. If $n < N$ go to 1;
4. Utilize N $\boldsymbol{R}_{s.cell}$ components to construct a generic space mapping surrogate $\boldsymbol{R}_{s.g}(\boldsymbol{x},\boldsymbol{P})$ of an unconventional microwave/RF circuit;
5. Utilize SBO scheme for the determination of desired functionality at the output space mapping level.

3.3 Surrogate Model Construction

Multipoint parameter extraction utilized for a construction of a reliable local model of SWRS component requires a combination of various space mapping techniques to

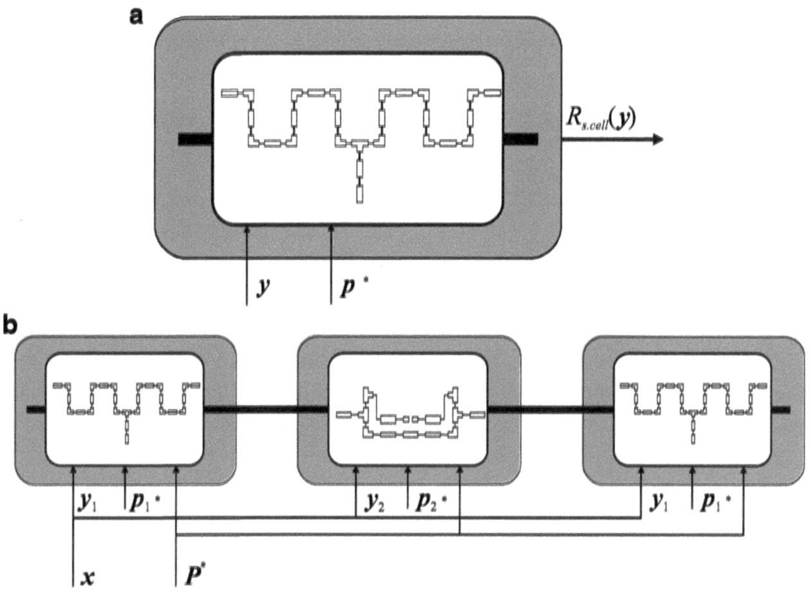

Fig. 6 The concept of nested space mapping (NSM): (**a**) local space mapping model of SWRS component. Extractable parameters $\boldsymbol{p}^*$ are optimized to match the circuit $\boldsymbol{R}_{s.cell}(\boldsymbol{y})$ and $\boldsymbol{R}_{f.cell}(\boldsymbol{y})$ within solution space; (**b**) global space mapping model composed of two SWRS components. Once extractable parameters $\boldsymbol{p}^*_{(1,2)}$ are set they are reused in global model. Vector of design parameters $\boldsymbol{x}$ is optimized to obtain desired specification

achieve desired generalization. The inner space mapping layer is constructed using the following relation:

$$\mathbf{R}_{s.g.cell}\,(\mathbf{y},\mathbf{p}) = \mathbf{R}_{c.F}\,(\mathbf{B}\cdot\mathbf{x}+\mathbf{c},\mathbf{p}_I) \tag{12}$$

where $\boldsymbol{B}$ and $\boldsymbol{c}$ are input space mapping parameters (a diagonal matrix $\boldsymbol{B}$ is utilized), $\boldsymbol{p}_I$ are implicit space mapping (ISM) parameters (a substrate parameters of individual microstrip subsection of SWRS component, specifically, dielectric permittivity and the substrate heights). Additionally, a frequency scaling $\boldsymbol{R}_{c.F}$ of low-fidelity model aimed at evaluation of the $\boldsymbol{R}_c$ model across the frequency band of interest, is performed. The frequency-scaled model $\boldsymbol{R}_{c.F}(\boldsymbol{y})$ corresponding to $\boldsymbol{R}_c(\boldsymbol{y}) = [R_c(\boldsymbol{y},\omega_1)\ R_c(\boldsymbol{y},\omega_2)\ \ldots\ R_c(\boldsymbol{y},\omega_m)]^T$ is defined as flows:

$$\mathbf{R}_{c.F}\,(\mathbf{y},\mathbf{p}) = [R_c\,(\mathbf{y},\,f_0+\omega_1\cdot f_1)\quad R_c\,(\mathbf{y},\,f_0+\omega_2\cdot f_1)\ \ldots\ R_c\,(\mathbf{y},\,f_0+\omega_m\cdot f_1)]^T \tag{13}$$

where f_0, f_1 are extractable scaling parameters. Technique is especially useful for correction of frequency misalignment between low- and high-fidelity models of the SWRS component [51]. An exemplary inner space mapping layer of SWRS component $\boldsymbol{R}_{s.cell}(\boldsymbol{y}, \boldsymbol{p})$ and its corresponding high-fidelity model $\boldsymbol{R}_{f.cell}(\boldsymbol{y})$ are shown in Fig. 7.

3.4 Generalization Capability of NSM

The most notable advantage of NSM technique lies in good generalization capability of the global NSM surrogate $\boldsymbol{R}_{s.g}$, which is achieved by global accuracy of each local SWRS model $\boldsymbol{R}_{s.cell}$ utilized for a construction of unconventional microwave/RF circuit. A typical modeling accuracy of an exemplary SWRS cell (see Fig. 7) after multipoint parameter extraction is illustrated in Fig. 8. A comparison of global NSM surrogate for an exemplary structure—in the form of unconventional

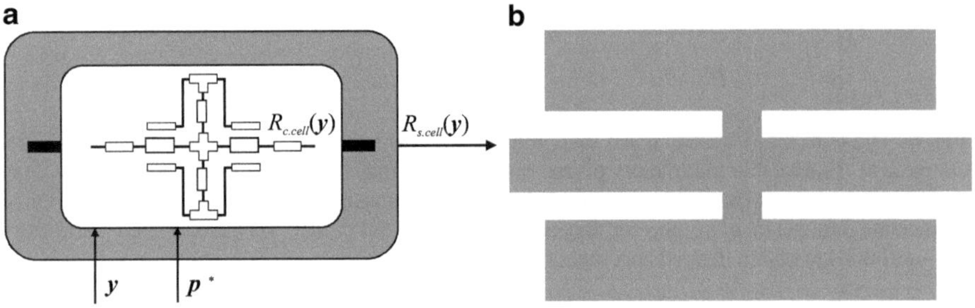

Fig. 7 An exemplary SWRS component: (**a**) coarse model $\boldsymbol{R}_{c.cell}(\boldsymbol{y})$ within inner space mapping layer $\boldsymbol{R}_{s.cell}(\boldsymbol{y})$; (**b**) high-fidelity EM model $\boldsymbol{R}_{f.cell}(\boldsymbol{y})$

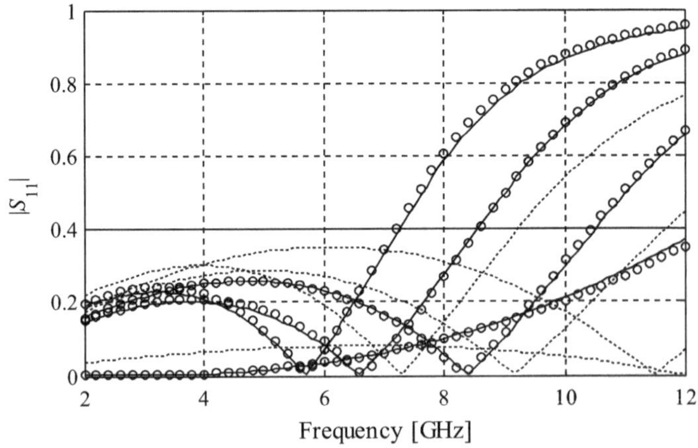

Fig. 8 NSM modeling of SWRS component. Responses at the selected test designs: coarse model (*dotted line*), fine model (*solid line*), NSM surrogate after multipoint parameter extraction (*open circle*). The plots indicate very good approximation capability of the surrogate

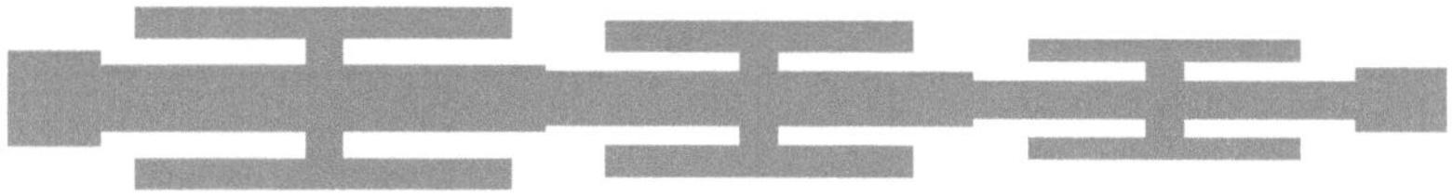

Fig. 9 Geometry of an exemplary matching transformer composed by cascading three SWRS sections of Fig. 7

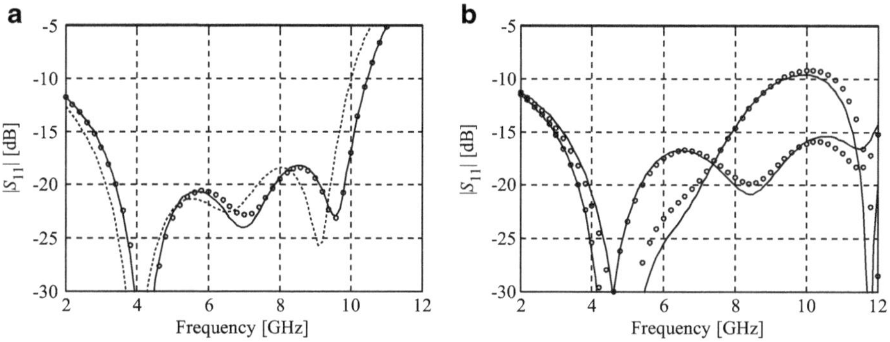

Fig. 10 NSM modeling of a matching transformer of Fig. 9: (**a**) high-fidelity model response (*solid line*), NSM surrogate before parameter extraction (*dotted line*) and NSM surrogate after parameter extraction (*open circle*); (**b**) visualization of excellent generalization of NSM surrogate (*open circle*) extracted at random designs. Corresponding high-fidelity model responses (*solid line*) are also provided

matching transformer of Fig. 9—before and after parameter extraction step (10) is shown in Fig. 10. One should emphasize that $\boldsymbol{R}_{s.g}$ model matches its high-fidelity counterpart even before parameter extraction step. Therefore, parameter extraction is aimed only at addressing the interactions (i.e., couplings) between respective SWRS components that are not accounted by the $\boldsymbol{R}_{s.cell}$ models.

For a comparison purpose, an exemplary transformer of Fig. 9 is also designed using conventional space mapping modeling (i.e., correction of its low-fidelity model $\boldsymbol{R}_c$). The process of circuit design in such a setup is considerably more complex because (1) the low-fidelity model of the entire unconventional microwave/RF circuit is much less accurate than $\boldsymbol{R}_{s.g}$ (cf. Fig. 11a), (2) a large number of space mapping parameters with considerable range of variation is required, (3) the parameter extraction process is more challenging and time consuming, and (4) generalization capability of the model is poor (cf. Fig. 11b).

Sufficient accuracy of the underlying low-fidelity model and good generalization capability of the surrogate are essential for fast convergence of the SBO optimization process (6) [51]. The NSM model exhibits both aforementioned features (here, the global model $\boldsymbol{R}_{s.g}$ is formally a low-fidelity model for (6)), which results in not only rapid, but also accurate design of complex microwave/RF circuits with enhanced functionality.

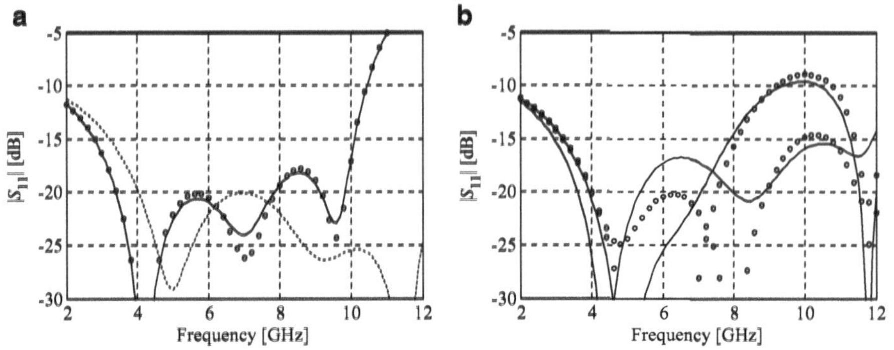

Fig. 11 Conventional space mapping modeling of a matching transformer of Fig. 9: (**a**) high-fidelity model response (*solid line*), SM surrogate before (*dotted line*) and after parameter extraction (*open circle*); (**b**) visualization of poor generalization capability of a surrogate (*open circle*) extracted at random designs. Corresponding high-fidelity model responses (*solid line*) are also provided

4 Case Studies

In this section, we present verification examples for the design of complex microwave/RF circuits using NSM methodology of Sect. 3. The technique is validated using a four-section ultra-wideband matching transformer with 16 independent design variables, a 15-variable broadband three section matching transformer, and a miniaturized rat-race coupler with a total of ten designable parameters. Unconventional properties of illustrative circuits are obtained by implementation of SWRS components using both the database approach of Sect. 2.2 and knowledge-based design of complementary SWRS explained in Sect. 2.3. A comparison of the NSM technique with sequential space mapping (SSM) and implicit space mapping (ISM) methods is also provided.

4.1 Design of Ultra-Wideband Four-Section Matching Transformer

Consider a four-section microstrip matching transformer (MT). The structure is aimed to mimic the functionality of a conventional MT, i.e., (1) match a 50 Ω line to a 130 Ω load, and (2) provide a reflection coefficient $|S_{11}| \leq -15$ dB within 3.1–10.6 GHz frequency band of interest (ultra-wideband structure). A circuit is considered to operate on Taconic RF-35 dielectric substrate ($\varepsilon_r = 3.5$, $\tan\delta = 0.0018$, $h = 0.762$).

A prototype circuit satisfying design specifications regarding reflection and matching properties is constructed using binomial design expressions [17] resulting

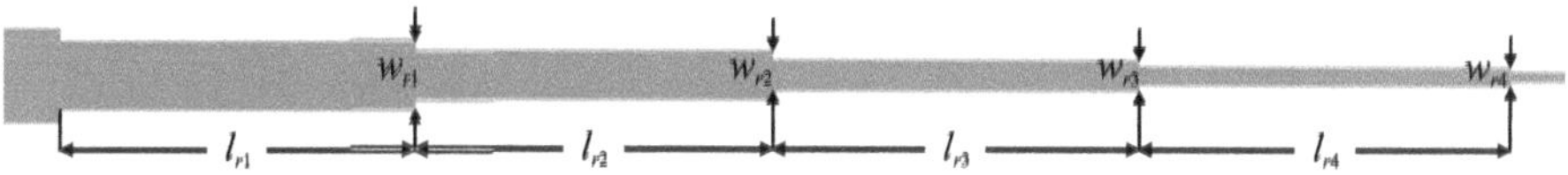

Fig. 12 Geometry of a conventional four-section MT

in a cascade of four quarter-wavelength ($\theta = 90°$ at $f_0 = 6.65$ GHz) TL sections described by the following vector of characteristic impedances $Z_C = [53.1\ 67.4\ 96.4\ 122.5]^T\ \Omega$. Subsequently, physical dimensions of the MT: $\boldsymbol{x}_r = [w_{r1}\ w_{r2}\ w_{r3}\ w_{r4}\ l_{r1}\ l_{r2}\ l_{r3}\ l_{r4}]^T$ are calculated using general equations for microstrip lines and slightly tuned. The design parameters of the reference structure (see Fig. 12) are $\boldsymbol{x}_r = [1.2\ 0.9\ 0.5\ 0.3\ 6.6\ 6.8\ 6.7\ 7.0]^T$. Moreover, variables $l_{r0} = 10$, $w_{ri0} = 1.7$, and $w_{ro0} = 0.18$ denote size of 50 and 130 Ω lines (all dimensions in mm).

The reference MT is characterized by a very simple geometry that allows for its decomposition into four TL sections (cf. Sect. 2.1). Furthermore, similarities between consecutive cells indicate that they might be substituted with one versatile SWRS component. An appropriate structure may be found using database approach described in Sect. 2.2. A desired SWRS is intended to mimic a range of characteristic impedances Z_C with preservation of electrical length around 90°. A double-T-type SWRS constituted by a vector of four independent design parameters: $\boldsymbol{y} = [l_1\ l_2\ w_1\ w_2]^T$ (dimensions $l_3 = 0.2$ and $w_3 = 0.2$ are fixed) is sufficient to fulfill our needs [73]. One should emphasize that due to technology limitations (i.e., minimum feasible width of the SWRS lines and the gaps between them equal to 0.1 mm), acceptable structure dimensions are defined by the following lower/upper $\boldsymbol{l}/\boldsymbol{u}$ bounds: $\boldsymbol{l} = [0.1\ 1\ 0.1\ 0.1]^T$ and $\boldsymbol{u} = [1\ 5\ 1\ 1]^T$. The high-fidelity model $\boldsymbol{R}_{f.cell}$ of the double-T-type SWRS (~200,000 mesh cells and evaluation time of 60 s) is implemented in CST Microwave Studio [84], whereas its low-fidelity model $\boldsymbol{R}_{c.cell}$ is constructed in Agilent ADS circuit simulator [85].

The inner layer model $\boldsymbol{R}_{s.cell}$ of the chosen double-T-type SWRS component is constituted by 16 space mapping parameters, including: eight input space mapping (four $\boldsymbol{B} = diag([B_1\ B_2\ B_3\ B_4]^T)$ and four $\boldsymbol{c} = [c_1\ c_2\ c_3\ c_4]^T$ parameters), six ISM (three various substrate heights and permittivity parameters $\boldsymbol{p}_I = [h_1\ h_2\ h_3\ \varepsilon_1\ \varepsilon_2\ \varepsilon_3]^T$), and two frequency scaling (f_0 and f_1) ones. A multipoint parameter extraction based on star-distribution scheme (c.f. Sect. 3.2) has been conducted to achieve $\boldsymbol{R}_{s.cell}$ model generalization within predefined lower/upper bounds. A comparison of the component characteristics before and after multipoint parameter extraction is shown in Fig. 8, whereas a double-T-type structure with highlighted geometrical and space mapping parameters is shown in Fig. 13.

A global model $\boldsymbol{R}_{s.g}$ of the unconventional MT has been constructed using a cascade connection of $\boldsymbol{R}_{s.cell}$ models of the double-T-type component. Subsequently, the high-fidelity model of the structure has been prepared in CST Microwave Studio (~1,060,000 mesh cells and average simulation time 10 min). The initial set of parameters is $\boldsymbol{x} = [0.55\ 3.75\ 0.65\ 0.35\ 0.55\ 3.75\ 0.65\ 0.35\ 0.55\ 3.75\ 0.65\ 0.35\ 0.55\ 3.75\ 0.65\ 0.35]^T$. Subsequently, the circuit has been optimized using

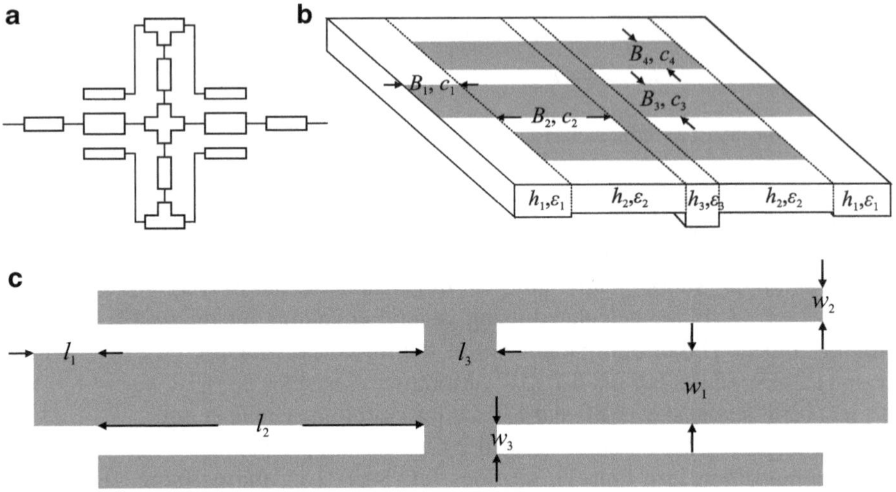

Fig. 13 A double-T-type SWRS component: (**a**) low-fidelity $\boldsymbol{R}_{c.cell}$ model; (**b**) conceptual visualization of the inner layer model $\boldsymbol{R}_{s.cell}$ with highlighted extractable parameters; (**c**) topology of $\boldsymbol{R}_{f.cell}$ model with highlighted geometrical parameters

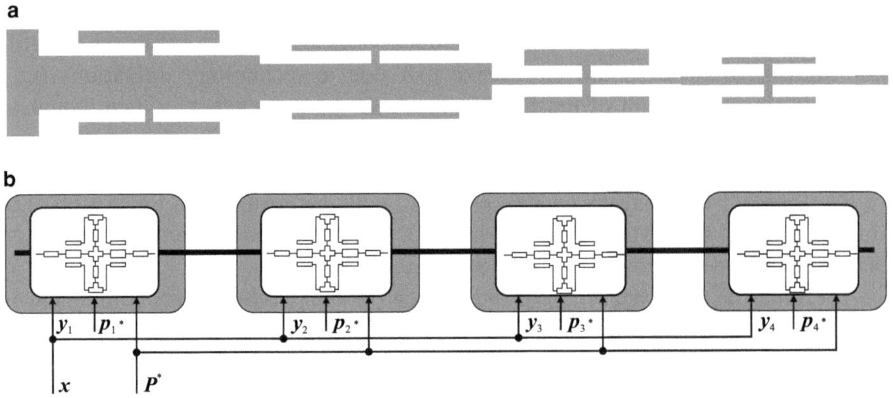

Fig. 14 An unconventional MT composed of cascade connection of double-T-type SWRS components [51]: (**a**) geometry of an optimized structure; (**b**) schematic diagram of SWRS interconnections

the NSM technique of Sect. 3. The final design of complex MT is denoted by vector $\boldsymbol{x} = [1.0\ 3.52\ 0.85\ 0.2\ 0.8\ 4.1\ 0.58\ 0.1\ 0.8\ 3.09\ 0.1\ 0.25\ 1\ 2.32\ 0.13\ 0.1]^T$. Figure 14 illustrates geometry of the structure as well as schematic diagram of $\boldsymbol{R}_{s.cell}$ interconnections.

The final design of the unconventional MT is obtained after three iterations of the NSM algorithm. The structure fulfills all assumed design specifications: (1) it provides 50–130 Ω matching as well as (2) $|S_{11}| \leq -16.2$ dB within band of interest.

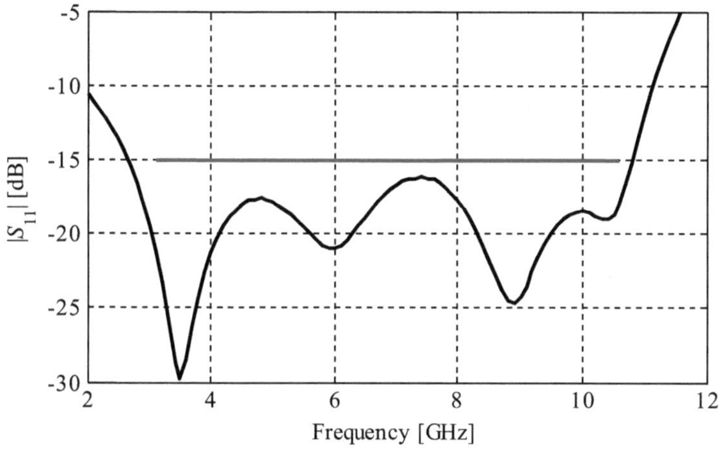

Fig. 15 Reflection response of the optimized unconventional MT

Moreover, the operational bandwidth of the circuit for reflection below −15 dB is 2.7–10.8 GHz, which is 15 % broader than assumed one. Figure 15 presents simulated characteristics of the unconventional MT. One should emphasize that the number of the outer layer SM parameters $\boldsymbol{P}$ is much smaller than the combined set of SM parameters for the inner layer (14 vs. 64 for the considered structure) as only frequency scaling and selected implicit SM parameters are used. Reduction of the number of parameters (introduced by excellent generalization capability of NSM) considerably speeds up the design process.

The cost of inner space mapping model preparation corresponds to nine $\boldsymbol{R}_{f.cell}$ model evaluations for multi parameter extraction step, while determination of the final design required only three evaluations of the $\boldsymbol{R}_f$ model. The accumulated cost of $\boldsymbol{R}_{s.cell}$ and $\boldsymbol{R}_{s.g}$ models evaluations corresponds to about 0.2 $\boldsymbol{R}_f$ evaluations, thus the total aggregated cost of the unconventional MT design using the NSM technique is about 40 min. For the sake of comparison, the design process of MT has been also conducted by means of ISM [47] and SSM [27] techniques resulting in considerably higher computational cost or failure of algorithms. Additionally, a direct optimization of the transformer using pattern search method [80] was carried out. The algorithm failed at seeking for desired circuit dimensions and has been terminated after 500 iterations. A detailed comparison of the computational costs of mentioned optimization techniques is provided in Table 1.

4.2 Design of a Broadband Three Section Matching Transformer

Consider a microstrip MT composed of three sections. The structure is aimed to mimic the functionality of a conventional TL-based MT, i.e., (1) match a 50 Ω

Table 1 Four-section unconventional MT: design and optimization cost

	Optimization algorithm			
Model evaluations	NSM	ISM	SSM	Direct search
SWRS $\boldsymbol{R}_{s.cell}$	$0.1 \times \boldsymbol{R}_f$	N/A	N/A	N/A
SWRS $\boldsymbol{R}_{f.cell}$	$0.6 \times \boldsymbol{R}_f$	N/A	N/A	N/A
MT low-fidelity $\boldsymbol{R}_s$	$0.1 \times \boldsymbol{R}_f$	$5.1 \times \boldsymbol{R}_f$	$1.7 \times \boldsymbol{R}_f$	N/A
MT high-fidelity $\boldsymbol{R}_f$	3	7	10[a]	500[b]
Total cost	$3.8 \times \boldsymbol{R}_f$	$12.1 \times \boldsymbol{R}_f$	$11.7 \times \boldsymbol{R}_f$	500[b] $\times \boldsymbol{R}_f$
Total cost [min]	38	121	N/A	5,000

[a]The algorithm started diverging and was terminated after ten iterations

[b]The algorithm failed to find a geometry satisfying performance specifications

line to 130 Ω load and (2) obtain reflection $|S_{11}| \leq -15$ dB within 1–3.5 GHz frequency band of interest. Additionally, circuit is intended to offer at least 50 % length diminution in comparison to the conventional one. A structure is considered to work on Taconic RF-35 dielectric substrate ($\varepsilon_r = 3.5$, $\tan\delta = 0.0018$, $h = 0.762$).

A binomial design expressions are utilized for a construction of a sufficient prototype circuit in the form of a cascade connection of three TL sections with $\theta = 90°$ for the given operating frequency of $f_0 = 2.25$ GHz. The characteristic impedances of the prototype circuit are represented by vector $Z_C = [56.3\ 80.6\ 115.4]^T$ Ω, while physical dimensions: $\boldsymbol{x}_r = [w_{r1}\ w_{r2}\ w_{r3}\ l_{r1}\ l_{r2}\ l_{r3}]^T$ of the MT are calculated using general microstrip equations. Subsequently, the circuit is tuned to match the design requirements resulting in the following variables: $\boldsymbol{x}_r = [1.09\ 0.64\ 0.34\ 20\ 21.7\ 23.6]^T$. Moreover, $l_{r0} = 10$, $w_{rl0} = 1.7$, and $w_{ro0} = 0.18$ denote dimensions of 50 and 130 Ω lines (all dimensions in mm).

The reference structure has been decomposed into three TL sections (cf. Sect. 2.1). Moreover, a database approach of Sect. 2.2 has been utilized for the determination of SWRS component that is versatile enough to substitute all TL sections. Therefore, a desired SWRS is intended to mimic their Z_C parameters for electrical length θ being around 90°. A SWRS in the form of C-type component is suitable to fulfill the specification [73]. The structure is represented by the following vector of geometrical parameters: $\boldsymbol{y} = [w_1\ w_2\ w_3\ l_1\ l_2]^T$. The lower/upper bounds imposed by technology limitations of circuit realization in microstrip technology (i.e., minimal feasible width of SWRS lines and gaps between them equal to 0.1) are set to: $\boldsymbol{l} = [0.1\ 0.1\ 0.1\ 5\ 0.1]^T$ and $\boldsymbol{u} = [2\ 2\ 0.5\ 10\ 0.5]^T$. The high-fidelity model $\boldsymbol{R}_{f.cell}$ of C-type SWRS component is implemented in CST Microwave Studio. An average evaluation time of the model is 60 s (~330,000 mesh cells). The low-fidelity model $\boldsymbol{R}_{c.cell}$ of the structure is prepared in Agilent ADS circuit simulator.

Eighteen space mapping parameters of the inner layer model $\boldsymbol{R}_{s.cell}$ include: ten input space mapping (five $\boldsymbol{B} = \mathrm{diag}([B_1\ B_2\ B_3\ B_4\ B_5]^T)$ and five $\boldsymbol{c} = [c_1\ c_2\ c_3\ c_4\ c_5]^T$ parameters), six ISM (three various substrate heights and three permittivity

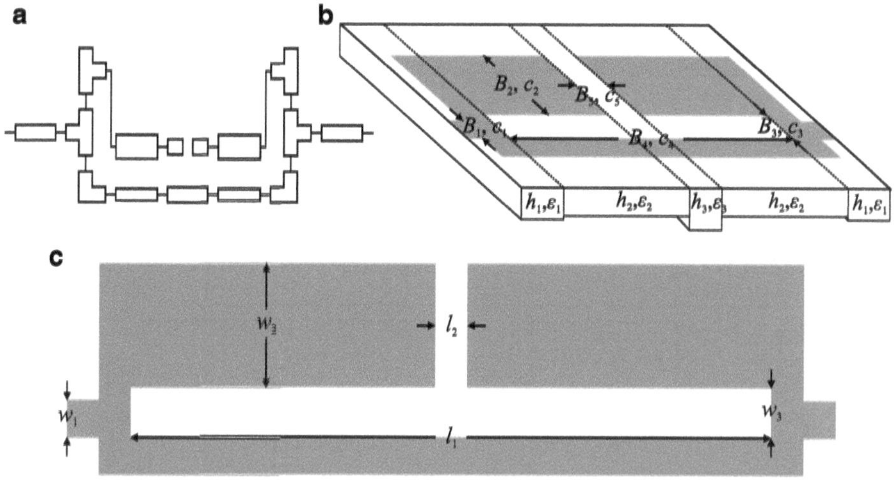

Fig. 16 A C-type SWRS component: (**a**) low-fidelity $\boldsymbol{R}_{c.cell}$ model; (**b**) conceptual visualization of inner layer model $\boldsymbol{R}_{s.cell}$ with highlighted 16 extractable parameters (except frequency scaling); (**c**) topology of $\boldsymbol{R}_{f.cell}$ model with highlighted five geometrical parameters

parameters $\boldsymbol{p}_I = [h_1\ h_2\ h_3\ \varepsilon_1\ \varepsilon_2\ \varepsilon_3]^T$), and two frequency scaling (f_0 and f_1) ones. A star-distribution scheme of Sect. 3.2 has been utilized for multipoint parameter extraction of the $\boldsymbol{R}_{s.cell}$ model. Figure 16 illustrates the structure with the emphasis on its geometrical and extractable parameters.

A cascade connection of three $\boldsymbol{R}_{s.cell}$ models of the C-type SWRS component has been utilized for a construction of a global $\boldsymbol{R}_{s.g}$ model of unconventional MT. A high-fidelity model $\boldsymbol{R}_f$ of the entire structure has been implemented in CST Microwave Studio (~1,400,000 mesh cells and average simulation time 18 min). The initial set of parameters is: $\boldsymbol{x} = [0.2\ 1\ 0.2\ 4\ 0.2\ 0.2\ 1\ 0.2\ 4\ 0.2\ 0.2\ 1\ 0.2\ 4\ 0.2]^T$. Subsequently, NSM methodology of Sect. 3 has been utilized for optimization of the structure resulting in the following vector of design parameters: $\boldsymbol{x} = [1.21\ 0.18\ 10\ 0.3\ 0.5\ 1.5\ 0.2\ 8.98\ 0.3\ 0.19\ 0.97\ 0.15\ 10\ 0.3\ 0.15]^T$. Geometry of unconventional structure and a schematic diagram of $\boldsymbol{R}_{s.cell}$ interconnections are shown in Fig. 17.

The unconventional MT that offers over 51 % length reduction (length of 31.9 mm) in comparison to conventional structure (length of 65.3 mm) and reflection $|S_{11}| \leq -17$ within band of interest is obtained using only three iterations of NSM algorithm. Furthermore, $|S_{11}| \leq -15$ dB of an abbreviated circuit is obtained within 0.9–3.6 GHz frequency band. Similarly to the results of Sect. 4.1, the width of an unconventional MT is slightly greater than of conventional one. Figure 18 shows comparison of the reflection characteristics of the conventional and the abbreviated MT.

A total design and optimization cost of the structure, including eleven $\boldsymbol{R}_{f.cell}$ model evaluations during generation of the inner space mapping layer, three $\boldsymbol{R}_f$ model simulations for optimization of unconventional MT circuit, and evaluations

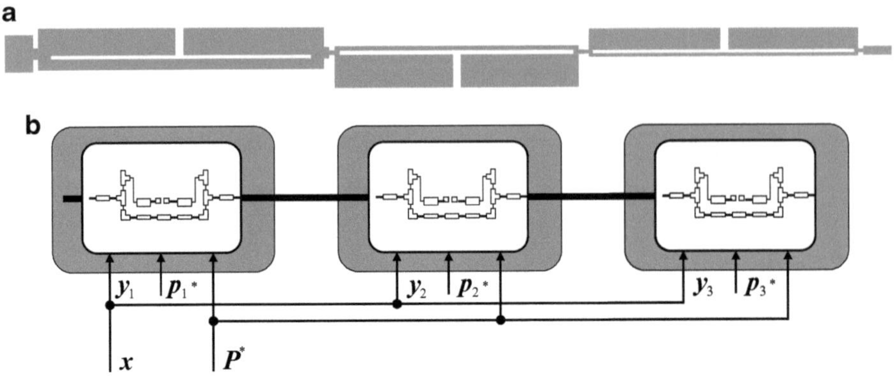

Fig. 17 An abbreviated MT composed of four C-type SWRS components: (**a**) geometry of an optimized structure; (**b**) schematic diagram of SWRS cascade connection

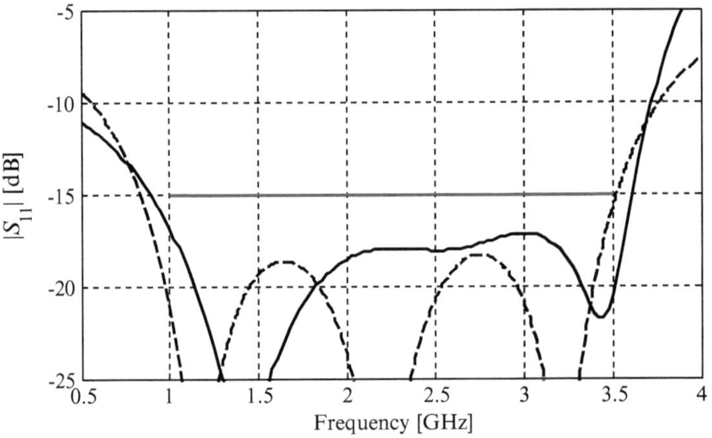

Fig. 18 Reflection characteristics of the conventional (*dashed line*) and abbreviated (*solid line*) MT

of surrogate models ($\boldsymbol{R}_{s.g}$ and $\boldsymbol{R}_{s.cell}$) corresponds to about 0.5 $\boldsymbol{R}_f$ model simulations (~1.2 h). Alternative techniques including ISM and SSM as well as direct optimization driven by pattern search algorithm have been also utilized for the design and optimization of abbreviated MT. SSM and ISM requires twice as many iterations as NSM approach, while direct optimization required 520 iterations to complete, which turns the method virtually impractical for such circuits. A detailed comparison of the methods in terms of iterations is collected in Table 2.

Table 2 An abbreviated three section MT: design and optimization cost

	Optimization algorithm			
Model evaluations	NSM	ISM	SSM	Direct search
SWRS $\boldsymbol{R}_{s.cell}$	$0.2\times\boldsymbol{R}_f$	N/A	N/A	N/A
SWRS $\boldsymbol{R}_{f.cell}$	$0.6\times\boldsymbol{R}_f$	N/A	N/A	N/A
MT low-fidelity $\boldsymbol{R}_s$	$0.3\times\boldsymbol{R}_f$	$3\times\boldsymbol{R}_f$	$1.3\times\boldsymbol{R}_f$	N/A
MT high-fidelity $\boldsymbol{R}_f$	3	7	6	520
Total cost	$4.1\times\boldsymbol{R}_f$	$10\times\boldsymbol{R}_f$	$7.3\times\boldsymbol{R}_f$	$520\times\boldsymbol{R}_f$
Total cost [h]	1.2	3.2	2.3	149.8

4.3 Design of a Compact Rat-Race Coupler

Our last example is an unconventional rat-race coupler (RRC). The structure is desired to fulfill the following design specifications: (1) at least 20 % bandwidth defined for both isolation $|S_{41}|$ and reflection coefficients $|S_{11}|$ below −20 dB and (2) −3 dB coupling. Both goals are considered for the given operating frequency $f_0 = 1$ GHz. Moreover, the design is intended to achieve at least 80 % of footprint reduction in comparison with conventional rectangle-based RRC. The circuit is desired to operate on Taconic RF-35 dielectric substrate ($\varepsilon_r = 3.5$, $\tan\delta = 0.0018$, $h = 0.762$).

A conventional, equal-split RRC may be constructed using well-known even-odd mode analysis [70]. The reference circuit is composed of six TL sections of characteristic impedance $Z_C = \sqrt{2}\ Z_0\,\Omega$ and electrical length $\theta = 90°$, where $Z_0 = 50\ \Omega$ is the characteristic impedance of feed lines. Moreover, TL sections are interconnected through tee junctions and microstrip bends (see Fig. 2b). Physical dimensions of the conventional structure are represented by a vector: $\boldsymbol{x}_r = [w_{r1}\ l_{r1}]^T$. The dimensions of reference design are: $\boldsymbol{x}_r = [0.87\ 45.8]^T$, while parameters $l_0 = 10$, $w_0 = 1.7$ are fixed to ensure 50 Ω feed (all dimensions in mm). One should emphasize that the design is characterized by a considerable footprint of ~4,536 mm^2 (47.5 × 95.5 mm^2).

A conventional RRC may be decomposed into six TL sections characterized by the same electrical parameters. Unfortunately, determination of SWRS component for TL replacement by means of database approach prevents sufficient circuit miniaturization. For that reason, we perform a knowledge-based construction of SWRSs (cf. Sect. 2.3) aimed at manual forming of components to maximally utilize interior of the coupler. To satisfy miniaturization requirements, $n = 2$ complementary SWRS components based on T-type topology have been constructed. Both cells are represented by the following vectors of design parameters: $\boldsymbol{y}^{(1)} = [w_{11}\ l_{11}\ l_{21}\ l_{31}\ l_{41}]^T$ and $\boldsymbol{y}^{(2)} = [w_{12}\ l_{12}\ l_{22}\ l_{32}\ l_{42}]^T$. Technology limitations impose the lower/upper bounds of each structure dimensions: $\boldsymbol{l}^{(1)} = [0.2\ 0.2\ 0.2\ 0.2\ 0.2]^T$, $\boldsymbol{u}^{(1)} = [0.5\ 0.5\ 4\ 0.5\ 4]^T$, and $\boldsymbol{l}^{(2)} = [0.2\ 0.2\ 0.2\ 0.2\ 0.2]^T$, $\boldsymbol{u}^{(2)} = [0.5\ 1\ 7\ 0.5\ 7]^T$. The high-fidelity models $\boldsymbol{R}_{f.cell}{}^{(n)}$ of SWRS are both prepared in CST Microwave Studio (~340,000

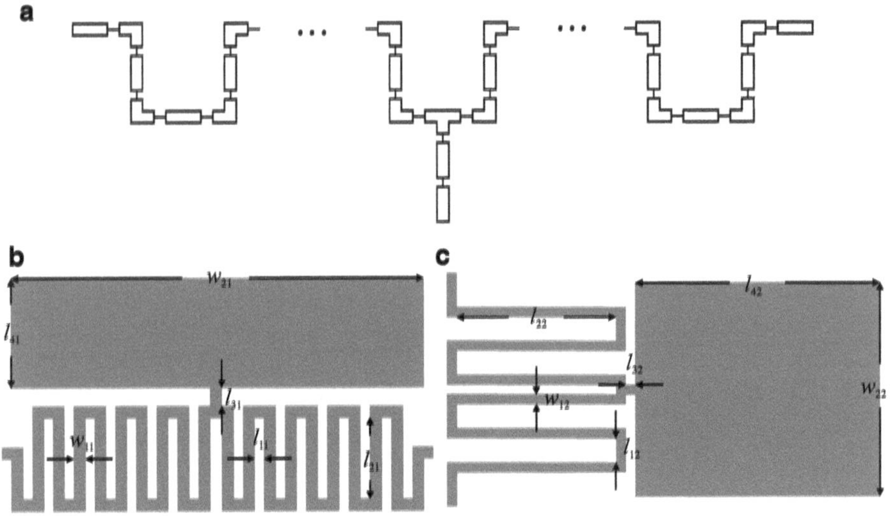

Fig. 19 A T-type SWRS components designed by means of knowledge-based approach: (**a**) general visualization of low-fidelity $\boldsymbol{R}_{c.cell}{}^{(1)}$ and $\boldsymbol{R}_{c.cell}{}^{(2)}$ models; (**b**) topology of $\boldsymbol{R}_{f.cell}{}^{(1)}$ model with highlighted six geometrical parameters ($w_{21} = 18l_{11} + 21w_{11}$); (**c**) topology of $\boldsymbol{R}_{f.cell}{}^{(2)}$ model 6 geometrical parameters are highlighted ($w_{22} = 6l_{12} + 7w_{12}$)

and ~400,000 mesh cells, as well as 4 and 4.5 min evaluation time for $\boldsymbol{R}_{f.cell}{}^{(1)}$ and $\boldsymbol{R}_{f.cell}{}^{(2)}$ respectively). The low-fidelity models $\boldsymbol{R}_{c.cell}{}^{(n)}$ of both structures are constructed in Agilent ADS circuit simulator. Designs of both components with highlighted geometrical parameters as well as visualization of their circuit models are illustrated in Fig. 19.

A star-distribution design of experiments scheme for training data allocation (cf. Sect. 3.2) has been utilized for a multipoint parameter extraction. Inner space mapping layers of $\boldsymbol{R}_{s.cell}{}^{(1)}$ and $\boldsymbol{R}_{s.cell}{}^{(2)}$ models have been both composed of 18 parameters. The surrogate model responses of both SWRS components before and after multipoint parameter extraction are shown in Fig. 20.

A global model $\boldsymbol{R}_{s.g}$ of compact RRC has been constructed using two $\boldsymbol{R}_{s.cell}{}^{(1)}$ and four $\boldsymbol{R}_{s.cell}{}^{(2)}$ models that substitute TL sections of a conventional circuit. The coupler dimensions are represented by the following vector: $\boldsymbol{x} = [w_{11}\ l_{11}\ l_{21}\ l_{31}\ l_{41}\ w_{12}\ l_{12}\ l_{22}\ l_{32}\ l_{42}]^T$, whereas $w_{10} = 0.75$, $l_{10} = 4.3$, $l_{20} = 0.4$ remain fixed. Moreover, $w_{21} = 18l_{11} + 21w_{11}$ and $w_{22} = 6l_{12} + 7w_{12}$. A high-fidelity model $\boldsymbol{R}_f$ of the structure has been prepared in CST Microwave Studio (~800,000 mesh cells and average simulation time 75 min per design). The initial design parameters are: $\boldsymbol{x} = [0.2\ 0.2\ 2.5\ 0.2\ 2.5\ 0.2\ 0.2\ 5.0\ 0.2\ 5.0]^T$. A NSM methodology of Sect. 3 has been utilized for optimization of the structure resulting in the following vector of design parameters: $\boldsymbol{x} = [0.24\ 0.26\ 3.35\ 0.28\ 2.04\ 0.25\ 0.29\ 6.52\ 0.29\ 5.63]^T$. Geometry of the unconventional structure and a schematic diagram of $\boldsymbol{R}_{s.g}$ are shown in Fig. 21.

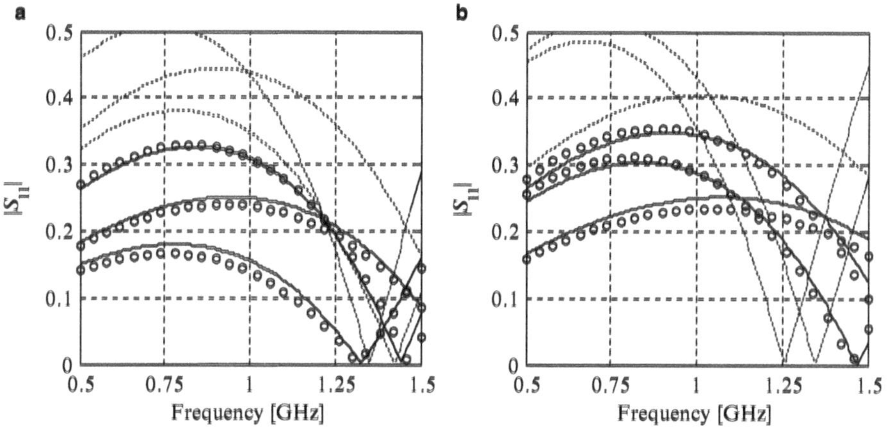

Fig. 20 NSM modeling of T-type SWRS components. Responses at the selected test designs—coarse model (*dotted line*), fine model (*solid line*), NSM surrogate after multipoint parameter extraction (*open circle*): (**a**) $\boldsymbol{R}_{s.cell}^{(1)}$; (**b**) $\boldsymbol{R}_{s.cell}^{(2)}$

The final design of compact RRC has been obtained after four iterations of the NSM algorithm. The footprint of the miniaturized structure is only $17 \times 27.3 = 464$ mm^2 and thus it offers a considerable miniaturization of almost 90 % with respect to the conventional circuit (4,536 mm^2).

The actual obtained −20 dB bandwidth is 23.5 %, which is broader than the one assumed in the specifications. The lower and upper operating frequencies are 0.915 and 1.150 GHz, respectively. A very slight shift of the operational frequency may be observed. What is also important, low-pass properties of SWRS components [73] introduced attenuation of harmonic frequencies up to 4 GHz. Narrow-band transmission characteristics of miniaturized RRC as well as a comparison of broadband responses of compact and conventional RRC is shown in Fig. 22.

A total design and optimization cost of the structure (~8.6 h), including 11 $\boldsymbol{R}_{f.cell}^{(1)}$ and 11 $\boldsymbol{R}_{f.cell}^{(2)}$ model evaluations during generation of the inner space mapping layers, four $\boldsymbol{R}_f$ model simulations for RRC optimization, and evaluations of surrogate models ($\boldsymbol{R}_{s.g}$, $\boldsymbol{R}_{s.cell}^{(1)}$ and $\boldsymbol{R}_{s.cell}^{(2)}$) corresponds to about 1.2 simulations of $\boldsymbol{R}_f$ model. Additionally, the circuit has been optimized using alternative SBO techniques and direct optimization driven by pattern search algorithm. The NSM algorithm clearly outperforms the benchmark techniques. A detailed comparison of the methods in terms of iterations is provided in Table 3.

5 Conclusions

In this chapter, a technique for fast design and optimization of computationally expensive microwave/RF circuits with enhanced functionality has been discussed. The design procedure is based on a decomposition of conventional passive structure

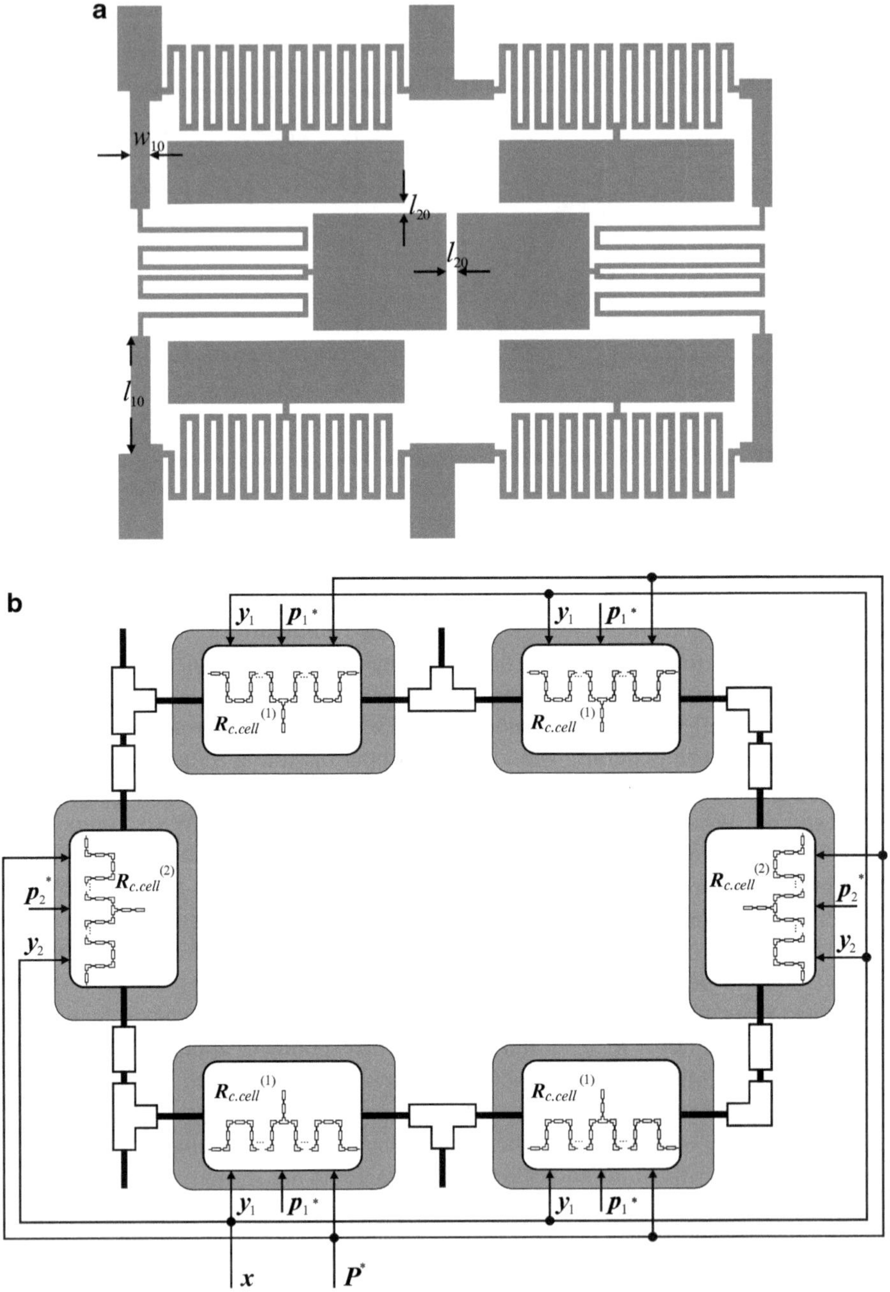

Fig. 21 A compact rat-race coupler composed of four $R_{s.cell}^{(1)}$ and two $R_{s.cell}^{(2)}$ T-type SWRS components: (**a**) geometry of an optimized structure; (**b**) schematic diagram of SWRS interconnections

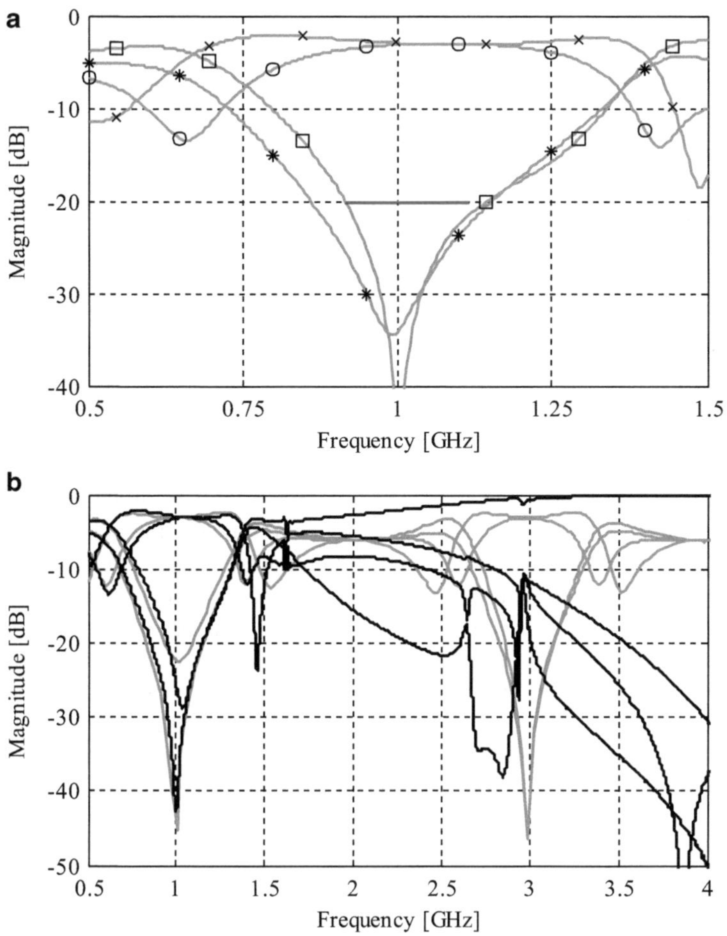

Fig. 22 A compact rat-race coupler: (**a**) transmission characteristics—(*asterisk*) reflection, (*open circle*) transmission, (*cross*) coupling, (*open square*) isolation; (**b**) attenuation of harmonic frequencies of compact RRC (*black line*) in comparison with conventional one (*gray line*)

Table 3 A compact rat-race coupler: design and optimization cost

	Optimization algorithm			
Model evaluations	NSM	ISM	SSM	Direct search
SWRS $\boldsymbol{R}_{s.cell}$	$0.9 \times \boldsymbol{R}_f$	N/A	N/A	N/A
SWRS $\boldsymbol{R}_{f.cell}$	$1.3 \times \boldsymbol{R}_f$	N/A	N/A	N/A
MT low-fidelity $\boldsymbol{R}_s$	$0.3 \times \boldsymbol{R}_f$	$3.9 \times \boldsymbol{R}_f$	$3.2 \times \boldsymbol{R}_f$	N/A
MT high-fidelity $\boldsymbol{R}_f$	4	16	12	286
Total cost	$6.5 \times \boldsymbol{R}_f$	$18.9 \times \boldsymbol{R}_f$	$15.2 \times \boldsymbol{R}_f$	$286 \times \boldsymbol{R}_f$
Total cost [h]	8.6	26.5	18.7	363.6

into a set of transmission line sections and their substitution with respective slow-wave resonant structures. Two distinct approaches for a determination of SWRS for transmission line replacement based on selection of a proper component from a predefined database and knowledge-based construction of the cell have been described.

Direct design of unconventional microwave/RF circuits constituted by SWRS components is a computationally expensive problem that may be efficiently handled only by means of surrogate-based optimization. Here, we discuss a NSM methodology, which is a two-stage design and optimization approach utilizing inner space mapping layer prepared at the level of each SWRS component, and the outer layer constructed at the level of the entire unconventional circuit. Representing of the design in such a way allows for a construction of a SWRS component model that exhibits good generalization capability and may be reused for the design various unconventional circuits.

The introduced technique allows for a fast and reliable design of computationally expensive circuits constituted of SWRS components. It is illustrated using three exemplary planar structures: a 16-variable four-section UWB matching transformer, a 15-variable three section broadband matching transformer, and a 10-variable rat-race coupler. All structures are successfully designed in a timeframe being only a fraction in comparison conventional design based on direct optimization scheme. Despite promising results, the discussed technique requires a considerable engineering knowledge to perform circuit decomposition and determine sufficient SWRS components. This is somehow problematic from the point of view of full automation of the design process. Expanding of the presented methods with the aim of design automation will be the subject of the future research.

References

1. Chen, S.-B., Jiao, Y.-C., Wang, W., Zhang, F.-S.: Modified T-shaped planar monopole antennas for multiband operation. IEEE Trans. Microw. Theory Tech. **54**, 3267–3270 (2006)
2. Xu, J., Miao, C., Cui, L., Ji, Y.-X., Wu, W.: Compact high isolation quad-band bandpass filter using quad-mode resonator. Electron. Lett. **48**, 28–30 (2012)
3. Liu, H.-W., Wang, Y., Wang, X.-M., Lei, J.-H., Xu, W.-Y., Zhao, Y.-L., Ren, B.-P., Guan, X.-H.: Compact and high selectivity tri-band bandpass filter using multimode stepped-impedance resonator. IEEE Microw. Wirel. Compon. Lett. **23**, 536–538 (2013)
4. Rodenbeck, C.T., Sang-Gyu, K., Wen-Hua, T., Coutant, M.R., Seungpyo, H., Mingyi, L., Kai, C.: Ultra-wideband low-cost phased-array radars. IEEE Trans. Microw. Theory Tech. **53**, 3697–3703 (2005)
5. Kuo, T.-N., Lin, S.-C., Chen, C.H.: Compact ultra-wideband bandpass filters using composite microstrip-coplanar-waveguide structure. IEEE Trans. Microw. Theory Tech. **54**, 3772–3778 (2006)
6. An-Shyi, L., Huang, T.-Y., Wu, R.-B.: A dual wideband filter design using frequency mapping and stepped-impedance resonators. IEEE Trans. Microw. Theory Tech. **56**, 2921–2929 (2008)
7. Zhang, X.-Y., Xue, Q.: High-selectivity tunable bandpass filters with harmonic suppression. IEEE Trans. Microw. Theory Tech. **58**, 964–969 (2010)

8. Sun, S., Zhu, L.: Periodically nonuniform coupled microstrip-line filters with harmonic suppression using transmission zero reallocation. IEEE Trans. Microw. Theory Tech. **53**, 1817–1822 (2005)
9. Ngoc-Anh, N., Ahmad, R., Yun-Taek, I., Yong-Sun, S., Seong-Ook, P.: A T-shaped wide-slot harmonic suppression antenna. IEEE Antennas Wirel. Propag. Lett. **6**, 647–650 (2007)
10. Deng, C., Li, P., Cao, W.: A high-isolation dual-polarization patch antenna with omnidirectional radiation patterns. IEEE Antennas Wirel. Propag. Lett. **11**, 1273–1276 (2012)
11. Zeng, S.-J., Wu, J.-Y., Tu, W.-H.: Compact and high-isolation quadruplexer using distributed coupling technique. IEEE Microw. Wirel. Compon. Lett. **21**, 197–199 (2011)
12. Chappell, W.J., Little, M.P., Katehi, L.P.B.: High isolation, planar filters using EBG substrates. IEEE Microw. Wirel. Compon. Lett. **11**, 246–248 (2001)
13. Hee-Ran, A., Itoh, T.: New isolation circuits of compact impedance-transforming 3-dB baluns for theoretically perfect isolation and matching. IEEE Trans. Microw. Theory Tech. **58**, 3892–3902 (2010)
14. Hee-Ran, A., Sangwook, N.: Compact microstrip 3-dB coupled-line ring and branch-line hybrids with new symmetric equivalent circuits. IEEE Trans. Micro. Theory Tech. **61**, 1067–1078 (2013)
15. Tao, Y., Pei-Ling, C., Itoh, T.: Compact quarter-wave resonator and its applications to miniaturized diplexer and triplexer. IEEE Trans. Microw. Theory Tech. **59**, 260–269 (2011)
16. Milligan, T.A.: Modern Antenna Design, 2nd edn. Wiley, New York (2005)
17. Pozar, D.M.: Microwave Engineering, 4th edn. Wiley, New York (2012)
18. Azadegan, R., Sarabandi, K.: Bandwidth enhancement of miniaturized slot antennas using folded, complementary, and self-complementary realizations. IEEE Trans. Antennas Propag. **55**, 2435–2444 (2007)
19. Ruiz-Cruz, J.A., Yunchi, Z., Zaki, K.A., Piloto, A.J., Tallo, J.: Ultra-wideband LTCC ridge waveguide filters. IEEE Microw. Wirel. Compon. Lett. **17**, 115–117 (2007)
20. Hou, J.-A., Wang, Y.-H.: Design of compact 90° and 180° couplers with harmonic suppression using lumped-element bandstop resonators. IEEE Trans. Microw. Theory Tech. **58**, 2932–2939 (2010)
21. Chen, W.-L., Wang, G.-M.: Exact design of novel miniaturised fractal-shaped branch-line couplers using phase-equalising method. IET Microw. Antennas Propag. **2**, 773–780 (2008)
22. Kaymaram, F., Shafai, L.: Enhancement of microstrip antenna directivity using double-superstrate configurations. Can. J. Electr. Comput. Eng. **32**, 77–82 (2007)
23. Opozda, S., Kurgan, P., Kitlinski, M.: A compact seven-section rat-race hybrid coupler incorporating PBG cells. Microw. Opt. Technol. Lett. **51**, 2910–2913 (2009)
24. Kurgan, P., Kitlinski, M.: Novel doubly perforated broadband microstrip branch-line couplers. Microw. Opt. Technol. Lett. **51**, 2149–2152 (2009)
25. Tseng, C.-H., Chen, H.-J.: Compact rat-race coupler using shunt-stub-based artificial transmission lines. IEEE Microw. Wirel. Compon. Lett. **18**, 734–736 (2008)
26. Kurgan, P., Bekasiewicz, A., Pietras, M., Kitlinski, M.: Novel topology of compact coplanar waveguide resonant cell low-pass filter. Microw. Opt. Technol. Lett. **54**, 732–735 (2012)
27. Bekasiewicz, A., Kurgan, P., Kitlinski, M.: A new approach to a fast and accurate design of microwave circuits with complex topologies. IET Microw. Antennas Propag. **6**, 1616–1622 (2012)
28. Nocedal, J., Wright, S.: Numerical Optimization, 2nd edn. Springer, New York (2006)
29. Rios, L.M., Sahinidis, N.V.: Derivative-free optimization: a review of algorithms and comparison of software implementations. J. Glob. Optim. **56**, 1247–1293 (2013)
30. Bakr, M.H., Nikolova, N.K.: An adjoint variable method for time domain TLM with wideband Johns matrix boundaries. IEEE Trans. Microw. Theory Tech. **52**, 678–685 (2004)
31. Chung, Y.S., Cheon, C., Park, I.H., Hahn, S.Y.: Optimal design method for microwave device using time domain method and design sensitivity analysis-part II: FDTD case. IEEE Trans. Magn. **37**, 3255–3259 (2001)
32. El Sabbagh, M.A., Bakr, M.H., Bandler, J.W.: Adjoint higher order sensitivities for fast full-wave optimization of microwave filters. IEEE Trans. Microw. Theory Tech. **54**, 3339–3351 (2006)

33. Koziel, S., Mosler, F., Reitzinger, S., Thoma, P.: Robust microwave design optimization using adjoint sensitivity and trust regions. Int. J. RF Microw. Comput. Aid. Eng. **22**, 10–19 (2012)
34. Deb, K.: Multi-Objective Optimization Using Evolutionary Algorithms. Wiley, Chichester (2001)
35. Talbi, E.-G.: Metaheuristics – From Design to Implementation. Wiley, Chichester (2009)
36. Bandler, J.W., Cheng, Q.S., Dakroury, S.A., Mohamed, A.S., Bakr, M.H., Madsen, K., Søndergaard, J.: Space mapping: the state of the art. IEEE Trans. Microw. Theory Tech. **52**, 337–361 (2004)
37. Echeverría, D., Hemker, P.W.: Manifold mapping: a two-level optimization technique. Comput. Vis. Sci. **11**, 193–206 (2008)
38. Koziel, S., Leifsson, L., Ogurtsov, S.: Reliable EM-driven microwave design optimization using manifold mapping and adjoint sensitivity. Microw. Opt. Technol. Lett. **55**, 809–813 (2013)
39. Koziel, S.: Shape-preserving response prediction for microwave design optimization. IEEE Trans. Microw. Theory Tech. **58**, 2829–2837 (2010)
40. Leifsson, L., Koziel, S.: Multi-fidelity design optimization of transonic airfoils using physics-based surrogate modeling and shape-preserving response prediction. J. Comput. Sci. **1**, 98–106 (2010)
41. Koziel, S., Cheng, Q.S., Bandler, J.W.: Space mapping. IEEE Microw. Magazine **9**, 105–122 (2008)
42. Koziel, S., Bandler, J.W., Madsen, K.: A space mapping framework for engineering optimization: theory and implementation. IEEE Trans. Microw. Theory Tech. **54**, 3721–3730 (2006)
43. Koziel, S., Bekasiewicz, A., Zieniutycz, W.: Expedited EM-driven multi-objective antenna design in highly-dimensional parameter spaces. IEEE Antennas Wirel. Propag. Lett. **13**, 631–634 (2014)
44. Redhe, M., Nilsson, L.: Optimization of the new Saab 9-3 exposed to impact load using a space mapping technique. Struct. Multidiscip. Optim. **27**, 411–420 (2004)
45. Crevecoeur, G., Dupre, L., Van de Walle, R.: Space mapping optimization of the magnetic circuit of electrical machines including local material degradation. IEEE Trans. Magn. **43**, 2609–2611 (2007)
46. Encica, L., Makarovic, J., Lomonova, E.A., Vandenput, A.J.A.: Space mapping optimization of a cylindrical voice coil actuator. IEEE Trans. Ind. Appl. **42**, 1437–1444 (2006)
47. Bandler, J.W., Cheng, Q.S., Nikolova, N.K., Ismail, M.A.: Implicit space mapping optimization exploiting preassigned parameters. IEEE Trans. Microw. Theory Tech. **52**, 378–385 (2004)
48. Cheng, Q.S., Bandler, J.W., Koziel, S.: Combining coarse and fine models for optimal design. IEEE Microw. Magazine **9**, 79–88 (2008)
49. Koziel, S., Bandler, J.W., Cheng, Q.S.: Constrained parameter extraction for microwave design optimisation using implicit space mapping. IET Microw. Antennas Propag. **5**, 1156–1163 (2011)
50. Kurgan, P., Bekasiewicz, A.: A robust design of a numerically demanding compact rat-race coupler. Microw. Opt. Technol. Lett. **56**, 1259–1263 (2014)
51. Koziel, S., Bekasiewicz, A., Kurgan, P.: Rapid EM-driven design of compact RF circuits by means of nested space mapping. IEEE Microw. Wirel. Compon. Lett. **24**(6), 364–366 (2014)
52. Awida, M.A., Safwat, A.M.E., El-Hennawy, H.: Compact rat-race hybrid coupler using meander space-filling curves. Microw. Opt. Technol. Lett. **48**, 606–609 (2006)
53. Eccleston, K.W., Ong, S.H.M.: Compact planar microstrip line branch-line and rat-race couplers. IEEE Trans. Microw. Theory Tech. **51**, 2119–2125 (2003)
54. Kurgan, P., Kitlinski, M.: Doubly miniaturized rat-race hybrid coupler. Microw. Opt. Technol. Lett. **53**, 1242–1244 (2011)
55. Kurgan, P., Kitlinski, M.: Novel microstrip low-pass filters with fractal defected ground structures. Microw. Opt. Technol. Lett. **51**, 2473–2477 (2009)
56. Wen, W., Lu, Y., Fu, J.S., Yong, Z.X.: Particle swarm optimization and finite-element based approach for microwave filter design. IEEE Trans. Magnetics **41**, 1800–1803 (2005)

57. Lai, M.-I., Jeng, S.-K.: Compact microstrip dual-band bandpass filters design using genetic-algorithm techniques. IEEE Trans. Microw. Theory Tech. **54**, 160–168 (2006)
58. Chen, C.-F., Lin, C.-Y., Weng, J.-H., Tsai, K.-L.: Compact microstrip broadband filter using multimode stub-loaded resonator. Electron. Lett. **49**, 545–546 (2013)
59. Meissner, P., Kitlinski, M.: A 3-dB multilayer coupler with UC-PBG structure. IEEE Microw. Wirel. Compon. Lett. **15**, 52–54 (2005)
60. Lin, B.-Q., Zheng, Q.-R., Yuan, N.-C.: A novel planar PBG structure for size reduction. IEEE Microw. Wirel. Compon. Lett. **16**, 269–271 (2006)
61. Hong, J.-S., Lancaster, M.J.: Theory and experiment of novel microstrip slow-wave open-loop resonator filters. IEEE Trans. Microw. Theory Tech. **45**, 2358–2365 (1997)
62. García-García, J., Bonache, J., Gil, I., Martín, F., Marqués, R., Falcone, F., Lopetegi, T., Laso, M.A.G., Sorolla, M.: Comparison of electromagnetic band gap and split-ring resonator microstrip lines as stop band structures. Microw. Opt. Technol. Lett. **44**, 376–379 (2005)
63. Zhang, F.: High-performance rat-race hybrid ring for RF communication using MEBE-on-microstrip technology. Microw. Opt. Technol. Lett. **51**, 1539–1542 (2009)
64. Nanbo, J., Rahmat-Samii, Y.: Hybrid real-binary particle swarm optimization (HPSO) in engineering electromagnetics. IEEE Trans. Antennas Propag. **58**, 3786–3794 (2010)
65. Lai, M.-I., Jeng, S.-K.: A microstrip three-port and four-channel multiplexer for WLAN and UWB coexistence. IEEE Trans. Microw. Theory Tech. **53**, 3244–3250 (2005)
66. Nishino, T., Itoh, T.: Evolutionary generation of microwave line-segment circuits by genetic algorithms. IEEE Trans. Microw. Theory Tech. **50**, 2048–2055 (2002)
67. Smierzchalski, M., Kurgan, P., Kitlinski, M.: Improved selectivity compact band-stop filter with Gosper fractal-shaped defected ground structures. Microw. Opt. Technol. Lett. **52**(1), 227–232 (2010)
68. Zhang, C.F.: Planar rat-race coupler with microstrip electromagnetic bandgap element. Microw. Opt. Technol. Lett. **53**, 2619–2622 (2011)
69. Bekasiewicz, A., Kurgan, P.: A compact microstrip rat-race coupler constituted by nonuniform transmission lines. Microw. Opt. Technol. Lett. **56**, 970–974 (2014)
70. Matthaei, G., Jones, E.M.T., Young, L.: Microwave Filters, Impedance-Matching Networks, and Coupling Structures. Artech House, Norwood (1980)
71. Hong, J.-S., Lancaster, M.J.: Microstrip Filters for RF/Microwave Applications. Wiley, Hoboken (2001)
72. Kurgan, P., Kitlinski, M.: Slow-wave fractal-shaped compact microstrip resonant cell. Microw. Opt. Technol. Lett. **52**, 2613–2615 (2010)
73. Kurgan, P., Filipcewicz, J., Kitlinski, M.: Development of a compact microstrip resonant cell aimed at efficient microwave component size reduction. IET Microw. Antennas Propag. **6**, 1291–1298 (2012)
74. Kurgan, P., Filipcewicz, J., Kitlinski, M.: Design considerations for compact microstrip resonant cells dedicated to efficient branch-line miniaturization. Microw. Opt. Technol. Lett. **54**, 1949–1954 (2012)
75. Bekasiewicz, A., Koziel, S.: Local–global space mapping for rapid EM-driven design of compact RF structures. Int. Conf. Microw. Radar Wirel. Commun. **1**, 313–316 (2014)
76. Koziel, S., Kurgan, P.: Low-cost optimization of compact branch-line couplers and its application to miniaturized Butler matrix design. Eur. Microw. Conf. Rome, Italy, Oct. 5–10, (2014)
77. Koziel, S., Echeverría-Ciaurri, D., Leifsson, L.: Simulation-driven design in microwave engineering: methods. In: Koziel, S., Yang, X.S. (eds.) Computational Optimization, Methods and Algorithms. Series: Studies in Computational Intelligence, pp. 33–60. Springer, New York (2011)
78. Cheng, Q.S., Rautio, J.C., Bandler, J.W., Koziel, S.: Progress in simulator-based tuning—the art of tuning space mapping [application notes]. IEEE Microw. Magazine **11**, 96–110 (2010)
79. Bandler, J.W., Cheng, Q.S., Hailu, D.M., Nikolova, N.K.: A space-mapping design framework. IEEE Trans. Microw. Theory Tech. **52**, 2601–2610 (2004)

80. Kolda, T.G., Lewis, R.M., Torczon, V.: Optimization by direct search: new perspectives on some classical and modern methods. SIAM Rev. **45**, 385–482 (2003)
81. Goldberg, D.E.: Genetic Algorithms in Search, Optimization and Machine Learning. Addison-Wesley, New York (1989)
82. Koziel, S., Echeverría-Ciaurri, D., Leifsson, L.: Surrogate-based methods. In: Koziel, S. Yang, X.S. (eds.) Computational Optimization, Methods and Algorithms. Series: Studies in Computational Intelligence, pp. 33–60. Springer, New York (2011)
83. Koziel, S., Leifsson, L., Ogurtsov, S.: Space mapping for electromagnetic-simulation-driven design optimization. In: Koziel, S. Leifsson, L. (eds.) Surrogate-Based Modeling and Optimization: Applications in Engineering, pp. 1–25. Springer, New York (2013)
84. CST Microwave Studio, ver. 2013.: CST AG, Bad Nauheimer Str. 19, D-64289 Darmstadt, Germany (2013)
85. Agilent ADS, ver. 2011.10: Agilent Technologies, 1400 Fountaingrove Parkway, Santa Rosa, CA 95403-1799, (2011)

Automated Low-Fidelity Model Setup for Surrogate-Based Aerodynamic Optimization

Leifur Leifsson, Slawomir Koziel, and Piotr Kurgan

Abstract Computational fluid dynamics (CFD) simulations are a fundamental tool in aerodynamic design. Unfortunately, accurate, high-fidelity CFD models may be computationally too expensive to conduct the design using numerical optimization procedures. Recently, variable-fidelity optimization algorithms have attracted attention for their ability to reduce high CPU-cost related to the design process solely based on accurate CFD models. Low-fidelity simulation models are the most critical components of such algorithms. They normally employ the same CFD solver as the high-fidelity model but with reduced discretization density and reduced number of flow solver iterations. Typically, the selection of the appropriate model parameters has only been guided by the designer experience. In this chapter, an automated low-fidelity model selection technique is described. By defining the model setup task as a constrained nonlinear optimization problem, suitable grid and flow solver parameters are obtained. The approach is compared to two conventional methods of generating a family of variable-fidelity models. Comparison of the standard and the proposed approach is carried out in the context of aerodynamic design of transonic airfoils using a multi-level optimization algorithm. The results obtained for several test cases indicate that the automated model generation may lead to significant computational savings of the CFD-based airfoil design process. Illustration of the entire optimization cycle involving automated low-fidelity model preparation and B-spline-parameterized airfoil design using space mapping algorithm is also provided.

Keywords Aerodynamic optimization • Design automation • CFD • Transonic airfoils • Low-fidelity modeling

---

L. Leifsson (✉) • S. Koziel • P. Kurgan
Engineering Optimization & Modeling Center, School of Science and Engineering, Reykjavik University, Menntavegur 1, 101 Reykjavik, Iceland
e-mail: leifurth@ru.is; koziel@ru.is; kurgan@ru.is

S. Koziel et al. (eds.), *Solving Computationally Expensive Engineering Problems*, Springer Proceedings in Mathematics & Statistics 97,
DOI 10.1007/978-3-319-08985-0_4

1 Introduction

Shape optimization is one of the fundamental procedures of aerodynamic design [1, 2]. Modern optimization methodologies exploit computational fluid dynamics (CFD) simulations for reliable evaluation of engineering systems and components such as aircraft wings and turbine blades. Accurate high-fidelity CFD simulations are computationally expensive. Consequently, performing CFD-based design may be unrealistic when using conventional numerical optimization techniques, even with adjoint sensitivity information [3–6]. Surrogate-based optimization (SBO) methods [3–5] provide a promising way to reduce the computational cost of aerodynamic design, particularly the ones using physics-based surrogates [6–10]. Those methods are often referred to as variable- (or multi-) fidelity optimization methods. In these approaches, the optimization burden is shifted from an expensive model over to a suitably corrected low-fidelity model.

The low-fidelity models are the most critical part of variable-fidelity design techniques. In CFD-based design, the low-fidelity models are, typically, exploiting the same flow solver as the high-fidelity one, but with coarser discretization and a reduced number of flow solver iterations. One of the main challenges here is the selection of the discretization parameters and the flow solver convergence criteria for a fast and reliable CFD analysis of the low-fidelity model [11]. Currently, this process is hands-on and guided by engineering experience.

This chapter describes an automated procedure for setting up low-fidelity CFD models. The model setup task is formulated as a constrained nonlinear optimization problem which is solved numerically to find appropriate values of the parameters of the CFD model discretization and convergence criteria. The process is carried out for multiple designs simultaneously to ensure consistency of the low-fidelity model performance (both simulation time and accuracy) across the entire design space. The technique replaces an ad hoc method of constructing the low-fidelity model by hand to automate the process. Applications of the technique to the design of transonic airfoil shapes as well as comparisons with conventional approaches are provided.

2 Multi-Level CFD-Based Aerodynamic Shape Optimization

This section provides a formulation of the aerodynamic shape design problem. We also describe a variable-fidelity optimization approach, as well as a so-called multi-level optimization algorithm [12] as an exemplary design technique. Multi-level optimization exploits a family of CFD models of increasing discretization density so that a proper selection of the models is critical for the algorithm performance.

2.1 Problem Formulation

Airfoil shape optimization aims, in general, at finding the best geometry which maximizes the aerodynamic performance at a certain operating condition(s) for a given set of constraints. The problem can be formulated as follows:

$$\begin{aligned} &\min_{\mathrm{x}} f(x) \\ &s.t.\ g_j(\mathrm{x}) \leq 0,\ j = 1, \ldots, M \\ &\quad h_k(\mathrm{x}) = 0,\ k = 1, \ldots, N \\ &\quad l \leq \mathrm{x} \leq u \end{aligned} \tag{1}$$

where $f(\boldsymbol{x})$ is the objective function, $\boldsymbol{x}$ is the design variable vector, $g_j(\boldsymbol{x})$ are the inequality constraints, M is the number of the inequality constraints, $h_k(\boldsymbol{x})$ are the equality constraints, N is the number of the equality constraints, and $\boldsymbol{l}$ and $\boldsymbol{u}$ are the design variables lower and upper bounds, respectively.

The objective and constraint functions are assumed to be obtained through high-fidelity CFD simulation. The objective can be written as, for example, $f(\boldsymbol{x}) = C_d(\boldsymbol{x})$ with the constraints $g_1(\boldsymbol{x}) = C_{l.\min} - C_l(\boldsymbol{x}) \leq 0$ and $g_2(\boldsymbol{x}) = A_{\min} - A(\boldsymbol{x}) \leq 0$, where C_d is the drag coefficient, C_l is the lift coefficient, $C_{l.min}$ is a minimum lift coefficient, A is the cross-sectional area, and A_{min} is a minimum cross-sectional area.

2.2 Optimization Algorithm

A multi-level optimization algorithm is used here to solve the problem (1). The algorithm was first introduced in the area of microwave engineering [13], and later extended and applied to airfoil shape optimization [12]. The algorithm exploits a family of low-fidelity models denoted as $\{c_j\}$, $j = 1, \ldots, K$, all evaluated by the same CFD solver as the one used for the high-fidelity model f. Discretization of the model c_{j+1} is finer than that of the model c_j, which results in higher accuracy and, unfortunately, a longer evaluation time. In practice, $K = 2$ or 3. The discretization density may be controlled by solver-dependent parameters (e.g., the grid parameters).

The multi-level optimization works as follows. Starting from the initial design $\boldsymbol{x}^{(0)}$, the coarsest model c_1 is optimized to produce a first approximation of the high-fidelity model optimum, $\boldsymbol{x}^{(1)}$. The vector $\boldsymbol{x}^{(1)}$ is used as a starting point to find the subsequent approximation of the high-fidelity model optimum, $\boldsymbol{x}^{(2)}$, which is obtained by optimizing the next model, c_2. The process continues until the optimum $\boldsymbol{x}^{(K)}$ of the last low-fidelity model c_K is found.

Having $\boldsymbol{x}^{(K)}$, we evaluate the model c_K at all perturbed designs around $\boldsymbol{x}^{(K)}$, i.e., at $\boldsymbol{x}_k{}^{(K)} = [\boldsymbol{x}_1{}^{(K)} \ldots\ \boldsymbol{x}_k{}^{(K)} + \mathrm{sign}(k) \cdot d_k\ \ldots\ \boldsymbol{x}_n{}^{(K)}]^T$, $k = -n, -n+1, \ldots, n{-}1, n$. We use the notation $c^{(k)} = c_K(\boldsymbol{x}_k{}^{(K)})$. This data is used to refine the final design without directly optimizing the high-fidelity model f. More specifically, we set up an approximation model involving $c^{(k)}$ and optimize it in the vicinity of $\boldsymbol{x}^{(K)}$ defined as $[\boldsymbol{x}^{(K)} - d,\ \boldsymbol{x}^{(K)} + d]$, where $d = [d_1\ d_2\ \ldots\ d_n]^T$. The size of the area, i.e., the parameter d, can be selected based on sensitivity analysis of c_1 (the cheapest of the low-fidelity models); usually d equals 2–5 % of $\boldsymbol{x}^{(K)}$.

Here, approximation is performed using a reduced quadratic model $q(\boldsymbol{x}) = [q_1\ q_2\ \ldots\ q_m]^T$, defined as

$$q_j(x) = q_j\left([x_1 \cdots x_n]^T\right) = \lambda_{j.0} + \lambda_{j.1}x_1 + \cdots + \lambda_{j.n}x_n + \lambda_{j.n+1}x_1^2 + \cdots + \lambda_{j.2n}x_n^2 \tag{2}$$

The coefficients $\lambda_{j,r}$, $j = 1, \ldots, m$, $r = 0, 1, \ldots, 2n$, are uniquely obtained by solving the linear regression problems

$$\begin{bmatrix} 1 & x_{-n.1}^{(K)} & \cdots & x_{-n.n}^{(K)} & \left(x_{-n.1}^{(K)}\right)^2 & \cdots & \left(x_{-n.n}^{(K)}\right)^2 \\ \vdots & \vdots & & \vdots & \vdots & & \vdots \\ 1 & x_{0.1}^{(K)} & \cdots & x_{0.n}^{(K)} & \left(x_{0.1}^{(K)}\right)^2 & \cdots & \left(x_{-n.n}^{(K)}\right)^2 \\ \vdots & \vdots & & \vdots & \vdots & & \vdots \\ 1 & x_{n.1}^{(K)} & \cdots & x_{n.n}^{(K)} & \left(x_{n.1}^{(K)}\right)^2 & \cdots & \left(x_{-n.n}^{(K)}\right)^2 \end{bmatrix} \cdot \begin{bmatrix} \lambda_{j.0} \\ \lambda_{j.1} \\ \vdots \\ \lambda_{j.2n} \end{bmatrix} = \begin{bmatrix} c_j^{(-n)} \\ \vdots \\ c_j^{(0)} \\ \vdots \\ c_j^{(n)} \end{bmatrix} \tag{3}$$

where $\boldsymbol{x}_{k.j}^{(K)}$ is a jth component of the vector $\boldsymbol{x}_k^{(K)}$, and $c_j^{(k)}$ is a jth component of the vector $c^{(k)}$. In our case, the components of the response vector consist of the lift and drag coefficients, as well as the cross-section area.

In order to account for unavoidable misalignment between c_K and f, instead of optimizing the quadratic model q, it is recommended to optimize a corrected model $q(\boldsymbol{x}) + [f(\boldsymbol{x}^{(K)}) - c_K(\boldsymbol{x}^{(K)})]$ that ensures a zero-order consistency [6] between c_K and f. The refined design can be then found as

$$\mathbf{x}^* = \arg \min_{\mathbf{x}^{(K)} - d \le \mathbf{x} \le \mathbf{x}^{(K)} + d} H\left(q(\mathbf{x}) + \left[f\left(\mathbf{x}^{(K)}\right) - c_K\left(\mathbf{x}^{(K)}\right)\right]\right) \tag{4}$$

This type of correction is also known as output space mapping [14]. If necessary, the step (4) can be performed a few times starting from a refined design, i.e., $\boldsymbol{x}^{*} = \text{argmin}\{\boldsymbol{x}^{(K)} - d \le \boldsymbol{x} \le \boldsymbol{x}^{(K)} + d{:}\ H(q(\boldsymbol{x}) + [f(\boldsymbol{x}^{*}) - c_K(\boldsymbol{x}^{*})])\}$. It should be noted that the high-fidelity model is not evaluated until executing the refinement step (4). Also, each refinement-iteration requires only a single evaluation of f.

The optimization procedure can be summarized as follows (where K is the number of models):

1. Set $j = 1$;
2. Select the initial design $\boldsymbol{x}^{(0)}$;
3. Starting from $\boldsymbol{x}^{(j-1)}$ find $\boldsymbol{x}^{(j)} = \arg\min\{\boldsymbol{x}{:}\ H(c_j(\boldsymbol{x}))\}$;
4. Set $j = j + 1$; if $j < K$ go to 3;
5. Obtain a refined design according to (4).

The main benefit of using several models of varying fidelity is that starting from a less accurate but faster model allows us to quickly find an approximate location of the optimum design. Switching to finer models at the later stages allows us to locate the optimum more accurately without excessive computational effort because each algorithm-iteration starts from an already reasonable approximation of the optimum. Another benefit of this procedure is that, aside from the refinement stage, no enhancement/correction of the low-fidelity models is necessary, which is in contrast to most of other SBO techniques. Therefore, the multi-level approach is less dependent on the low-fidelity model quality.

3 High-Fidelity Model

In this section we consider a two-dimensional CFD model describing transonic flow past airfoil sections. The steady compressible Euler equations are taken to be the governing fluid flow equations.

3.1 Computational Grid

The computational grids are of structured curvilinear body-fitted C-topology, as shown in Fig. 1, with elements clustering around the airfoil and growing in size with distance from the airfoil surface. The free-stream Mach number, static pressure, and angle of attack are prescribed at the farfield boundary. The solution domain boundaries are placed at 25 chord lengths in front of the airfoil, 50 chord lengths behind it, and 25 chord lengths above and below it.

The grid density is controlled by the following parameters (shown in Fig. 1a): m_1 = number grid points on the upper and lower airfoil surfaces, m_2 = number grid points on the horizontal line behind the airfoil from the trailing edge to the farfield boundary, m_3 = number of grid points on the vertical line from the airfoil surface to one-fourth of the distance to the farfield, m_4 = number of grid points on the three-fourths of the vertical line from the farfield down to the airfoil surface, and m_5 = distance from the airfoil surface to the first grid point. Local clustering on the airfoil surface is also controlled, but not parameterized. The computer code ICEM CFD [15] is used for the grid generation. An example grid is shown in Fig. 1b.

3.2 Flow Solver

The flow solver is of implicit density-based formulation and the inviscid fluxes are calculated by an upwind-biased second-order spatially accurate Roe flux scheme. Asymptotic convergence to a steady state solution is obtained in each case.

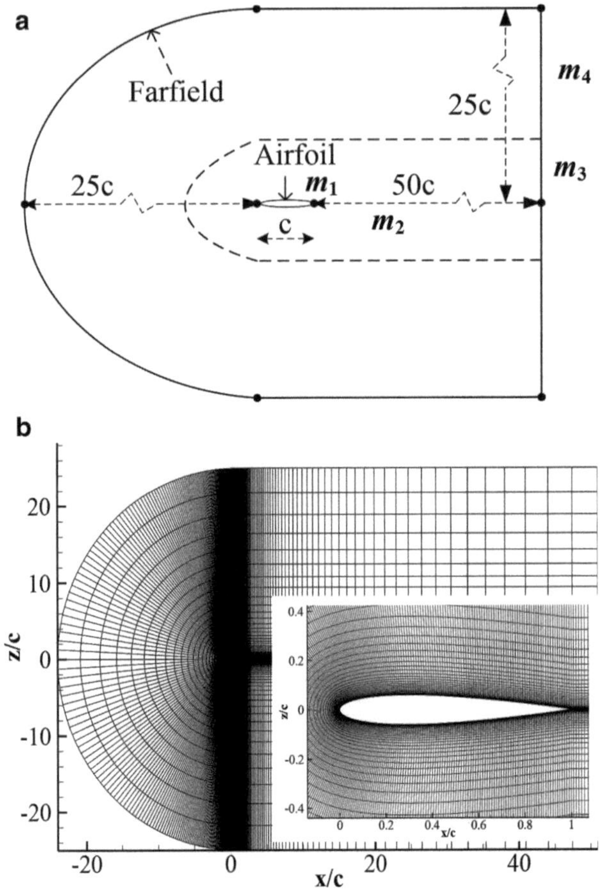

Fig. 1 Structured curvilinear body-fitted C-topology grid for transonic airfoil flow simulation, (**a**) the computational domain, and (**b**) an example grid for the NACA 0012 airfoil

The iterative convergence of each solution is examined by monitoring the overall residual, which is the sum (over all the cells in the computational domain) of the L^2 norm of all the governing equations solved in each cell. The solution convergence criterion for the high-fidelity model is the one that occurs first of the following: a reduction in the residual by six orders, or a maximum number of iterations of 1,000. Numerical fluid flow simulations are performed using the computer code FLUENT [16].

3.3 Grid Independence

A grid independence study was performed using the NACA 0012 airfoil at Mach number $M_\infty = 0.75$ and an angle of attack $\alpha = 1°$. The results of the study, shown in Fig. 2a, reveal that ~400,000 grid cells are needed for mesh convergence.

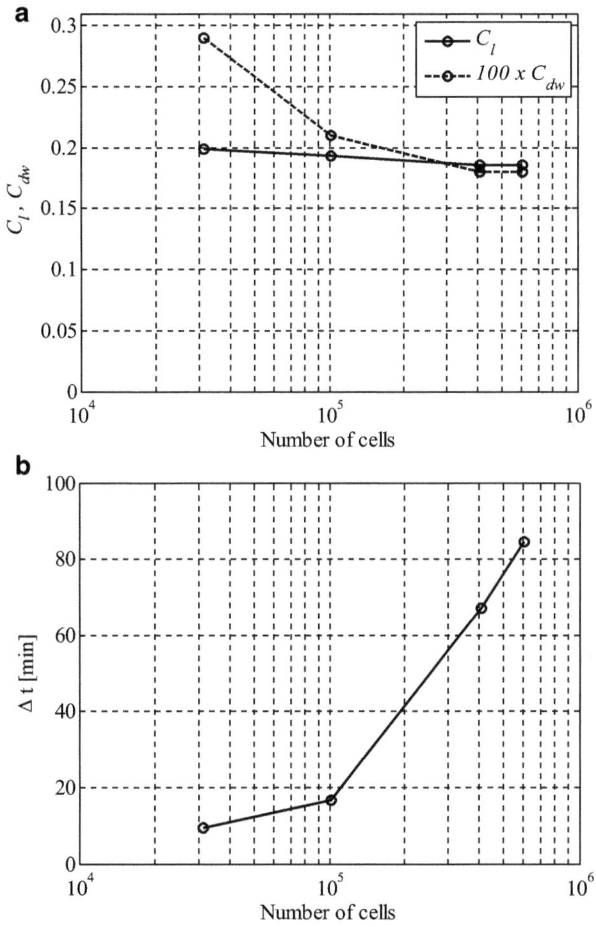

Fig. 2 Grid independence study using the NACA 0012 airfoil at Mach number $M_\infty = 0.75$ and an angle of attack $\alpha = 1°$; (**a**) lift and drag coefficients versus the number mesh cells, and (**b**) the simulation time versus the number of mesh cells

The high-fidelity CFD model is based on that particular grid. The overall simulation time for the case considered is roughly 67 min (Fig. 2b). The flow solver reached a converged solution after 352 iterations. The other grids required around 350–500 iterations to converge, except the coarsest one, which terminated after 1,000 iterations, with the overall simulation time around 9.5 min.

4 Low-Fidelity Models

The low-fidelity models are evaluated using the same CFD solver as the one utilized for the high-fidelity model, but with a coarser computational mesh and relaxed convergence criteria. As explained in Sect. 2, the optimization algorithm presented here may exploit several models of different fidelity. The setup of the low-fidelity CFD models is essential for robust performance and high efficiency of the multi-fidelity optimization algorithm.

The grid density is controlled by five parameters, m_i, $i = 1, \ldots, 5$, as described in Sect. 3.1. The number of flow solver iterations is denoted by N. The vector of all combined parameters will be referred to as $\mathbf{z} = [m_1\ m_2\ m_3\ m_4\ m_5\ N]^T$. We consider three different ways of defining the grid parameters in order to setup the low-fidelity models for the multi-level optimization algorithm. In Sect. 5, we demonstrate the influence of the low-fidelity model families set up with these approaches on the performance of the multi-level optimization algorithm.

4.1 Low-Fidelity Model Setup Based on Grid Independence Study

The most common strategy for setting up the low-fidelity models is by using the results of a grid independence study. The process is typically done in "reverse," meaning a high-fidelity grid is developed by an experienced engineer and then the number of grid points in each direction is reduced by half. The distance to the grid point closest to the surface is doubled as well. Figure 2 is an example of such a study.

The flow solution history, shown in Fig. 3a, for a low-fidelity model indicates that the lift and drag coefficients are nearly converged after 80–100 iterations. Therefore, the maximum number of iterations is set to 100 for the low-fidelity model and, thereby, reducing the simulation time further.

A comparison of the pressure distributions of the high- and the low-fidelity models, shown in Fig. 3b, indicates that the low-fidelity model, in spite of being based on much coarser mesh and reduced flow solver iterations, captures the main features of the high-fidelity model pressure distribution quite well. The biggest discrepancy in the distributions is around the shock on the upper surface, leading to an over estimation of both lift and drag (Fig. 2a).

4.2 Low-Fidelity Model Setup Based on Insight

An alternative way of setting up the low-fidelity models is by modifying the grid parameters based on the insight of the engineer. The objective would be to reduce the simulation time, but at the same time retain the accuracy of the high-fidelity model. For example, regions with large gradients need to be resolved better than other regions. With that in mind, one can reduce the number of grid points in the outer regions more than in the region close to the surface. In our case, the grid parameters m_2 and m_4 can be reduced more rapidly than the other grid parameters. The number of flow solver iterations is set in the same way as described in Sect. 4.1.

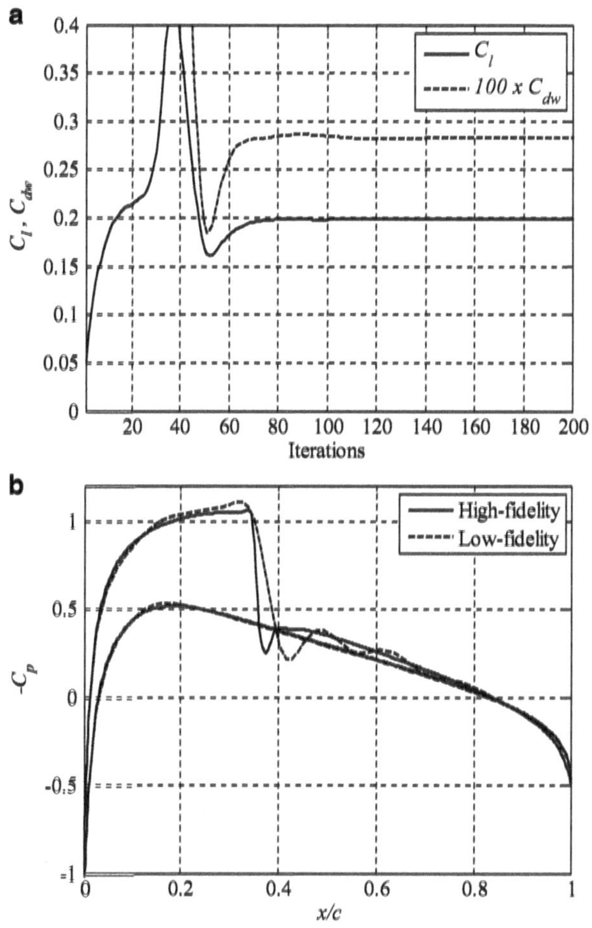

Fig. 3 Simulation results for NACA 0012 at Mach number $M_\infty = 0.75$ and angle of attack $\alpha = 1°$; (**a**) evolution of the lift and drag coefficients obtained by the "coarser" low-fidelity model in Fig. 2; (**b**) comparison of the pressure distributions obtained by the high- and low-fidelity models

4.3 Low-Fidelity Model Setup Based on Numerical Optimization

The last low-fidelity model setup methodology considered here exploits numerical optimization. More specifically, the grid parameters as well as the number of iterations N, i.e., the vector $\mathbf{z}$, are optimized in order to reduce the discrepancy between the drag coefficients predicted by the low-fidelity model and the high-fidelity one, assuming given simulation time ratios between the models.

Let us denote the drag coefficient predicted by a CFD model simulated using the grid/iteration parameters z as $C_d(x,M,\alpha,z)$, where x represents the airfoil geometry, whereas M and α are operating conditions for a reference airfoil for which the low-fidelity model is being set up. The optimization problem is defined as follows

$$\mathbf{z}^* = \arg\min_{z} H_z\left(C_d\left(\mathbf{x}, M, \alpha, \mathbf{z}\right)\right) \tag{5}$$

with the objective function defined as

$$H_z(C_d(\mathbf{x}, M, \alpha, \mathbf{z})) = \left(\frac{C_d(\mathbf{x}, M, \alpha, \mathbf{z}) - C_d\left(\mathbf{x}, M, \alpha, \mathbf{z}_f\right)}{C_d\left(\mathbf{x}, M, \alpha, \mathbf{z}_f\right)}\right)^2 + \gamma\left(\frac{t\left(\mathbf{z}_f\right)}{t(\mathbf{z})} - R_{\text{target}}\right)^2 \tag{6}$$

Here, z_f are the grid parameters of the high-fidelity model, $C_d(x,M,\alpha,z_f)$ is the drag coefficient predicted by the high-fidelity model for the reference airfoil and operating conditions, whereas $t(z)$ represents the CFD model simulation time for given grid parameters z. The objective function contains a penalty factor with the proportionality coefficient γ (here, we use $\gamma = 1{,}000$) that forces the optimization process to obtain the given simulation time ratio R_{target}.

The low-fidelity model setup through the optimization process (5), (6) allows us to obtain the best possible grid setup for a required simulation time ratio that can be controlled much more precisely than for typical methods of Sects. 4.1 and 4.2.

5 Numerical Results

This section presents the effects of various low-fidelity model setup strategies on the performance of the CFD-simulation-based airfoil design. In particular, we consider the multi-level optimization algorithm of Sect. 2 to the design of airfoils in transonic flow using the NACA airfoil parameterization. The high-fidelity CFD model is described in Sect. 3 and the setup of the low-fidelity CFD models is described in Sect. 4. Additionally, design of transonic airfoils exploiting a B-spline parameterization is considered with the low-fidelity model set up using the approach of Sect. 4.3 and a space mapping algorithm utilized as an optimization engine.

5.1 Airfoil Shape Design with NACA Parameterization

Three test cases involving both drag minimization and lift maximization are considered. The performance of the algorithm (with respect to the quality of the final design, as well as the computational complexity of the design process) is compared for different setups of the low-fidelity models (as described in Sect. 4).

5.1.1 NACA Airfoils

The airfoil shape is parameterized with the NACA four-digit method [17]. The NACA airfoils are constructed by combining a thickness function $z_t(x)$ with a mean camber line function $z_c(x)$. The x-coordinates are

$$x_{u,l} = x \mp z_t \sin\theta \tag{7}$$

and the z-coordinates are

$$z_{u,l} = z_c \pm z_t \cos\theta \tag{8}$$

where u and l refer to the upper and lower surfaces, respectively, $z_t(x)$ is the thickness function, $z_c(x)$ is the mean camber line function, and

$$\theta = \tan^{-1}\left(\frac{dz_c}{dx}\right) \tag{9}$$

is the mean camber line slope. The NACA four-digit thickness distribution is given by

$$z_t = t\left(a_0 x^{1/2} - a_1 x - a_2 x^2 + a_3 x^3 - a_4 x^4\right) \tag{10}$$

where $a_0 = 1.4845$, $a_1 = 0.6300$, $a_2 = 1.7580$, $a_3 = 1.4215$, $a_4 = 0.5075$, and t is the maximum thickness. The mean camber line is given by

$$z_c = \begin{cases} \dfrac{m}{p^2}\left(2px - x^2\right), & x < p \\ \dfrac{m}{(1-p)^2}\left(1 - 2p + 2px - x^2\right), & x \geq p \end{cases} \tag{11}$$

where m is the maximum ordinate of the camber line and p is the location of the maximum ordinate. Example NACA four-digit airfoils are shown in Fig. 4.

The NACA four-digit airfoil shapes are designed for low-speed applications and are therefore not suitable for transonic flow speeds. However, the NACA four-digit parameterization is a convenient approach for numerical experiments of the optimization algorithms since there are only three well defined parameters controlling the airfoil shape. The design variable vector is written as $\boldsymbol{x} = [m\ p\ t]^T$.

5.1.2 Setup of the Low-Fidelity Models

The low-fidelity models were configured using the NACA 0012 airfoil shape as a baseline and assuming the following operating conditions: $M_\infty = 0.75$ and $\alpha = 1°$. Three sets of the low-fidelity models were considered: LFM 1 = models created using linear variation of all grid parameters (cf. Sect. 4.1), LFM 2 = models created

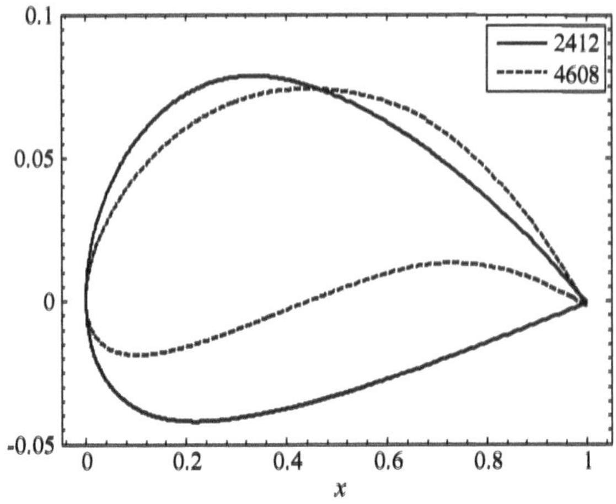

Fig. 4 Shown are two different NACA four-digit airfoil sections: NACA 2412 ($m = 0.02, p = 0.4, t = 0.12$) and NACA 4608 ($m = 0.04, p = 0.6, t = 0.08$)

using variation of a set of grid parameters (cf. Sect. 4.2), and LFM 3 = models obtained by numerical optimization of the grid parameters (cf. Sect. 4.3).

The characteristics of all three low-fidelity model sets are presented in Fig. 5. It can be observed that the lift coefficient values (Fig. 5a) diverge from the high-fidelity value upon reducing the grid cells for all the low-fidelity model families. The behavior of the drag coefficient values (Fig. 5b) is different. For the sets LFM 1 (constructed by a typical grid independence study) and LFM 2 (constructed by insight), a deviation from the high-fidelity model drag due to the reduction of the cell number is much more pronounced as compared to LFM 3 (constructed by numerical optimization). LFM 3 can be characterized by the drag coefficient being nearly constant with respect to the number of cells.

What is even more important, the time ratio (Fig. 5c) increases more rapidly with reduced number of cells for the LFM 3 family than the other two. It can be therefore concluded that constructing the low-fidelity model families using numerical optimization yields both more accurate and faster models, at least for the our baseline airfoil shape.

5.1.3 Shape Optimization

The multi-level optimization algorithm of Sect. 2.2 is applied to three different design cases. The pattern-search algorithm, a derivative-free optimization method (see, e.g., Koziel [13]), is used for the low-fidelity model optimization (Step 3 of the algorithm) with the maximum number of model evaluations set to 200. The stopping criterion on the argument is 10^{-4}.

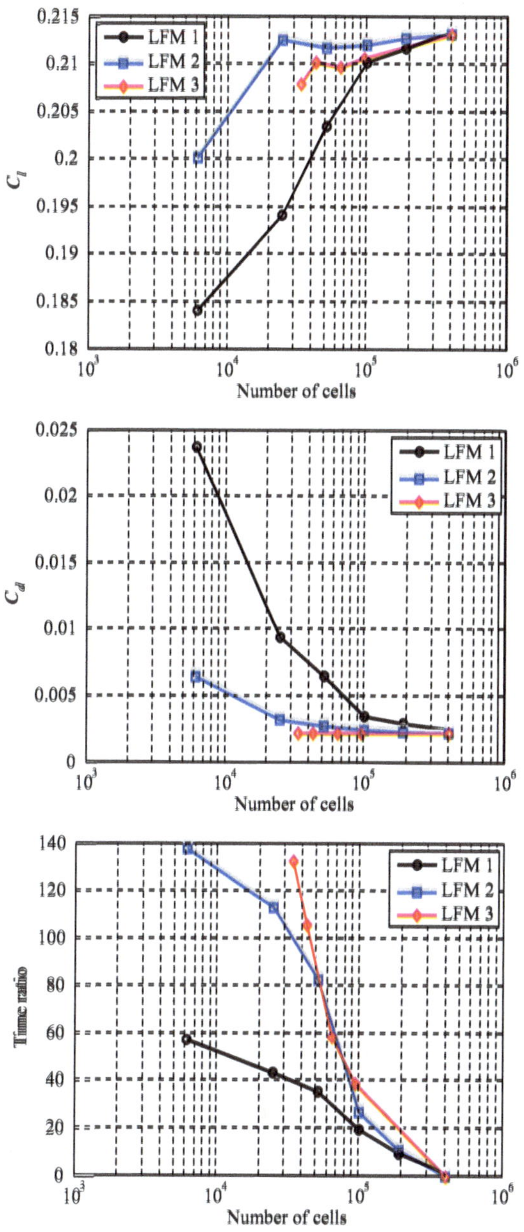

Fig. 5 Comparison of low-fidelity model set characteristics LFM 1 through LFM 3 for NACA 0012 at $M_\infty = 0.75$, $\alpha = 1^\circ$. Shown are the variations of (**a**) the lift coefficient, (**b**) drag coefficient, and (**c**) ratio of the high-to-low-fidelity simulation time with the number of grid cells

Table 1 Simulation time ratio for the high-to-low-fidelity models for each low-fidelity model used from different families

Model	LFM 1	LFM 2	LFM 3
c_1	25	84	100
c_2	3	12	40

Table 2 Design case formulations

Case	M_∞	α (°)	Objective	Constraint 1	Constraint 2
1	0.70	1	min C_d	$C_l \geq 0.6000$	$A \geq 0.075$
2	0.75	0	max C_l	$C_d \leq 0.0050$	$A \geq 0.075$
3	0.80	0	min C_d	$C_l \geq 0.5000$	$A \geq 0.065$

For each design case, three optimization studies are performed using the three different low-fidelity model sets constructed as described in the previous section. Two low-fidelity models are used by the multi-level optimization algorithm in each case. Table 1 shows the time ratio of the two low-fidelity models used from each family. Using faster models for LFM 1 and LFM 2 either resulted in failed simulations during the optimization run, i.e., the grids were simply to coarse for the flow solver to handle, or the optimizer would not yield improved designs.

The design cases are described in Table 2. In each case, the airfoil shape is parameterized using the NACA four-digit method as described in Sect. 5.1.1. The bounds on the design variables are $0 \leq m \leq 0.1$, $0.2 \leq p \leq 0.8$, and $0.05 \leq t/c \leq 0.2$. The Mach number is $M_\infty = 0.75$ and the angle of attack $\alpha = 1°$. Details of the optimization results are given in Table 2.

The results presented in Tables 3, 4, and 5 indicate that the performance of the multi-level algorithm is consistent throughout all the test cases. The results show that the optimized designs obtained by using the three different low-fidelity model families are similar. The thickness-to-chord ratio is nearly the same for all designs, but the maximum camber and the location of maximum camber differ slightly. All designs satisfy both constraints, but they differ slightly in the objective function.

The algorithm using the low-fidelity model set LFM 3 yields designs similar to those produced with the low-fidelity model sets LFM 1 and LFM 2, but at a considerably lower CPU cost (4–5 equivalent evaluations of the high-fidelity model on average). At the same time, the quality of the final designs produced with all sets is similar. It should be emphasized that because of the algorithm setup, the number of evaluations of particular models are similar (N_{c1}~50, N_{c2}~40, N_f~3) so that the computational benefit mostly comes from the fact that the models of LFM 3 are faster than those of LFM 1 and LFM 2. However, the fact that the LFM 3 models are

Table 3 Numerical results for Case 1

Variable	Initial	LFM 1	LFM 2	LFM 3
m	0.0200	0.0175	0.0166	0.0180
p	0.4000	0.5500	0.5800	0.5233
t	0.1200	0.1114	0.1164	0.1114
C_l	0.5963	0.6001	0.6000	0.6000
C_d	0.0047	0.0016	0.0018	0.0017
A	0.0808	0.0750	0.0784	0.0751
N_{c1}	–	51	54	52
N_{c2}	–	38	38	38
N_f	–	3	3	2
Cost	–	<18	<7	<4

Table 4 Numerical results for Case 2

Variable	Initial	LFM 1	LFM 2	LFM 3
m	0.0200	0.0152	0.0146	0.0151
p	0.4000	0.7433	0.7656	0.7661
t	0.1200	0.1140	0.1140	0.1120
C_l	0.4745	0.5676	0.5702	0.5900
C_d	0.0115	0.0050	0.0050	0.0050
A	0.0808	0.0767	0.0768	0.0754
N_{c1}	–	52	51	51
N_{c2}	–	38	38	38
N_f	–	3	4	3
Cost	–	<18	<8	<5

Table 5 Numerical results for Case 3

Variable	Initial	LFM 1	LFM 2	LFM 3
m	0.0000	0.0150	0.0153	0.0151
p	0.0000	0.5300	0.5150	0.5200
t	0.1000	0.0966	0.0965	0.0966
C_l	0.0006	0.5002	0.5004	0.4999
C_d	0.0016	0.0156	0.0162	0.0159
A	0.0673	0.0651	0.0650	0.0650
N_{c1}	–	51	51	51
N_{c2}	–	38	38	38
N_f	–	3	3	2
Cost	–	<18	<7	<4

also more accurate than LFM 1 and LFM 2 has its contribution to the computational savings: the average number of the refinement iterations of the multi-level algorithm (and, consequently, the number of high-fidelity model evaluations) equals 3 for LFM 1, 3.3 for LFM 2 but only 2.3 for LFM 3.

5.2 Airfoil Shape Design with B-Spline Parameterization

In this section, we consider optimization of the airfoil shape described by a B-spline parameterization. Formulation of the parameterization is followed by the results of the low-fidelity model setup, as well as a description of the optimization approach. The section is concluded with the numerical results and discussion.

5.2.1 Airfoil Shapes with B-Spline Curves

A B-spline is a piecewise polynomial function of the order k in a variable t defined over the range $t_0 \leq t \leq t_m$, $m = k + 1$, where the points $t = t_j$ are known as knots (or break-points). A spline function of order k on a given set of knots can be expressed as a linear combination of B-splines as [18]

$$p(t) = \sum_{l=0}^{n} N_{l,k}(t) P_l \tag{12}$$

where $N_{l,k}$ are the B-spline blending functions, also referred to as the basis functions, and P_l, $i = 0, 1, \ldots n$, are control points. The basis functions are calculated from the Cox-DeBoor recursion relation as

$$N_{l,1} = \begin{cases} 1 & if \quad t_l \leq t \leq t_{l+1} \\ 0 & \text{elsewhere} \end{cases} \tag{13}$$

and

$$N_{l,k}(t) = \frac{t - t_l}{t_{l+k-1} - t_l} N_{l,k-1}(t) + \frac{t_{l+k} - t}{t_{l+k} - t_{l+1}} N_{l+1,k-1}(t) \tag{14}$$

for $k = 2, 3, \ldots,$ K, as well as for all needed values of i. The basis functions $N_{l,k}$ can be polynomials of the order one, two, or higher.

Figure 6 shows an example of two B-spline curves of the order 3. We can observe that the curves have the same control points, aside from control point 5. The two curves, however, are very similar, except locally near the perturbed control point. This highlights the local control capability of the B-spline curves.

To obtain a typical airfoil shape, a few control points need to be set at specific locations:

- A control point is fixed at the leading edge, typically set at $(x, z) = (0,0)$.
- A control point is fixed at the trailing edge, i.e., at $x = 1$. If the trailing edge is fixed on the x-axis, then it is set at $(x,z) = (0,1)$, and
- To ensure a rounded leading edge, two control points are allocated on the line $x = 0$.

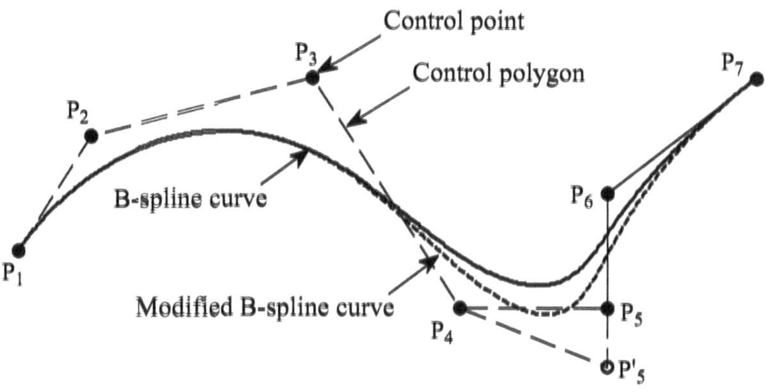

Fig. 6 Example B-spline curves with one different control point

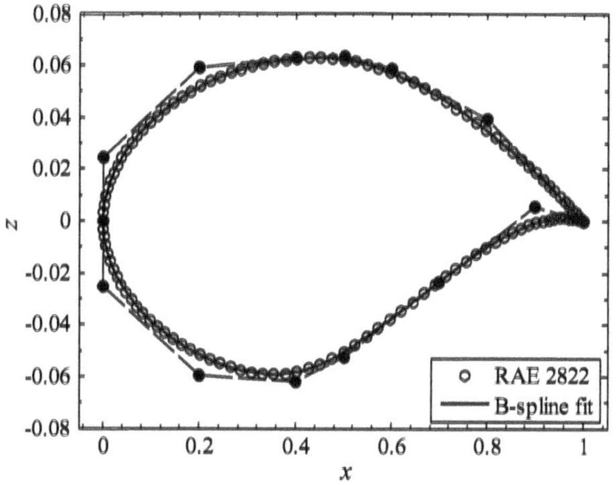

Fig. 7 B-spline curve approximation to the RAE 2822 airfoil (see coordinates in Selig [19]). The free control points can only move vertically

With these control points in place, the upper and lower surfaces are determined by the remaining control points. In general, the remaining control points need not be bounded. However, often in practice, e.g., within an optimization run, the control points are set at fixed x-locations and allowed to move in the z-direction. This is done to prevent unrealistic airfoil shapes or shapes which cannot be handled by the fluid flow solver.

Figure 7 shows an example of a B-spline curve approximation to the RAE 2822 supercritical airfoil (see coordinates in Selig [19]). The airfoil shape is found by minimizing the norm of the difference between the target airfoil shape and the curve shapes generated by (12).

5.2.2 Setup of the Low-Fidelity Models

The low-fidelity models are configured by the numerical optimization procedure described in Sect. 4.3. The RAE 2822 airfoil shape (Fig. 7) is used as a baseline at $M_\infty = 0.734$ and $\alpha = 1.944°$. The target time ratios are 20, 40, 60, 80, and 100.

The results are shown in Fig. 8. We can see that the optimizer sets up grids which have time ratios of 20 and 40 rather precisely. However, it turns out to be more challenging to reach the remaining time ratios, i.e., 60, 80, and 100. The grids with the target time ratios of 60 and 80 can still be obtained, but the optimizer is unable to find a grid with a time ratio close to 100. This essentially means that time evaluation ratios higher than 80 are unreachable for the considered model.

5.2.3 Model Validation

The low-fidelity model setups are validated by performing CFD simulations for (1) airfoils shapes other than the baseline, and (2) for a different operating condition. In particular, the new shapes are generated by applying small random perturbations to the control point locations of the baseline airfoil shape. Thus, generating shapes which are located close to the baseline in the design space, but with a significantly different shape. The CFD simulations are performed at $M_\infty = 0.75$ and $\alpha = 1.0°$.

Figure 9 shows the obtained time ratios of the CFD simulations of seven different shapes with three different low-fidelity model setups, i.e., for ratios of 40, 60, and 80. We can see that the results for time ratios of 40 and 60 are slightly lower than originally required, i.e., the mean time ratio is 36.7 and 56.8, respectively, but slightly higher for the time ratio of 80, i.e., the mean time ratio is 80.4.

5.2.4 Shape Optimization

The initial shape is set as the RAE 2822. The Mach number is $M_\infty = 0.734$ and the angle of attack is $\alpha = 1.944°$. The objective is to minimize the drag coefficient subject to constraints on the lift coefficient and the cross-sectional area. The minimum lift coefficient is $C_{l.min} = 65.9$ l.c., where l.c. = lift count $= 0.01 \times C_l$, and the minimum cross-sectional area is $A_{min} = 0.0779$.

The optimization problem is solved using the space mapping (SM) algorithm [20]. The SM algorithm produces a sequence $\boldsymbol{x}^{(i)}$, $i = 0, 1, \ldots,$ of approximate solutions to (1) (here, $\boldsymbol{x}^{(0)}$ is the initial design) as

$$\mathbf{x}^{(i+1)} = \arg\min_{\mathbf{x}} H(s^{(i)}(\mathbf{x})) \tag{15}$$

where $s^{(i)}(\boldsymbol{x}) = [C_{l.s}{}^{(i)}(\boldsymbol{x})\ C_{d.s}{}^{(i)}(\boldsymbol{x})\ A_s{}^{(i)}(\boldsymbol{x})]^T$ is a surrogate model at iteration i. Here $C_{l.s}$, $C_{d.s}$, and A_s denote the lift and drag coefficients, as well as the cross-sectional area for the surrogate. The surrogate model is a composition of the low-fidelity

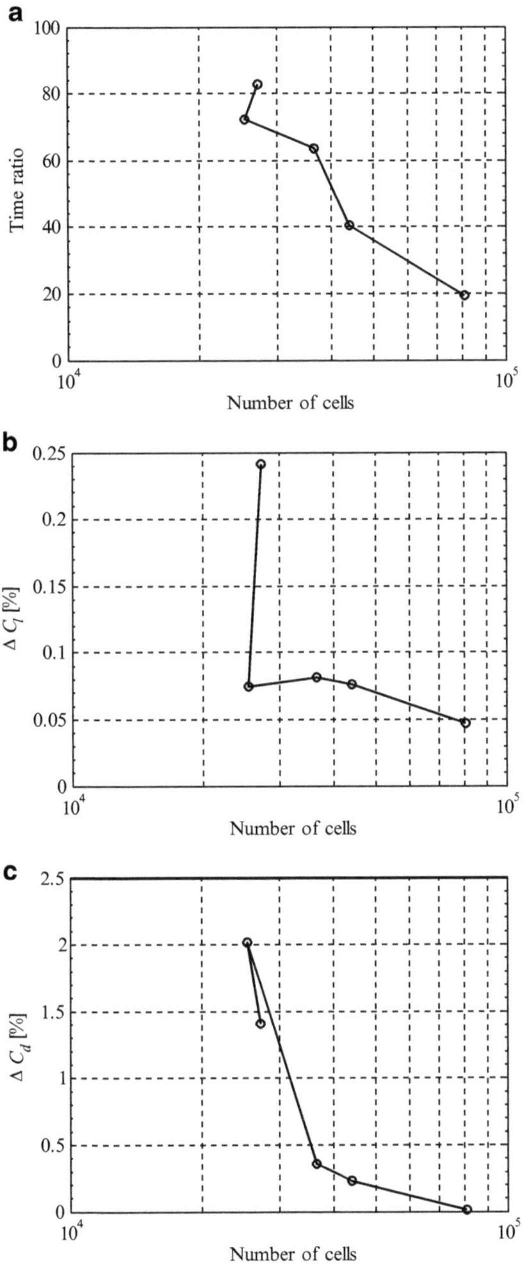

Fig. 8 Setup of low-fidelity models for several target time ratios using the RAE 2822 airfoils shape: (**a**) time ratio, (**b**) lift absolute error, and (**c**) drag absolute error

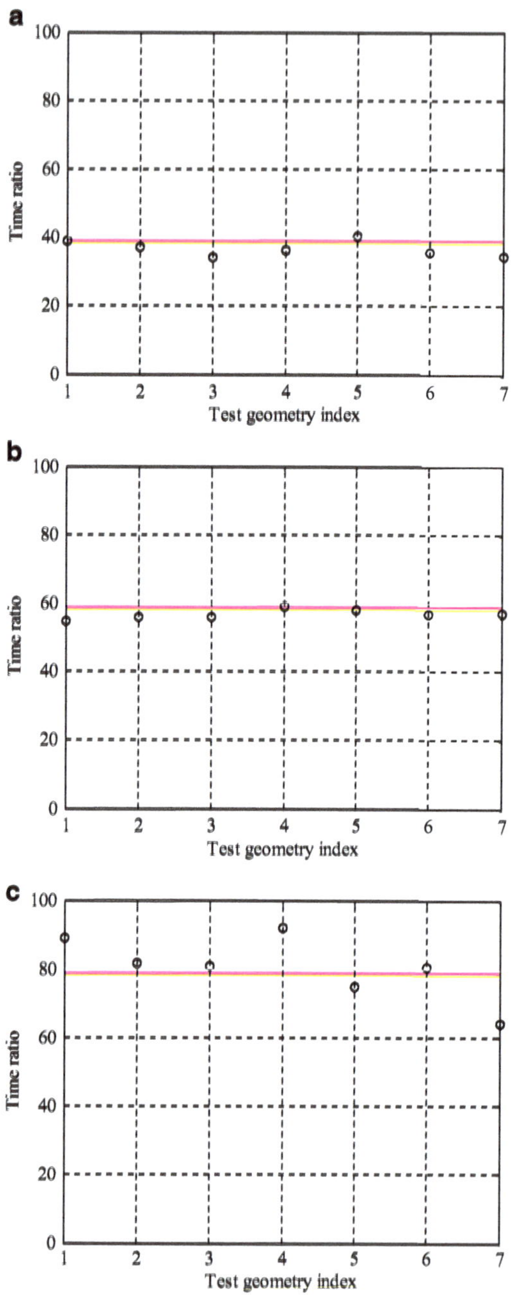

Fig. 9 Time ratios of airfoil shapes randomly perturbed (locally) from the RAE 2822 airfoil evaluated with grids setup by numerical optimization for time ratios of 40, 60, and 80. (**a**) Mean time ratio = 36.7, (**b**) Mean time ratio = 56.8, and (**c**) Mean time ratio = 80.4

model and simple, usually linear, transformations (or mappings). Here, we utilize so-called output SM [20] of the form:

$$s^{(i)}(\mathbf{x}) = \mathbf{A}^{(i)} \circ c(\mathbf{x}) + \mathbf{D}^{(i)} + \mathbf{q}^{(i)} =$$

$$= \left[a_l^{(i)} C_{l.c}(\mathbf{x}) + d_l^{(i)} + q_l^{(i)} \quad a_d^{(i)} C_{d.c}(\mathbf{x}) + d_d^{(i)} + q_d^{(i)} \quad A_c(\mathbf{x})\right]^T \quad (16)$$

where ° denoted component-wise multiplication. $C_{l.c}$, $C_{d.c}$, and A_c denote the lift and drag coefficients, as well as the cross-sectional area for the low-fidelity model. Note that there is no need to map A_c because $A_c(\boldsymbol{x}) = A_f(\boldsymbol{x})$ for all $\boldsymbol{x}$. Response correction parameters $\boldsymbol{A}^{(i)}$ and $\boldsymbol{D}^{(i)}$ are obtained as

$$\left[\mathbf{A}^{(i)}, \mathbf{D}^{(i)}\right] = \arg\min_{[\mathbf{A},\mathbf{D}]} \sum_{k=0}^{i} \| f(\mathbf{x}^{(k)}) - \mathbf{A} \circ c(\mathbf{x}^{(k)}) + \mathbf{D} \|^2 \quad (17)$$

i.e., the response scaling is supposed to (globally) improve the matching for all previous iteration points. The additive response correction term $\boldsymbol{q}^{(i)}$ is defined as

$$\mathbf{q}^{(i)} = f(\mathbf{x}^{(i)}) - \left[\mathbf{A}^{(i)} \circ c(\mathbf{x}^{(i)}) + \mathbf{D}^{(i)}\right] \quad (18)$$

i.e., it ensures perfect matching between the surrogate and the high-fidelity model at the current design $\boldsymbol{x}^{(i)}$, $s^{(i)}(\boldsymbol{x}^{(i)}) = f(\boldsymbol{x}^{(i)})$ (so-called zero-order consistency [21]). All mapping parameters can be found analytically as shown in [20].

The problem has been solved several times, each with a different maximum number of low-fidelity model evaluations ($N_{c.max}$) allowed for the surrogate model optimization (here, executed using the pattern-search algorithm). The stopping criterion on the argument is set to 10^{-3}. The low-fidelity model with the time ratio of 40 is used in all instances. The results are shown in Table 6.

One can observe that both constraints, the lift and the area constraints, are active, but not violated, in all the cases. The drag coefficient is reduced as the maximum number of low-fidelity model evaluations is increased. The difference between the highest and the lowest drag coefficient values is 1.7 d.c., where d.c. = drag count = 0.0001 × C_d, which issignificant.

Table 6 Optimization results for varying number of maximum low-fidelity model evaluations

$N_{c.max}$	65	100	150	200	250	300	350	400	450
C_l (l.c.)	65.9	66.1	65.9	65.9	66.0	65.9	65.9	65.9	65.9
C_d (d.c.)	22.0	22.0	21.5	21.0	20.6	20.7	20.9	20.9	20.3
A (×10^4)	78.9	78.4	78.4	78.6	78.6	78.6	78.9	78.9	78.7
N_f	3	3	3	3	3	3	3	3	3
N_{total}	5.29	6.23	7.83	8.82	8.78	11.32	11.89	11.58	13.75

The design iteration cost, i.e., the number of high-fidelity model evaluations, is constant for all the cases, $N_f = 3$. The total optimization cost (N_{tot}) increases with the maximum number of low-fidelity model evaluations, with the lowest being less than six equivalent high-fidelity model evaluations and the highest being less than 14. It can be observed that allowing more function evaluations in the surrogate model optimization results in slight improvement of the final design quality (i.e., lower drag coefficient).

The evolution of the optimization for the case with $N_{c.max} = 200$ is shown in Fig. 10. One can see how the optimization quickly reaches the stopping criteria (Fig. 10a) and how the lift coefficient reaches the minimum value (Fig. 10b).

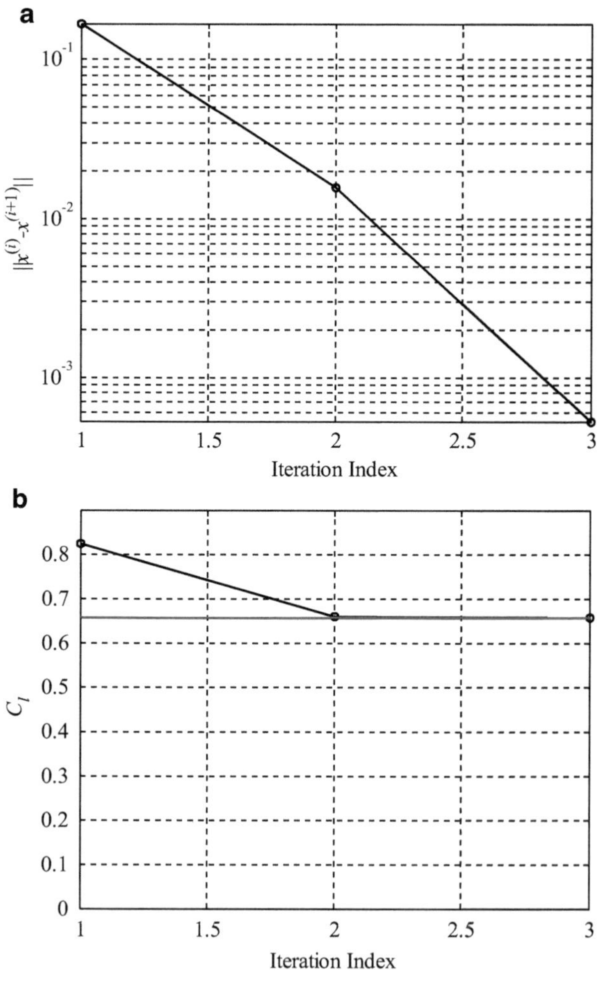

Fig. 10 Optimization history: (**a**) convergence plot, (**b**) evolution of the drag coefficient. The constraint value is indicated by the *straight solid line*

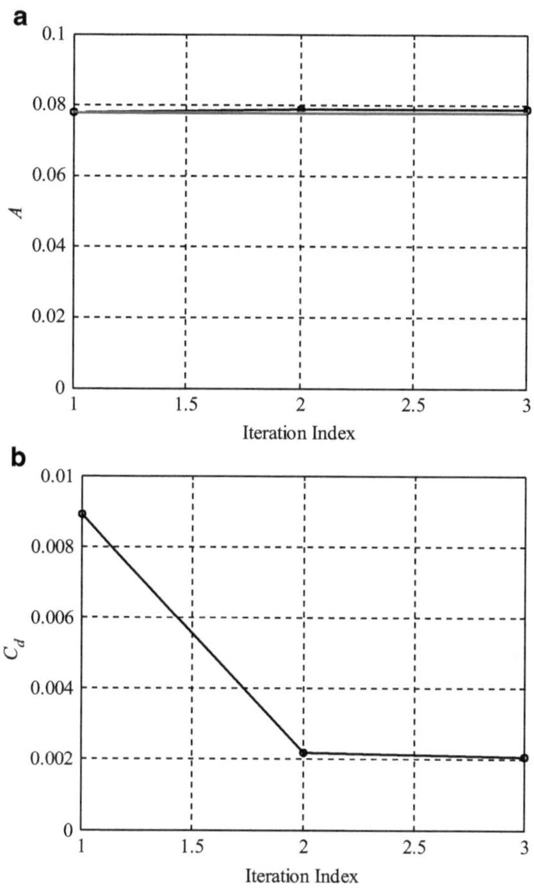

Fig. 11 Optimization history: (**a**) evolution of the lift coefficient, (**b**) evolution of the cross-sectional area. The constraint value is indicated by the *straight solid line*

The cross-sectional area stays close to the constraint value (Fig. 11a). The drag coefficient is reduced significantly in the first iteration (Fig. 11b). Compared to the initial design, the optimized shape has a thinner leading edge and a thicker trailing edge (Fig. 12).

6 Conclusion

A technique for automated low-fidelity model setup in the context of variable-resolution SBO is described. The approach replaces a hands-on process guided by experience to construct accurate and reliable low-fidelity models by an

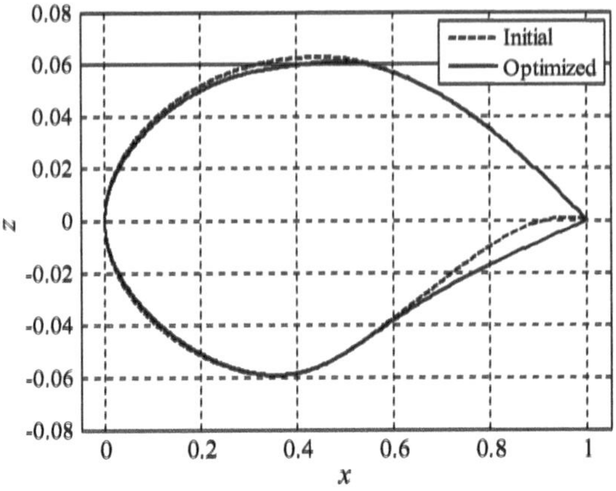

Fig. 12 Initial and optimized airfoil shapes

optimization-guided procedure, where the objective is to adjust the grid parameters so that an accurate model as possible is obtained assuming a given evaluation time in reference to the high-fidelity model. It has been demonstrated that the methodology leads to faster and more accurate low-fidelity models than those obtained using conventional approaches.

References

1. Leoviriyakit, K., Kim, S., Jameson, A.: Viscous aerodynamic shape optimization of wings including planform variables. In: 21st Applied Aerodynamics Conference, Orlando, 23–26 June 2003
2. Braembussche, R.A.: Numerical optimization for advanced turbomachinery design. In: Thevenin, D., Janiga, G. (eds.) Optimization and Computational Fluid Dynamics, pp. 147–189. Springer, Berlin (2008)
3. Queipo, N.V., Haftka, R.T., Shyy, W., Goel, T., Vaidyanathan, R., Tucker, P.K.: Surrogate-based analysis and optimization. Prog. Aerosp. Sci. **41**(1), 1–28 (2005)
4. Forrester, A.I.J., Keane, A.J.: Recent advances in surrogate-based optimization. Prog. Aerosp. Sci. **45**(1–3), 50–79 (2009)
5. Koziel, S., Echeverría-Ciaurri, D., Leifsson, L.: Surrogate-based methods. In: Koziel, S., Yang, X.S. (eds.) Computational Optimization, Methods and Algorithms. Series: Studies in Computational Intelligence, pp. 33–60. Springer, Berlin (2011)
6. Alexandrov, N.M., Lewis, R.M., Gumbert, C.R., Green, L.L., Newman, P.A.: Optimization with variable-fidelity models applied to wing design. In: 38th Aerospace Sciences Meeting & Exhibit, Reno, AIAA Paper 2000–0841, Jan 2000
7. Robinson, T.D., Eldred, M.S., Willcox, K.E., Haimes, R.: Surrogate-based optimization using multifidelity models with variable parameterization and corrected space mapping. AIAA J. **46**(11), 2814–2822 (2008)
8. Booker, A.J., Dennis Jr., J.E., Frank, P.D., Serafini, D.B., Torczon, V., Trosset, M.W.: A rigorous framework for optimization of expensive functions by surrogates. Struct. Optim. **17**(1), 1–13 (1999)

9. Barrett, T.R., Bressloff, N.W., Keane, A.J.: Airfoil design and optimization using multi-fidelity analysis and embedded inverse design. In: 47th AIAA/ASME/ASCE/AHS/ASC Structures, Structural Dynamics, and Materials Conference, Newport, AIAA Paper 2006-1820 (2006)
10. Leifsson, L., Koziel, S.: Multi-fidelity design optimization of transonic airfoils using physics-based surrogate modeling and shape-preserving response prediction. J. Comput. Sci. **1**(2), 98–106 (2010)
11. Leifsson, L., Koziel, S.: Variable-resolution shape optimization: low-fidelity model setup and algorithm scalability. In: 14th AIAA/ISSMO Multidisciplinary Analysis and Optimization Conference, Indianapolis, 17–19 Sept 2012
12. Koziel, S., Leifsson, L.: Multi-level surrogate-based airfoil shape optimization. In: 51st AIAA Aerospace Sciences Meeting including the New Horizons Forum and Aerospace Exposition, Grapevine, 7–10 Jan 2013
13. Koziel, S.: Multi-fidelity multi-grid design optimization of planar microwave structures with Sonnet. In: International Review of Progress in Applied Computational Electromagnetics, Tampere, pp. 719–724. 26–29 Apr 2010
14. Koziel, S., Cheng, Q.S., Bandler, J.W.: Space mapping. IEEE Microw. Mag. **9**(6), 105–122 (2008)
15. ICEM CFD, ver. 14.0, ANSYS Inc., Southpointe, 275 Technology Drive, Canonsburg, PA 15317 (2012)
16. FLUENT, ver. 14.0, ANSYS Inc., Southpointe, 275 Technology Drive, Canonsburg, PA 15317 (2012)
17. Abbott, I.H., Von Doenhoff, A.E.: Theory of Wing Sections. Dover, New York (1959)
18. Samareh, J.A.: Survey of shape parameterization techniques for high-fidelity multidisciplinary shape optimization. AIAA J. **39**(5), 877–884 (2001)
19. Selig, M., The University of Illinois at Urbana-Champaign Airfoil Coordinates Database. http://aerospace.illinois.edu/m-selig/ads/coord_database.html (2014)
20. Koziel, S., Leifsson, L.: Knowledge-based airfoil shape optimization using space mapping. In: 30th AIAA Applied Aerodynamics Conference, New Orleans, Louisiana, 25–28 June 2012
21. Alexandrov, N.M., Lewis, R.M.: An overview of first-order model management for engineering optimization. Optim. Eng. **2**(4), 413–430 (2001)

Design Space Reduction for Expedited Multi-Objective Design Optimization of Antennas in Highly Dimensional Spaces

Adrian Bekasiewicz, Slawomir Koziel, and Wlodzimierz Zieniutycz

Abstract A surrogate-based technique for efficient multi-objective antenna optimization is discussed. Our approach exploits response surface approximation (RSA) model constructed from low-fidelity antenna model data (here, obtained through coarse-discretization electromagnetic simulations). The RSA model enables fast determination of the best available trade-offs between conflicting design goals. The cost of RSA model construction for multi-parameter antennas is significantly lowered through initial design space reduction. Optimization of the response surface approximation model is carried out by a multi-objective evolutionary algorithm (MOEA). Additional response correction techniques are subsequently applied to improve selected designs at the level of high-fidelity electromagnetic antenna model. The refined designs constitute the final Pareto set representation. The presented multi-objective design approach is validated using three examples: a six-variable ultra-wideband dipole antenna, an eight-variable planar Yagi-Uda antenna, and an ultra-wideband monocone with 13 design variables.

Keywords Simulation-driven design • UWB antenna • Response surface approximation • Electromagnetic (EM) simulation • Space mapping • Multi-objective optimization • Multi-fidelity models • Kriging interpolation • Coarse model • Surrogate-based optimization • Design space reduction

A. Bekasiewicz (✉) • W. Zieniutycz
Faculty of Electronics, Telecommunications and Informatics, Gdansk University of Technology, Narutowicza 11/12, 80-233 Gdansk, Poland
e-mail: adrian.bekasiewicz@pg.gda.pl

S. Koziel (✉)
Engineering Optimization & Modeling Center, School of Science and Engineering, Reykjavik University, Menntavegur 1, 101 Reykjavik, Iceland
e-mail: koziel@ru.is

S. Koziel et al. (eds.), *Solving Computationally Expensive Engineering Problems*, Springer Proceedings in Mathematics & Statistics 97,
DOI 10.1007/978-3-319-08985-0_5

1 Introduction

Microwave/RF antennas belong to the key components of modern wireless communication systems. They have to fulfill stringent requirements upon their electrical and geometrical properties [1–4]. For the sake of reliability, the antenna design process requires a realistic structure setup that comprises not only the radiator together with its feeding network, but also the nearest environment of the structure, e.g., housing, connectors, or neighboring subsystems [5, 6]. Such a configuration cannot be evaluated using conventional empirical equations. Consequently, a computationally expensive analysis of antenna, based on high-fidelity electromagnetic (EM) simulations becomes a necessity [5, 7]. An important factor is also the fact that the relationships between adjustable parameters (both material and geometry) and the antenna performance parameters might be rather complex so that a conventional design (or tuning) procedures based on repetitive parameter sweeps driven by engineering experience are prone to failure [8, 9]. These difficulties make contemporary antenna design a very complicated and multifaceted problem that may be efficiently solved only by means of suitably developed optimization algorithms.

High demands related to both the antenna geometry and its field properties create the necessity of simultaneous account for several (often conflicting) objectives including not only the minimization of reflection characteristics within the frequency band of interest, but also reduction of the antenna footprint [7, 10], minimization of side-lobe level [11, 12], cross polarization [13, 14], or maximization of gain [12, 15], to name just a few. Coexistence of many objectives constitutes a multi-objective design problem that is significantly more challenging than conventional single-objective optimization. The complexity of such a setup lies in the occurrence of two solution spaces bonded by a unique and highly nonlinear mappings [16, 17]: (1) the decision variable space (so-called design space), whose dimensionality is determined by the number of design variables selected for the optimization process, and (2) the feature space (or a so-called objective space) that represents the responses of the designed structure with respect to given design objectives (its dimensionality is determined by the number of optimization goals). Due to a nature of the problem, a conventional definition of design priorities in multi-objective optimization scheme is not possible and a set of trade-off solutions between non-commensurable objectives have to be sought. Such solutions form a so-called Pareto optimal set (a representation of a Pareto front), where improvement with respect to single objective is impossible without degradation of others [17, 18]. The use of conventional algorithms (both gradient-based and derivative-free) for solving multi-objective optimization problems is not possible unless the objectives are aggregated into a scalar merit function [17–20]. In the latter case, only a single Pareto optimal solution may be generated at a time, and multiple algorithm runs with various objective aggregation parameters are necessary to yield a Pareto set representation [17].

High diversity of Pareto optimal solutions can be accounted for by population-based algorithms. Particularly, metaheuristic algorithms are attractive in such a setup, mostly due to their ability to process and outcome the entire Pareto set representation in a single algorithm run [10, 17]. Metaheuristics may be considered as simple and universal optimization strategies, usually imitating various biological or social phenomena (e.g., the swarm intelligence [21], genetic processes [22], behavior of cuckoos [23], etc.). In particular, they benefit from lack of restrictive assumptions upon model formulation. This is especially useful if complex problems that may be represented as a black box are considered [18]. Metaheuristic algorithms proved their usefulness in the context of seeking for globally optimal solutions for highly nonlinear and noisy functions with multiple discontinuities [17, 18], and therefore they tend to be very useful for design and optimization of contemporary antennas. Most common schemes applied for these structures are genetic algorithms (GA) [22, 24–26] and particle swarm optimizers (PSOs) [27–30]. Both have strong foundations in the context of antenna optimization [31–35], even in multi-objective sense [7, 10, 25, 36, 37]. Nonetheless, all the benefits of population-based metaheuristics come with a great drawback, which is a tremendous number of model evaluations needed to complete the optimization process. Unfortunately, single evaluation of a realistic antenna model may take even a few hours [38], which significantly hinders direct utilization of metaheuristics in the design process. These difficulties led to the development of various design strategies that aim at lowering the computational cost [39–41]. On the other hand, the problem of high computational cost may be partially addressed by the utilization of massive computational resources in the form of supercomputers with multiple CPU or GPU units together with multiple licenses for computer-aided design software (especially, electromagnetic solvers) [42]. Nevertheless, such hardware configurations are not widely available and they offer very poor speedup-to-cost ratio.

In this chapter, we discuss a fast multi-objective optimization technique that exploits population-based metaheuristic algorithm in the design process of numerically demanding antenna structures [7, 10, 42]. Our procedure expedites seeking for a trade-off solutions by the utilization of computationally cheap response surface approximation (RSA) model [43, 44] as an antenna evaluation engine. Moreover, we address difficulties related to the generation of a reliable RSA model of antennas with multiple independent design variables by initial design space reduction. The initial Pareto set representation is obtained by optimizing the RSA model using a multi-objective evolutionary algorithm (MOEA) [17, 18]. Subsequently, discrepancies between the RSA and EM antenna models are reduced by means of surrogate-based optimization (SBO) techniques. The presented design methodology allows us to perform antenna design at a cost being only a fraction of that corresponding to direct multi-objective optimization of the EM antenna model (without involving massive computational resources).

The chapter is organized as follows. In Sect. 2, we briefly discuss a multi-objective antenna design problem. We also introduce the concept of multi-fidelity antenna models that may be utilized for expedited optimization process, and we explain in detail the optimization algorithm. In Sect. 3, we describe the importance of design space reduction to limit a number of test samples for the generation of a reliable response surface approximation model in multi-dimensional parameter space. We also comment on the scaling properties of the RSA model in the reduced design space. Section 4 introduces the design space reduction algorithm based on identification of extreme designs that reside on a Pareto optimal set by means of single-objective optimizations. In Sect. 5, we explain the algorithm aimed at reduction of the solution space by analysis of a Pareto dominance relation between the designs obtained in consecutive iterations, whereas in Sect. 6, we discuss a scheme constituted by volume-wise restriction of design space by means of identification of extreme Pareto designs at two levels of fidelity. Moreover, Sects. 4, 5, and 6 are supplemented with illustrative antenna design examples. Section 7 concludes the chapter with discussion and recommendations for the future research related to multi-objective antenna design.

2 Multi-Objective Optimization: Methodology

Direct optimization of contemporary antennas in multi-objective setup is very troublesome, mostly due to a large number of computationally expensive EM simulations required to find a set of trade-off solutions [10, 17]. In such a setup, the overall optimization cost may correspond to several days of CPU time[13, 14], even if the antenna model is relatively simple (with the evaluation time a few minutes per design). However, many modern structures feature complex architectures with unconventional asymmetric geometries and multiple design parameters (e.g., [6, 45–47]), which does not only influence the cost of single EM evaluation but also increases the number of optimization algorithm iterations. As a matter of fact, multi-objective metaheuristic optimization of high-fidelity EM antenna model that simulates, say, in 1 h or more would be simply impractical: even a few thousands of antenna simulations necessary to yield a reasonable Pareto front approximation would take almost a year on a single PC machine.

Difficulties related to high computational cost of antenna simulation may be partially addressed by utilizing SBO methods that recently gained a considerable attention in various engineering fields [48–50]. Examples of such techniques include space mapping [51], manifold mapping [52], or shape preserving response prediction [53]. The attractiveness of SBO lies in its ability to iteratively correct/enhance a computationally cheap yet less accurate low-fidelity model using limited amount of data acquired from the simulations of a high-fidelity yet computationally expensive antenna model [54]. In such a setup, the antenna design variables are adjusted using a corrected low-fidelity model (referred to as a surrogate model). Subsequently, the optimal dimensions are then applied to a high-fidelity model for verification purposes. The utilization of SBO for antenna optimization is well described in literature [5, 7, 55]. SBO methods proved to be very efficient design tools, capable of yielding desired solutions at the cost of only a few simulations of respective high-fidelity antenna models. However, so far, SBO techniques have mostly been applied for solving single-objective antenna problems [56–58].

In this section, we formulate the multi-objective optimization problem with the emphasis on Pareto dominance relation. We also explain the differences between functional and physics-based surrogate models and their importance for expedited optimization of contemporary antennas. Finally, we formulate an SBO algorithm for fast antenna design and optimization in a multi-objective setup.

2.1 Multi-Objective Antenna Design Problem

Let $\boldsymbol{R}_f(\boldsymbol{x})$ denote a response of an accurate, high-fidelity model of the antenna structure under consideration (usually obtained using an EM solver). In particular, it may represent an antenna reflection coefficient [59], gain [10], directivity [24], isolation [60], etc. A vector $\boldsymbol{x}$ represents design variables, specifically, antenna dimensions. Here, $n = \dim(\boldsymbol{x})$ represents a dimensionality of the design space.

Let $F_k(\boldsymbol{x})$, where $k = 1, \ldots, N_{obj}$ denote a kth design objective. A typical performance objective is minimization of antenna reflection over a certain frequency band of interest, and to ensure that $|S_{11}| < -10$ dB over that band. There might be also some additional geometrical objectives such as minimization of antenna size defined in a convenient way (e.g., maximal lateral size, overall antenna footprint, the maximal dimension, antenna volume) [8, 42]. Similar objectives can be formulated with respect to antenna gain, radiation pattern, efficiency, etc.

If a number of design objectives $N_{obj} > 1$ then any two designs $\boldsymbol{x}^{(1)}$ and $\boldsymbol{x}^{(2)}$ for which $F_k(\boldsymbol{x}^{(1)}) < F_k(\boldsymbol{x}^{(2)})$ and $F_l(\boldsymbol{x}^{(2)}) < F_l(\boldsymbol{x}^{(1)})$ for at least one pair $k \neq l$, are not commensurable, i.e., none is better than the other in the multi-objective sense. We may define the Pareto dominance relation $\prec$ [17] saying that for the two designs $\boldsymbol{x}$ and $\boldsymbol{y}$, we have $\boldsymbol{x} \prec \boldsymbol{y}$ ($\boldsymbol{x}$ dominates $\boldsymbol{y}$) if $F_k(\boldsymbol{x}) < F_k(\boldsymbol{y})$ for all k. The goal of the multi-objective optimization is to find a representation of a Pareto optimal set X_P of the design space X, such that for any $\boldsymbol{x} \in X_P$, there is no $\boldsymbol{y} \in X$ for which $\boldsymbol{y} \prec \boldsymbol{x}$ [17]. A conceptual illustration of the relations between the solutions in a feature space with the emphasis on non-dominated designs is shown in Fig. 1.

2.2 Multi-Fidelity Antenna Models in Multi-Objective Setup

Determination of a Pareto optimal set in a multi-objective optimization setup requires a population-based algorithm that needs a very large number of model evaluations to converge. A restriction of population size, or the number of evaluations in such a scheme significantly degrades the diversity of optimal solutions and the quality of the overall Pareto front representation found by the algorithm. Therefore, the cost of multi-objective optimization can be effectively reduced primarily by modifications in the antenna model.

A high-fidelity model $\boldsymbol{R}_f$ of antenna is computationally too expensive to be directly utilized in a multi-objective optimization setup. Nonetheless, the lack of

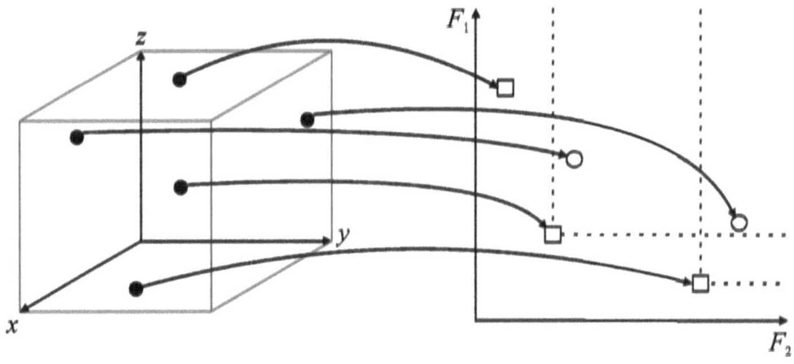

Fig. 1 Example designs in a three-dimensional ($n = 3$ independent design variables) design space (*filled circle*) mapped into a two-dimensional ($k = 2$ design objectives) feature space (*open square, open circle*). The goal of the multi-objective optimization is to find a set of non-dominated solutions (*open square*) that represents the Pareto optimal set X_P

analytical description of contemporary structures forces the utilization of their physics-based representations that may be evaluated only by means of EM solvers. In such a setup, a computational cost of single EM simulation may be reduced by replacement of the $\boldsymbol{R}_f$ model with its low-fidelity counterpart $\boldsymbol{R}_{cd}$ [10]. Generally, the $\boldsymbol{R}_{cd}$ model is constructed using certain simplifications with respect to $\boldsymbol{R}_f$. On the other hand, the low-fidelity model has to be carefully adjusted to ensure its decent accuracy. The most common $\boldsymbol{R}_{cd}$ simplifications include: sparse mesh, modeling of metallization as infinitely thin sheet ($h = 0$), neglecting losses of dielectric substrate ($tan\delta = 0$), or utilization of perfect electric conductor in place of metals with finite-conductivity ($\sigma = \infty$) [61]. Additionally, evaluation cost may be reduced by neglecting the nearest antenna environment (e.g., housing, connectors, or neighboring subsystems). One should emphasize that the utilization of a low-fidelity $\boldsymbol{R}_{cd}$ model usually decreases the cost of single simulation by a factor of 10–50 in comparison to its high-fidelity counterpart $\boldsymbol{R}_f$ [42].

Despite its low computational cost, the low-fidelity model $\boldsymbol{R}_{cd}$ is still too expensive to be directly utilized in multi-objective optimization. On the other hand, the accuracy of a well prepared model is sufficiently good to use it for construction of a response surface approximation model $\boldsymbol{R}_s$. The latter is very fast and thus well suited for direct multi-objective optimization using a population-based metaheuristic algorithm. Preparation of the RSA model $\boldsymbol{R}_s$ requires some computational effort due to data acquisition from the low-fidelity model $\boldsymbol{R}_{cd}$. The training samples necessary to set up the RSA model are allocated using appropriate design of experiments technique, here, Latin Hypercube Sampling (LHS) [62–65].

Because the model $\boldsymbol{R}_s$ prepared this way is merely a representation of $\boldsymbol{R}_{cd}$ rather than $\boldsymbol{R}_f$, it needs to be further refined using SBO methods. An illustration of three ways of modeling the antenna structure of interest utilized in the presented approach is shown in Fig. 2, whereas Fig. 3 illustrates the concept of RSA model preparation.

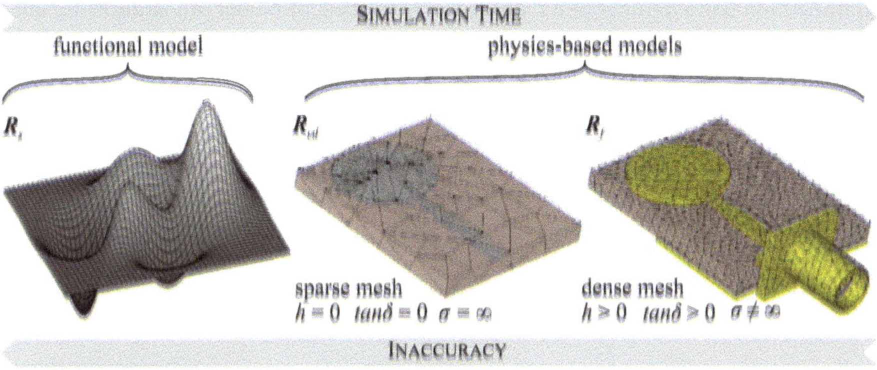

Fig. 2 Various representations of the same antenna structure. The functional model $\boldsymbol{R}_s$ is generated using data acquired from a simplified physics-based model $\boldsymbol{R}_{cd}$, here coarse-discretization EM simulation one. Simplifications introduced in $\boldsymbol{R}_{cd}$ with respect to $\boldsymbol{R}_f$ include, among others, lack of connector, zero thickness metallization ($h = 0$), coarse mesh, perfect conductivity of metallization ($\sigma = \infty$), and lossless dielectric ($tan\delta = 0$)

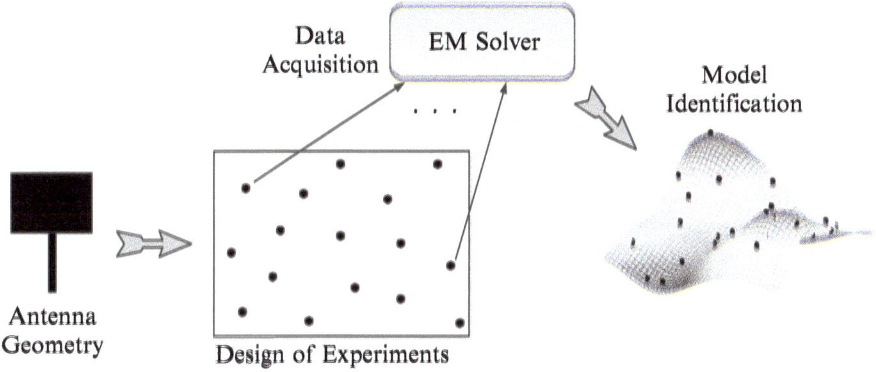

Fig. 3 Conceptual illustration of antenna modeling using RSA

A variety of RSA methods may be utilized for the construction of fast antenna models. These include polynomial approximation [66], neural networks [67–71] kriging [66, 72, 73], multi-dimensional Cauchy approximation [74], or support vector regression [75]. In this chapter, we discuss the utilization of a kriging interpolation technique for a construction of $\boldsymbol{R}_s$ surrogate. This method is attractive for multi-objective antenna optimization, especially due to very low evaluation cost, smoothness, and simple implementation provided through availability of MATLAB-based kriging toolboxes. In our implementation, we use a DACE toolbox [76]. For the sake of brevity, we omit details of kriging formulation. Interested reader is referred to the literature (e.g., [10, 43], or [66]).

Response surface approximation in general (and kriging interpolation in particular) as a method of generating the surrogate model for multi-objective optimization has a number of advantages: (1) the cost of model preparation depends only on the number of independent design variables; (2) there is no need for empirical-equivalent model of an antenna, and, consequently, no extra simulation software needs to be involved; (3) the RSA model may be applied for antenna structures that have no fast empirical representation; (4) initial design obtained through optimization of the coarse-mesh EM model is usually better than the initial design that could be possibly obtained by means of other methods.

2.3 Optimization Algorithm

The algorithm for expedited design and optimization of contemporary antennas in multi-objective setup utilizes variable-fidelity EM simulations of $\boldsymbol{R}_{cd}$ and $\boldsymbol{R}_f$ models as well as RSA model $\boldsymbol{R}_s$ in the form of a computationally cheap kriging interpolation model. Only the latter may be directly utilized as the fast evaluation engine for the metaheuristic algorithm. The RSA model is generated within the initially reduced design space using suitable design of experiments technique [65], followed by $\boldsymbol{R}_{cd}$ model data acquisition [77]. Moreover, a design space reduction scheme is also carried out at the $\boldsymbol{R}_{cd}$ model level, however $\boldsymbol{R}_f$ simulations may also be exploited at this step to ensure that high-fidelity representation of Pareto optimal solutions is contained within the reduced design space [7]. Finally, the $\boldsymbol{R}_f$ model data is utilized at the step of Pareto set refinement. The SBO engine is exploited here to reduce the misalignment between the $\boldsymbol{R}_s$ and $\boldsymbol{R}_f$ responses.

The design algorithm flow (see Fig. 4 for a detailed block diagram) can be summarized as follows:

1. Design space reduction using $\boldsymbol{R}_{cd}$ model (optionally refine the reduced space using the $\boldsymbol{R}_f$ model simulations in an SBO setup);
2. Sample the design space and acquire the $\boldsymbol{R}_{cd}$ data;
3. Construct the kriging interpolation model $\boldsymbol{R}_s$;
4. Obtain the Pareto set representation by optimizing $\boldsymbol{R}_s$ using MOEA;
5. Refine a set of designs selected from the initial Pareto set by means of SBO.

A construction of reliable RSA model requires a sufficient amount of training samples. Their number depends on the dimensionality and size of the design space, as well as the type (especially, nonlinearity) and the ranges of the training data outputs. In general, it cannot be determined beforehand. In our approach, the number of training samples within design space is iteratively increased until the RSA model reaches sufficient accuracy. For our purposes, the highest acceptable error is usually about 5 %. Cross-validation is performed for the accuracy verification, specifically, to estimate the generalization error of the surrogate [66].

Among various metaheuristic schemes available in the literature [21, 23, 27], MOEAs with the emphasis of genetic algorithms [19, 24, 35, 40], and PSOs [13, 29, 30, 32] belong to the most popular approaches in the context of antenna optimization. Here, we use the in-house implementation of an evolutionary

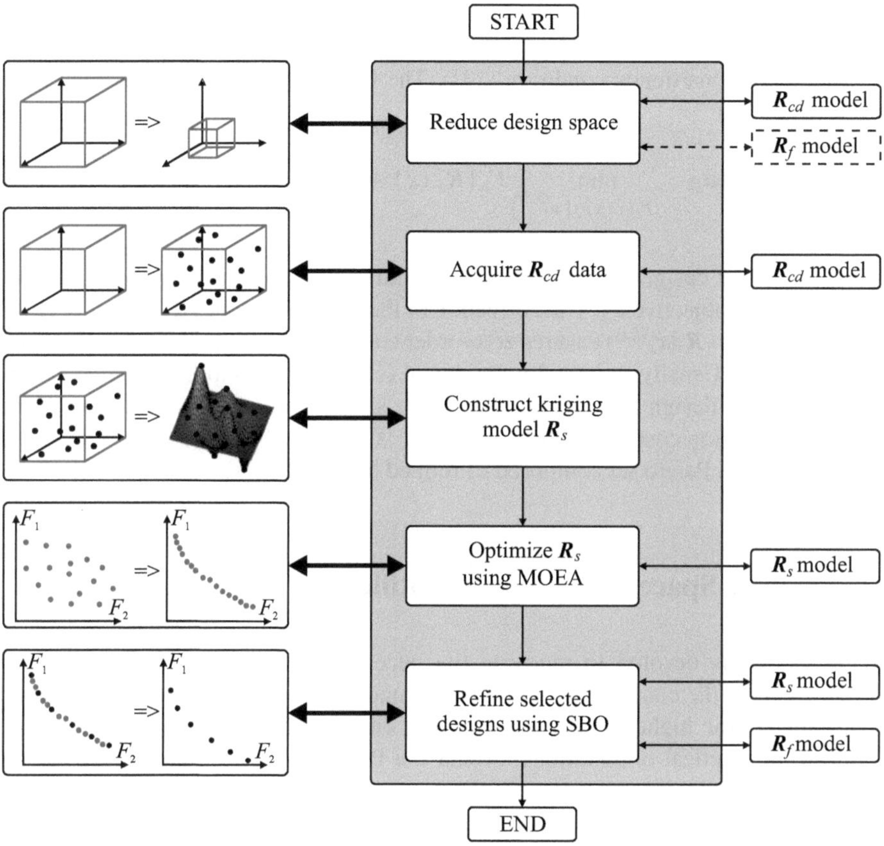

Fig. 4 Design flow of the multi-objective optimization technique. In the first stage, design variable ranges are reduced using a specialized algorithm and the $\boldsymbol{R}_{cd}$ model (optionally, $\boldsymbol{R}_f$ simulations may be involved in the process as well). Subsequently, the $\boldsymbol{R}_{cd}$ model data is acquired at the allocated training points, and a kriging interpolation model $\boldsymbol{R}_s$ is identified (here, using a DACE toolbox [76]). In the fourth stage, an in-house MOEA is used to optimize the $\boldsymbol{R}_s$ model and to determine the initial Pareto set. Finally, the SBO algorithm is utilized to refine the Pareto optimal designs by correcting the $\boldsymbol{R}_s$ model using data gained from $\boldsymbol{R}_f$ model simulations

algorithm with fitness sharing, mating restrictions, and Pareto dominance tournament selection [17, 18] as the optimization engine for identification of a Pareto optimal set.

The set of designs generated as an outcome of the MOEA-based optimization of the kriging model $\boldsymbol{R}_s$ is considered as an initial approximation of the Pareto optimal set. Subsequently, K designs selected from that initial set, i.e., $\boldsymbol{x}_s^{(k)}$, $k = 1, \ldots, K$, are refined using SBO to find a Pareto front representation at the high-fidelity $\boldsymbol{R}_f$ model level. Without loss of generality, we consider here two design objectives F_1 and F_2. The chosen $\boldsymbol{x}_s^{(k)}$ solutions are refined using output space mapping (OSM) [51, 54] which enhances the surrogate model by a correction term in the form of a difference between $\boldsymbol{R}_f$ and the original response of the $\boldsymbol{R}_s$ model at the current iteration point so that a perfect match between them is ensured (also referred to as a zero-order consistency condition [78]). The OSM algorithm used here is of the following form:

$$\boldsymbol{x}_f^{(k.i+1)} = \arg \min_{\boldsymbol{x},\ F_2(\boldsymbol{x}) \le F_2\left(\boldsymbol{x}_s^{(k.i)}\right)} F_1\left(\boldsymbol{R}_s(\boldsymbol{x}) + \left[\boldsymbol{R}_f\left(\boldsymbol{x}_s^{(k.i)}\right) - \boldsymbol{R}_s\left(\boldsymbol{x}_s^{(k.i)}\right)\right]\right) \quad (1)$$

The goal of design refinement is to minimize F_1 for each design $\boldsymbol{x}_f^{(k)}$ without degrading the objective F_2. The correction of the surrogate model $\boldsymbol{R}_s$ using the OSM term $\boldsymbol{R}_f(\boldsymbol{x}_s^{(k.i)}) - \boldsymbol{R}_s(\boldsymbol{x}_s^{(k.i)})$ ensures zero-order consistency between the models (here, $\boldsymbol{x}_f^{(k.0)} = \boldsymbol{x}_s^{(k)}$). Usually, only 2–3 iterations of (1) are required to find a refined high-fidelity model design $\boldsymbol{x}_f^{(k)}$, so the cost of $\boldsymbol{R}_f$ simulations is only a fraction of the total optimization cost. The OSM procedure is repeated for all K chosen samples, resulting in the Pareto set composed of refined high-fidelity solutions [7, 10].

3 Feature Space: Pareto Set Identification

This section is devoted to motivate the necessity of search space reduction in order to make the construction of the RSA model feasible in computational terms, particularly, for higher-dimensional design cases. The fundamental prerequisite (following practical observations) is that the Pareto set resides in a small part of the original design space, and only this very subset is of interest from the multi-objective optimization standpoint. Another important aspect is that narrowing down the ranges of design parameter variability which occurs during the space reduction process has important (and advantageous) consequences for scalability of the RSA model error with respect to the number of training samples.

3.1 Design Space Reduction as Prerequisite for Expedited Multi-Objective Antenna Optimization

Construction of the RSA model is a numerically demanding process involving multiple simulations of the $\boldsymbol{R}_{cd}$ model within a predefined solution space. In this chapter, we exploit uniform training point allocation realized through LHS [62–65]. LHS allows for generating any number of required samples for any number of design variables. The only restriction is that the sampled region of the design space should be a hypercube (i.e., a multi-dimensional interval). Here, we utilize a modified version of the algorithm LHS that allows to iteratively add any number of samples into the design space while retaining the LHS properties of the entire sample set [65].

Regardless of the design of experiments technique chosen for test samples generation, the numerical cost of RSA model preparation may be very high, especially if the number of adjustable parameters of the antenna structure of interest is large. In general, the cost of RSA model preparation—in terms of a number of training samples necessary to ensure usable accuracy—grows exponentially with the dimensionality of design space. For that reason, construction of RSA models is only practical for low-dimensional cases (up to a few independent parameters). For higher number of dimensions, the cost of a model preparation may surpass the number of evaluations needed for the direct determination of Pareto front by means of population-based metaheuristics. Unfortunately, contemporary antennas are often described by complex geometries with many (more than ten) design parameters. Setting up an accurate RSA model for such structures may be computationally prohibitive. This difficulty may be partially alleviated by decomposition of a structure into a sub-structures and generation of individual RSA models for each of them [10]. Nonetheless, the applicability of such a technique is restricted only to specific antenna structures, which may be divided into separate circuits (e.g., [59, 79]). Moreover, decomposition introduces additional inaccuracy into a model and complicates the design process [10].

On the other hand, the initial ranges of geometrical parameters of contemporary antennas are usually set rather wide to ensure that the optimum design (or, in case of multi-objective optimization, the Pareto optimal set) is allocated within the prescribed bounds. Setting up the RSA model in such redundant initial spaces, particularly for large number of adjustable parameters, is virtually impossible. Therefore instead of sampling the entire design space, a more practical approach is to determine the space region where Pareto optimal solutions reside. The reduction of initial solution space may alleviate the curse of dimensionality preventing the utilization of RSA model setup for a multi-parameter designs [7, 80]. It is also important that only a fraction of the Pareto optimal set representing the designs with reflection coefficient $|S_{11}| \leq -10$ dB within the frequency band of interest is considered relevant with respect to antenna applications [42]. Therefore, even higher restrictions for the solution space that accounts only the designs with acceptable reflection coefficient may be applied. Such an approach allows to decrease the volume of solution space by several orders, which significantly reduce the cost of RSA model preparation, especially for the structures with many independent variables. An illustration example of the location of the Pareto optimal set within redundant design space is shown in Fig. 5.

3.2 RSA Model Error Scaling in Reduced Solution Spaces

The ability to identify a relevant part of the Pareto optimal set within the pre-defined design space has a serious consequence that facilitates the RSA model setup in highly dimensional spaces. More specifically, Pareto optimal solutions obtained in a diminished solution space may be very similar in terms of some

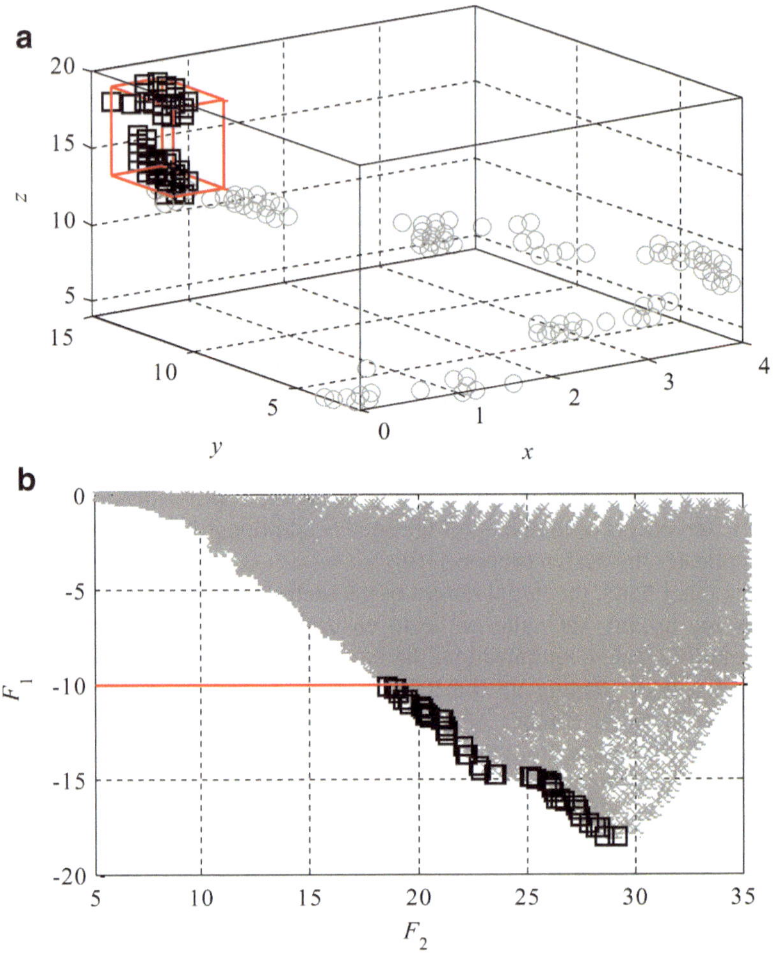

Fig. 5 (**a**) Visualization of the Pareto optimal set (*open circle*) in 3-dimensional solution space ($n = 3$). The portion of the design space X_r that contains the part of the Pareto set we are interested in (*red cuboid*, where $F_1 \leq -10$) is only a small fraction of the initial space X (*black cuboid*). (**b**) The Pareto set of interest (*open square*) versus the entire design space mapped to the feature space (*cross*). The set X_r is 456 times smaller (volume-wise) than the initial space X. The benefits of design space reduction are even more pronounced for higher-dimensional cases [7, 42, 80]

dimensions. This results in a flattening of the solution space with respect to the problem dimensionality. Moreover, some dimensions in the design space may be completely flattened, which allows excluding them from the process of solution space sampling [7].

Figure 6a shows a typical "thick" design space where increasing the number of training samples results in only slight reduction of the average minimum distance between the training points (proportional to $1/N^{1/n}$ with n being dimension of the

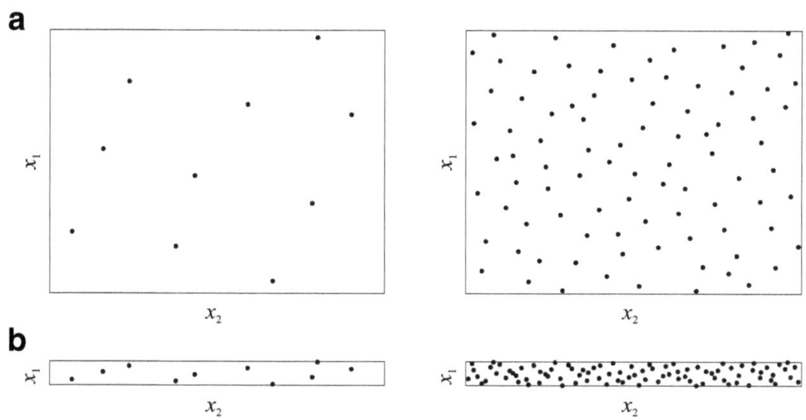

Fig. 6 Test sample allocation within two-dimensional design space. An increase of the number of samples from 10 to 90 in.: (**a**) "thick" design space; (**b**) "flat" design space. In the latter case, the increase of the number of training points results in larger reduction of the average minimum distance between the points along the "thick" dimension leading to better RSA model scalability

design space). The modeling error, on the other hand, is more or less proportional to that average distance. For "thin" space (Fig. 6b), the average minimum distance along "thick" (i.e., critical) dimensions decreases much faster than $1/N^{1/n}$ so that increasing the number of training points has more effect on the RSA model error. Of course, rigorous assessment of these effects is not possible in general because they are dependent on "flattening" effects and nonlinearity of the model along specific dimensions, which are both problem dependent.

4 Design Space Reduction Based on Sequential Single-Objective Optimizations

In the following sections we discuss different approaches to the design space reduction. These techniques allow for generating reliable RSA models even in multi-dimensional design spaces, which is critical for optimization of antennas in multi-objective setup. More specifically, design space reduction process is utilized to: (1) narrow down initial frontiers (lower/upper bounds for design variables), (2) enable preparation of the RSA model with desired accuracy, (3) reduce the number of necessary training samples, (4) identify the relevant fraction of the Pareto optimal set, and finally (5) speed-up generation of the RSA model. The design space reduction techniques require antenna model simulations, which increases the overall cost of antenna design. Notwithstanding, this overhead is mitigated by the identification of a relevant design space fraction using low-fidelity model simulations. Most importantly, the reduction step allows for a construction of a reliable RSA model within orders of magnitude (volume-wise) smaller design space using reasonably small number of samples [7, 42]. As a consequence, it enables the possibility of generating accurate models of antennas in multi-dimensional spaces. In this section, we discuss a reduction technique that utilizes sequential single-objective optimizations to find the extreme designs that determine the boundaries of the reduced search region. The approach is illustrated using a planar ultra-wideband dipole antenna.

4.1 Design Space Reduction Method

Here, we describe a simple scheme for design space reduction based on identification of the designs that determine the extreme points of a Pareto optimal set of interest. The designs are obtained by means of single-objective optimizations with respect to each objective, one at a time. Design parameter values corresponding to these extreme points may be utilized for the determination of the new frontiers of a refined solution space X_r, which is normally significantly smaller than the initial search space X. The design flow of the reduction algorithm is shown in Fig. 7. The algorithm operates as follows [42, 81]. Consider $\boldsymbol{l}$ and $\boldsymbol{u}$ as initially defined lower/upper bounds for the design parameters. Let

$$\boldsymbol{x}_{cd}^{*(k)} = \arg \min_{\boldsymbol{l} \le \boldsymbol{x} \le \boldsymbol{u}} F_k\left(\boldsymbol{R}_{cd}(\boldsymbol{x})\right) \tag{2}$$

where $k = 1, \ldots, N_{obj}$, be the optimum design of $\boldsymbol{R}_{cd}$ model with respect to the kth objective. The vectors $\boldsymbol{x}_{cd}^{*(k)}$ determine the extreme designs of the Pareto optimal set. The reduced solution space is then defined through the following lower/upper bounds:

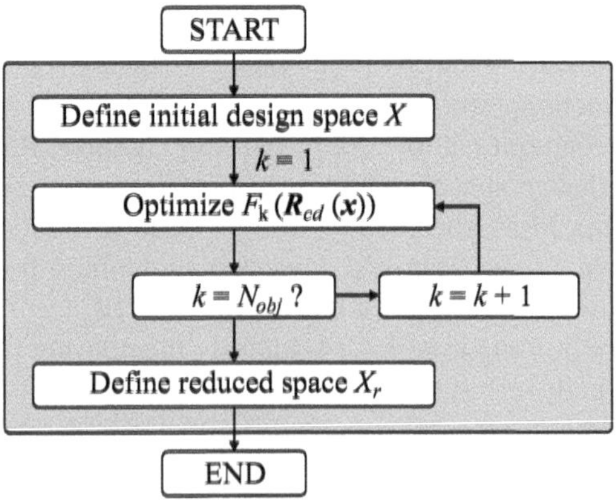

Fig. 7 The block diagram of design space reduction scheme based on sequential single-objective optimizations

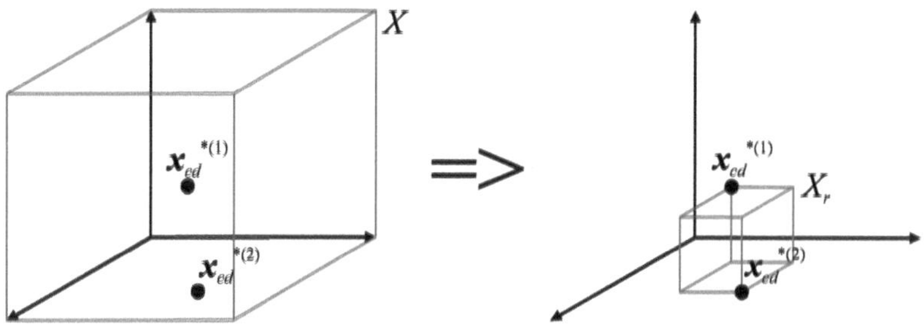

Fig. 8 Conceptual illustration of the design space reduction technique for $n = 3$ independent design variables (three-dimensional solution space) and $k = 2$ design objectives. The initial design space X is reduced by means of a sequential single-objective optimizations with respect to each objective, one at a time. The dimensions of the extreme designs (*filled circle*) $\boldsymbol{x}_{cd}{}^{*(k)}$ are used for the determination of refined solution space X_r

$$l^* = \min\left\{x^{*(1)}, x^{*(2)}, \ldots, x^{*(N_{obj})}\right\} \tag{3}$$

and

$$u^* = \max\left\{x^{*(1)}, x^{*(2)}, \ldots, x^{*(N_{obj})}\right\} \tag{4}$$

In practice, the refined space X_r is only a small subset of the initial one, which enables the generation of a reliable RSA model using reduced amount of data even if a design with multiple independent variables is considered. It should be emphasized that the above frontiers may not contain the entire Pareto optimal set, however the majority of it should be accounted for (including, of course, the extreme points). A conceptual illustration of the design space reduction scheme is shown in Fig. 8.

4.2 *Design Example: Planar Dipole Antenna*

Consider a planar, single layer dipole antenna shown in Fig. 9. The structure is composed of a main radiator with 50 Ω input impedance and two parasitic strips [82], with a total of six independent design variables. The antenna is designated to operate on a Rogers RT5880 dielectric substrate ($\varepsilon_r = 2.2$, $tan\delta = 0.0004$, $h = 1.58$ mm). Design variables considered for optimization are: $\boldsymbol{x} = [l_0\ w_0\ a_0\ l_p\ w_p\ s_0]^T$, whereas $a_1 = 0.5$ and $w_1 = 0.5$ remain fixed (all dimensions in mm). Both the high-fidelity model $\boldsymbol{R}_f$ (~12,510,000 mesh cells, average evaluation time: 20 min) and its low-fidelity counterpart $\boldsymbol{R}_{cd}$ (~167,900 mesh cells, average evaluation time: 30 s) are implemented in CST Microwave Studio [83]. The initial solution space X is defined by the following lower/upper bounds: $\boldsymbol{l} = [10\ 5\ 0.5\ 2\ 1\ 0.1]^T$ and $\boldsymbol{u} = [30\ 20\ 5\ 20\ 10\ 5]^T$. Two design objectives are considered: (1) minimization of antenna reflection within 3.1–10.6 GHz frequency band of interest (objective F_1),

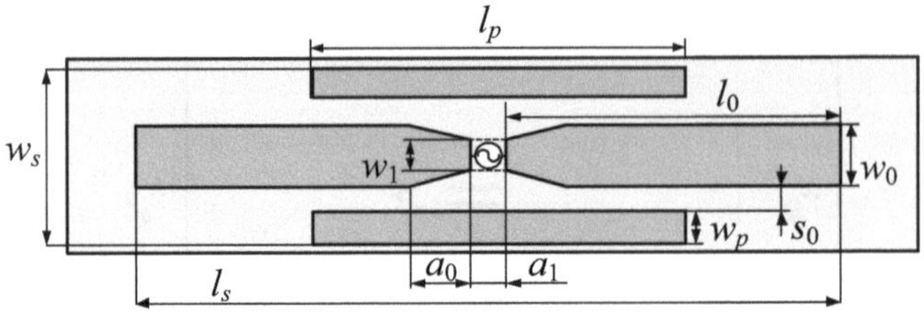

Fig. 9 Geometry of a considered planar ultra-wideband dipole antenna with six independent parameters [81]

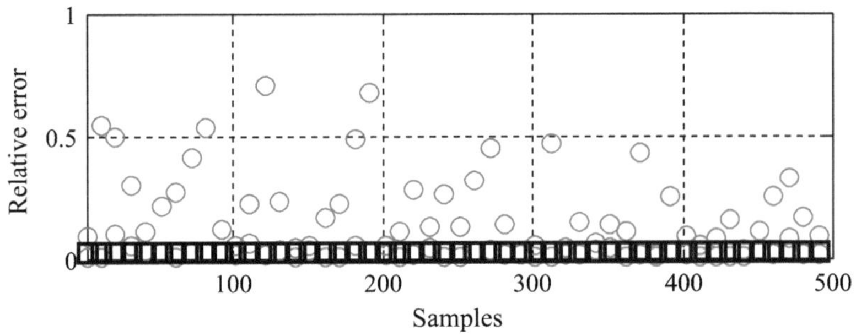

Fig. 10 Relative error of the kriging interpolation with respect to objective F_1. Model is constructed using 500 samples obtained in the initial design space (*open circle*) and in the refined design space (*open square*)

and (2) reduction of the antenna footprint defined by a rectangle $A = w_s \times l_s$, where $w_s = 2w_p + 2s_0 + w_0$ and $l_s = 2l_0 + 2a_0 + a_1$ (objective F_2).

An optimization scheme introduced in Sect. 2.3 and design space reduction technique of Sect. 4.1 are both utilized to perform multi-objective optimization of the structure [81]. The refined lower/upper bounds: $\boldsymbol{l}^* = [18\ 7.96\ 0.5\ 12.8\ 4.01\ 1.08]^T$ and $\boldsymbol{u}^* = [18.7\ 12.98\ 0.53\ 13.72\ 8.45\ 1.54]^T$ are obtained at a total cost of 250 evaluations of the $\boldsymbol{R}_{cd}$ model. We utilized pattern search algorithm [84] as single-objective optimization engine. The refined space X_r obtained by the utilization of discussed scheme is—volume-wise—six orders smaller than the initially defined one.

A kriging interpolation model $\boldsymbol{R}_s$ is constructed within a reduced solution space X_r using a base set composed of 500 $\boldsymbol{R}_{cd}$ samples (423 samples obtained using LHS scheme, supplemented with 64 design space corners, and 13 based on star-distribution [51]). The same set of samples is utilized for a construction of the RSA model in the initial solution space for comparison purposes. The average relative error of the $\boldsymbol{R}_s$ model generated in such a space is over 20 %, which makes it useless for the prediction of the antenna behavior. The error of the model obtained in refined space is only 2.2 % (see Fig. 10 for comparison). One should emphasize that the model generated within the initial space should be composed of well over 500,000 samples to achieve similar accuracy.

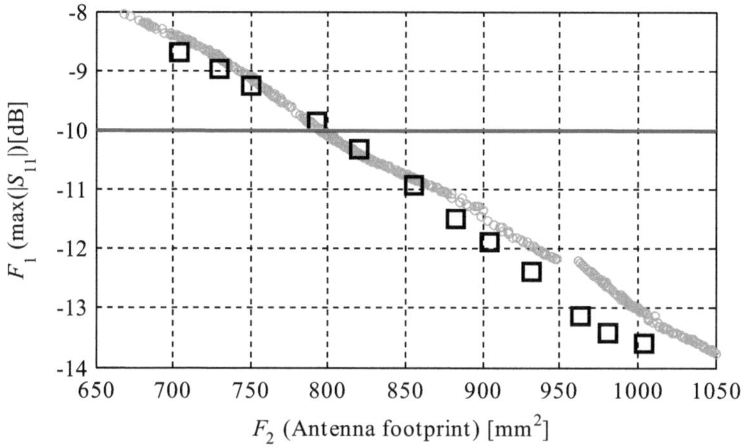

Fig. 11 Pareto optimal set of planar dipole antenna: RSA model optimized by means of MOEA (*open circle*) and fine model (*open square*) representation constituted by 12 samples obtained by OSM algorithm

The $\boldsymbol{R}_s$ model generated in the refined solution space is utilized as an evaluation engine for MOEA. In the next step, a set of 12 designs is chosen from the obtained Pareto optimal set and then refined using the OSM algorithm [54]. The smallest footprint of the antenna that fulfills the criteria upon reflection is 820 mm^2, while the lowest in-band reflection is −13.6 dB (corresponding antenna footprint: 1,004 mm^2). Note that the overall size of the structure with highest acceptable reflection (−10 dB) is over 18 % smaller than for the structure with the lowest reflection. The comparison of Pareto optimal sets composed of $\boldsymbol{R}_f$ and $\boldsymbol{R}_s$ models is shown in Fig. 11, while Table 1 gathers detailed dimensions of the selected high-fidelity antenna designs. Frequency characteristics of the selected Pareto optimal designs are shown in Fig. 12.

The design space reduction and generation of samples for RSA model preparation corresponds to 750 $\boldsymbol{R}_{cd}$ evaluations and about 37 $\boldsymbol{R}_f$ model simulations for MOEA-based optimization and refinement step. The detailed evaluation cost including number of each model evaluations is collected in Table 2. The total aggregated cost of multi-objective optimization of dipole antenna is about 19 h, which is negligible in comparison to cost of direct optimization being over 208 days (estimation based on a number of evaluations required by MOEA to yield initial front).

Table 1 Optimization results of planar dipole antenna

			Design variables [mm]					
$x_f^{(k)}$	F_1 [dB]	F_2 [mm^2]	l_0	w_0	a_0	l_p	w_p	s_0
1	−8.7	703	18.1	8.60	0.51	12.8	4.28	1.15
2	−9.0	730	18.0	8.74	0.52	12.8	4.58	1.16
3	−9.2	750	18.0	9.02	0.52	12.8	4.74	1.15
4	−9.9	792	18.0	9.44	0.52	12.8	5.11	1.17
5	−10.3	820	18.0	9.76	0.52	12.8	5.31	1.18
6	−10.9	856	18.2	10.2	0.50	12.8	5.60	1.08
7	−11.5	883	18.1	10.3	0.50	12.8	5.98	1.08
8	−11.9	905	18.0	10.6	0.50	12.8	6.15	1.10
9	−12.4	932	18.0	11.0	0.51	12.8	6.31	1.14
10	−13.1	964	18.1	12.6	0.50	12.8	5.52	1.47
11	−13.4	982	18.1	12.6	0.50	12.9	5.80	1.45
12	−13.6	1,004	18.2	12.4	0.51	12.8	6.25	1.36

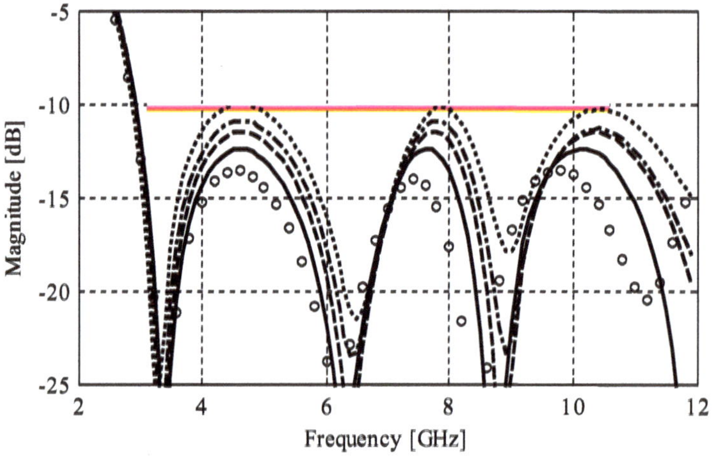

Fig. 12 Frequency responses of the selected designs from Table 1: $x_f^{(5)}$ (*dotted line*), $x_f^{(6)}$ (*dotted dashed line*), $x_f^{(7)}$ (*dashed line*), $x_f^{(9)}$ (*solid line*), $x_f^{(12)}$ (*open circle*)

Table 2 Planar dipole antenna: optimization cost

		CPU time	
Algorithm component	Number of model evaluations[a]	Absolute [min]	Relative to R_f
Evaluation of R_s	15,000	16	0.8
Evaluation of R_{cd}	750	375	18.8
Evaluation of R_f	36	720	36
Total cost[a]	N/A	1,111	**55.6**

[a]Excludes R_f and R_{cd} evaluation at the initial design

5 Pareto Dominance-Based Design Space Reduction for Multi-Objective Optimization of Antennas

In this section, we discuss another design space reduction technique based on identification and rejection of the most dominated solutions together with the design subspaces associated with them. A considered algorithm improves the design space reduction scheme referred in Sect. 4.1 by reducing the number of low-fidelity model evaluations required to seek the Pareto optimal set of interest. The operation of the algorithm is illustrated using a planar Yagi-Uda antenna with eight independent variables.

5.1 Design Space Reduction Algorithm

The concept of design space reduction explained in this section stems from identification of non-dominated designs within the initially defined frontiers. In practice, the region of solution space containing non-dominated designs is very small compared to the entire space, thus even rough identification of such a region could make the task of setting up the RSA model computationally feasible, even for high-dimensional problems. Here, we describe a two-stage design space reduction algorithm to achieve this goal [80]. The algorithm is based on a sparse sampling of the solution space X using low-fidelity $\boldsymbol{R}_{cd}$ model simulations and the utilization of Pareto ranking scheme [17] to cut out unpromising design space subsets, as a way toward seeking for the relevant fraction of the design space X_r that contains the Pareto optimal set of interest. In each iteration, a simple factorial design of experiments, a so-called star-distribution [51] is used to sample design space on the frontiers of X. The star-distribution scheme generates a test set X_t composed of $2n+1$ samples (where n is the number of independent design variables), referred to as nodes. The algorithm evaluates each node using the $\boldsymbol{R}_{cd}(\boldsymbol{x})$ model, and ranks them using Pareto dominance criteria (see Sect. 2.1). Then, the most dominated nodes are considered as irrelevant and they are rejected together with their corresponding design subspace. This step refines the lower/upper bounds defining the current approximation of the solution space $X_r^{\#}$ being of interest. The block diagram of the algorithm is shown in Fig. 13, whereas an exemplary workflow of the algorithm for the three-dimensional design space ($n = 3$) is detailed in Fig. 14.

In the course of the solution space reduction, all the nodes considered in the process are stored and then utilized for the determination of new lower/upper bounds of the temporary region of interest $X_r^{\#}$. This is realized by assigning to each node its Pareto rank; the nodes with rank being less or equal two (i.e., those that are dominated by at most one other design, cf. Fig. 14b, d, f), contribute to $X_r^{\#}$. The latter is defined as the smallest n-dimensional interval that contains all the contributing nodes. The algorithm stops if no further $X_r^{\#}$ reduction can be obtained in three consecutive iterations.

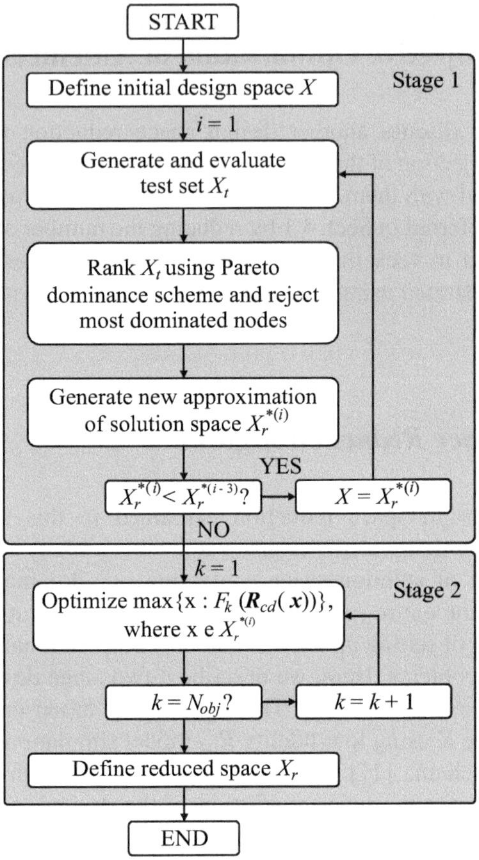

Fig. 13 The flow of the two-stage design space reduction scheme [80]. The first stage of the algorithm is executed until temporary region of interest X_r^* stays unchanged for three consecutive iterations. In the second, stage the solution space is extended by additional k designs separately optimized toward each $F_k(\boldsymbol{R}_{cd}(\boldsymbol{x}))$ objective

This restrictive approach for the rejection of dominated nodes can produce X_r^* that does not contain the entire Pareto front X_p. In order to prevent this, we expand the previously obtained solution space X_r^* by means of results of single-objective optimizations carried out with respect to each of the design objectives separately (cf. Fig. 14e–f). The starting points for the optimization of both nodes are max{$\boldsymbol{x}$: $F_k(\boldsymbol{R}_{cd}(\boldsymbol{x}))$}, $k = 1, 2$, where $\boldsymbol{x} \in X_r^*$. The region of interest X_r obtained using the explained algorithm is usually a few orders smaller (volume-wise) than the original design space, so that a sufficiently accurate RSA model can be generated in it using reasonable number of training samples. Procedure may be directly applied for larger number of objectives.

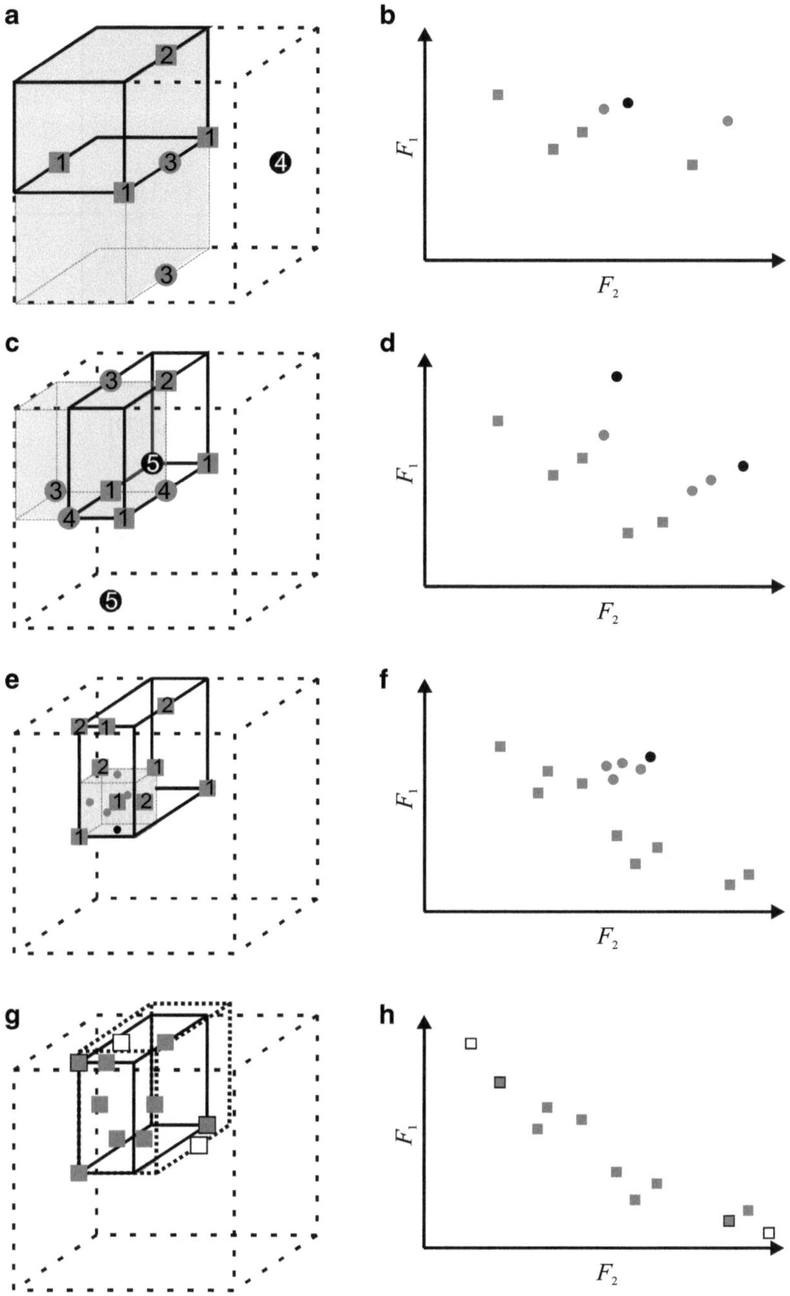

Fig. 14 Exemplary workflow of an algorithm for $n = 3$: (**a–f**) stage one—reduction of the design space—first, second, and last iteration; (**g–h**) stage two—expansion of X_r^* by single-objective optimization. *Squares* represent nodes with acceptable domination, *circles* are other nodes. *Black circles* are most dominated, rejected nodes. *White squares* are nodes refined by single-objective optimization. *Dashed, solid and dotted cuboid* represents X, X_r^*, and X_r, respectively. *Gray cuboid* is X^*

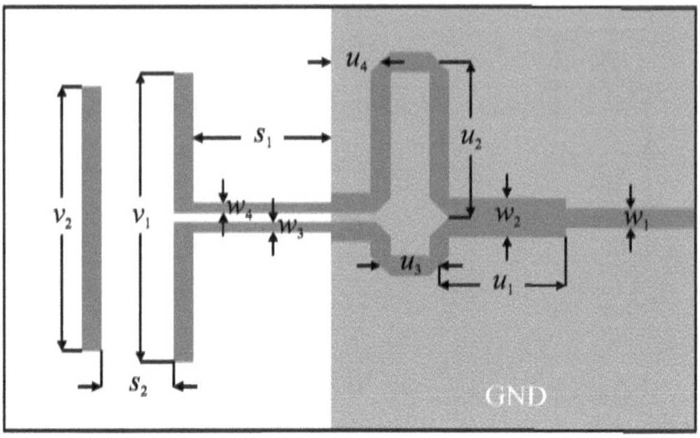

Fig. 15 Topology of the optimized eight-variable planar Yagi-Uda antenna [79]

5.2 Design Example: Planar Yagi-Uda Antenna

Consider a planar Yagi-Uda antenna (see Fig. 15) with eight independent design variables [79]. The antenna is designated to work on Rogers RO6010 dielectric substrate ($\varepsilon_r = 10.2$, $tan\delta = 0.0023$, $h = 0.635$ mm), and it is constituted by a driven element fed by a microstrip-to-coplanar strip transition, a director and an asymmetrical microstrip balun excited by a 50 Ω line. Design variables considered for optimization are: $\boldsymbol{x} = [s_1\ s_2\ v_1\ v_2\ u_1\ u_2\ u_3\ u_4]^T$. Other parameters, $w_1 = w_3 = w_4 = 0.6$, $w_2 = 1.2$, $u_5 = 1.5$, $s_3 = 3$, and $v_3 = 17.5$ remain fixed (all dimensions in mm). The high-fidelity model $\boldsymbol{R}_f$ (~1,512,000 mesh cells, average evaluation time: 18 min) and its low-fidelity counterpart $\boldsymbol{R}_{cd}$ (~85,680 mesh cells, average evaluation time: 110 s) are both prepared in CST Microwave Studio [83]. The initial solution space is defined by the following lower/upper bounds: $\boldsymbol{l} = [3.8\ 2.8\ 8.0\ 4.0\ 3.0\ 4.5\ 1.8\ 1.3]^T$, and $\boldsymbol{u} = [4.4\ 4.5\ 9.8\ 5.2\ 4.2\ 5.2\ 2.6\ 1.8]^T$. Two design objectives are considered: (1) minimization of the reflection coefficient (objective F_1), and (2) maximization of the antenna gain (objective F_2), both within 10–11 GHz frequency band of interest.

A multi-objective optimization of the structure follows the general design flow described in Sect. 2.3 and design space reduction scheme of Sect. 5.1. The first stage of the algorithm terminated after 153 evaluations of the $\boldsymbol{R}_{cd}$ model, while the expansion stage driven by a conventional gradient-based optimization scheme [85] needed 137 $\boldsymbol{R}_{cd}$ model evaluations to complete. The refined lower/upper bounds $\boldsymbol{l}^*/\boldsymbol{u}^*$ are: $\boldsymbol{l}^* = [4.1\ 3.63\ 8.11\ 4.27\ 3.6\ 4.67\ 1.8\ 1.3]^T$, $\boldsymbol{u}^* = [4.4\ 4.5\ 8.9\ 5.4\ 3.8\ 4.85\ 2.2\ 1.55]^T$, which resulted in reduction of design space by three orders [80].

Kriging interpolation model $\boldsymbol{R}_s$ is constructed within the reduced design space using 1,344 $\boldsymbol{R}_{cd}$ samples (1,000 LHS-allocated samples supplemented with 256 corners of X_r as well as those nodes considered in the space reduction stage that are within X_r). For the sake of comparison, the same set of 1,344 samples is used for the generation of kriging model within reduced and initial design space. The average

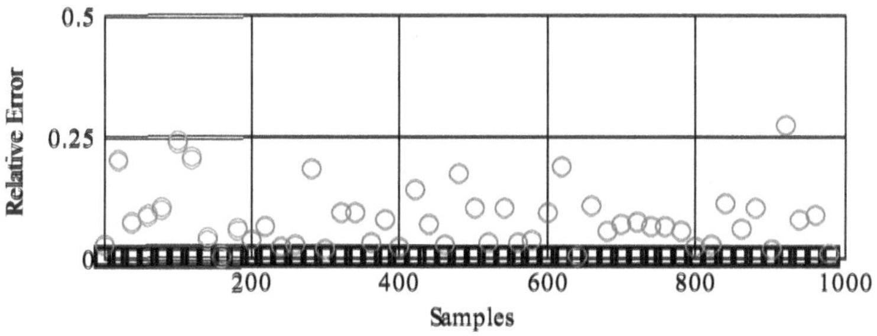

Fig. 16 Relative error of the kriging interpolation model constructed using 1,344 samples obtained using LHS scheme in the initial design space (*open circle*) and in the refined design space (*open square*) with respect to objective F_1

relative error of $\boldsymbol{R}_s$ model estimated using a cross-validation scheme [66] is 3 % for the former and 9 % for the latter. The error of the RSA model generated in the initial space excludes its utilization for prediction of the structure behavior. Due to space flattening by design space reduction algorithm the number of samples needed for the generation of an accurate RSA model should be increased at least by 3 orders, which is computationally infeasible. Relative errors of both models are graphically compared in Fig. 16.

The prepared RSA model has been used as an evaluation engine for optimization driven by MOEA. Subsequently, a set of ten design samples selected from the initial Pareto set has been refined using OSM technique. The results indicate that the best average gain of the antenna that still fulfills the requirements upon reflection ($|S_{11}| = -10.5$ dB) is almost 6.4 dB, while the minimum reflection coefficient is about -19.8 dB (with corresponding average gain being about 5.5 dB). Moreover, an average gain of the structure that satisfies requirements upon reflection is over 16 % greater in comparison with the antenna having the best reflection. The Pareto optimal sets constituted by $\boldsymbol{R}_s$ model and ten $\boldsymbol{R}_f$ samples refined by OSM algorithm are shown in Fig. 17, while detailed antenna dimensions of selected designs are gathered in Table 3. Moreover, a comparison of frequency responses of the selected antenna designs is presented in Fig. 18.

The total computational cost of multi-objective optimization of the considered Yagi-Uda antenna is about 57 h. The detailed data related to the number of each model evaluations and the corresponding numerical cost is gathered in Table 4. The total estimated cost of direct multi-objective optimization of $\boldsymbol{R}_f$ model based on number of evaluations required by MOEA to yield initial front is 50,000 (over 625 days), which is over 263 times longer in comparison with the described fast multi-objective optimization technique.

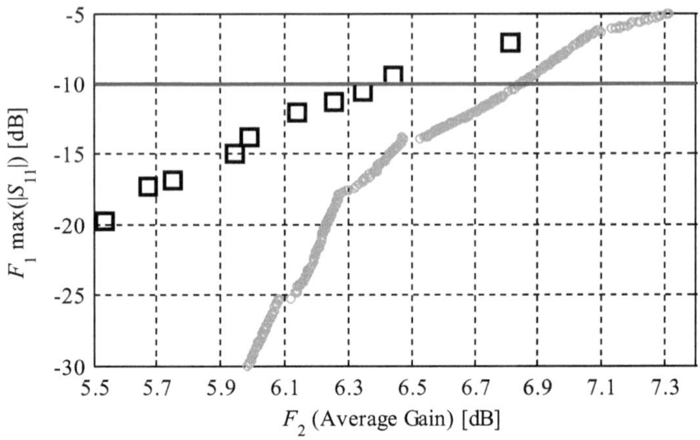

Fig. 17 Pareto optimal set of considered planar Yagi-Uda antenna obtained for low- (*open circle*) and high-fidelity (*open square*) model

Table 3 Planar Yagi-Uda antenna: multi-objective optimization results

			Design variables [mm]							
$x_f^{(k)}$	F_1 [dB]	F_2 [dB]	s_1	s_2	v_1	v_2	u_1	u_2	u_3	u_4
1	6.81	−7.14	4.39	4.43	8.11	5.40	3.78	4.84	2.20	1.55
2	6.44	−9.42	4.19	4.34	8.30	5.07	3.67	4.77	2.10	1.51
3	6.35	−10.55	4.19	4.34	8.29	4.98	3.67	4.78	2.10	1.51
4	6.26	−11.28	4.18	4.35	8.29	4.86	3.67	4.78	2.11	1.51
5	6.14	−12.04	4.17	4.34	8.31	4.73	3.67	4.78	2.10	1.51
6	5.99	−13.82	4.23	4.26	8.40	4.58	3.68	4.76	2.18	1.46
7	5.94	−14.95	4.20	4.28	8.48	4.49	3.69	4.76	2.18	1.46
8	5.75	−16.84	4.13	3.80	8.81	4.61	3.72	4.77	2.18	1.51
9	5.67	−17.27	4.12	3.71	8.83	4.54	3.73	4.76	2.17	1.51
10	5.54	−19.79	4.12	3.64	8.89	4.34	3.80	4.73	2.13	1.50

Table 4 Planar Yagi-Uda antenna: optimization cost

		CPU time	
Algorithm component	Number of model evaluations[a]	Absolute [min]	Relative to $\boldsymbol{R}_f$
Evaluation of $\boldsymbol{R}_s$	50,000	20	1.1
Evaluation of $\boldsymbol{R}_{cd}$	1,540	2,823	156.8
Evaluation of $\boldsymbol{R}_f$	30	540	30
Total cost[a]	N/A	3,383	**187.9**

[a]Excludes $\boldsymbol{R}_f$ and $\boldsymbol{R}_{cd}$ evaluation at the initial design

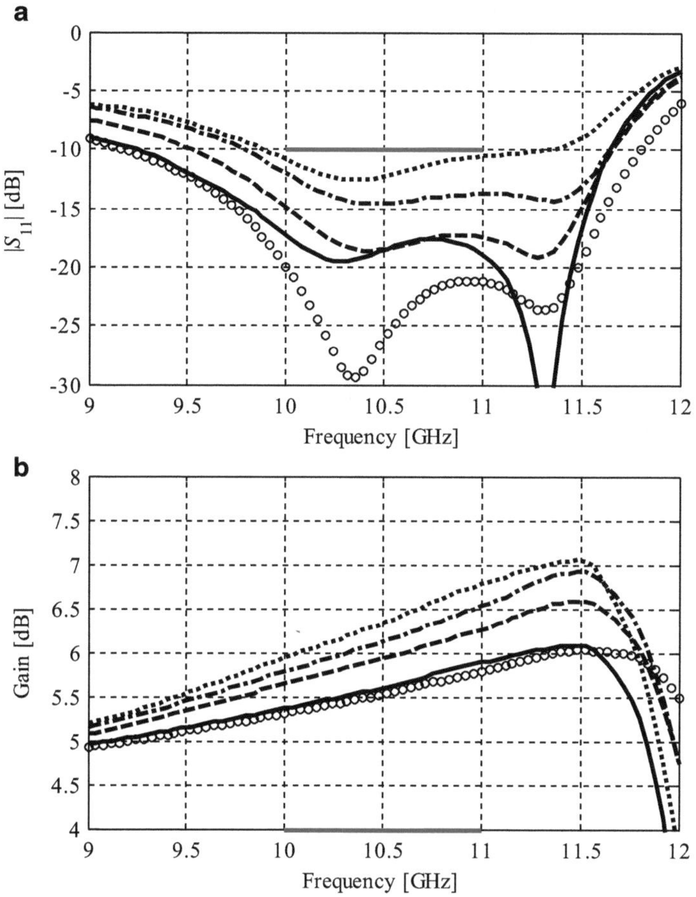

Fig. 18 Yagi-Uda antenna frequency responses: (**a**) reflection coefficient; (**b**) gain. Plots correspond to selected designs from Table 3, i.e., $\boldsymbol{x}_f^{(3)}$ *(dotted line)*, $\boldsymbol{x}_f^{(5)}$ *(dotted dashed line)*, $\boldsymbol{x}_f^{(7)}$ *(dashed line)*, $\boldsymbol{x}_f^{(9)}$ *(solid line)*, $\boldsymbol{x}_f^{(10)}$ *(open circle)*

6 Design Space Reduction Based on Sequential Single-Objective Optimizations Refined by SBO

In this section, we explain a modified design space reduction technique based on determination of extreme designs from the Pareto optimal set. Similarly to the method of Sect. 4.1, the technique exploits sequential single-objective optimizations. Additionally, optimal solutions based on low-fidelity model simulations are refined using SBO setup for the determination of their corresponding high-fidelity representations. Inclusion of the high-fidelity optimal extreme designs ensures that the reduced design space not only contains the majority of the high-fidelity Pareto front but also the low-fidelity one, which is important for subsequent creation of the RSA model as well as design refinement. An illustration example of antenna optimization using introduced scheme is presented.

6.1 Design Space Reduction Algorithm

In general, the responses of high-fidelity model $\boldsymbol{R}_f$ and its corresponding $\boldsymbol{R}_{cd}$ model are in reasonable agreement, which allows for the prediction of the refined design space X_r boundaries by performing simulations of the latter. Nonetheless, due to misalignment between the model responses, the refinement procedure may not be able to find some of the actual high-fidelity Pareto optimal designs, particularly those allocated close to the "ends" of the front. One should emphasize that the defined solution space X_r used for the generation of the RSA model cannot be simply expanded without $\boldsymbol{R}_{cd}$ model evaluations, and for that reason the subspace where Pareto front resides should be possibly accurate determined beforehand. In the approach detailed here [7], we estimate the boundaries of this region using the two-stage optimization procedure that involves both $\boldsymbol{R}_{cd}$ and $\boldsymbol{R}_f$ models (see Fig. 19

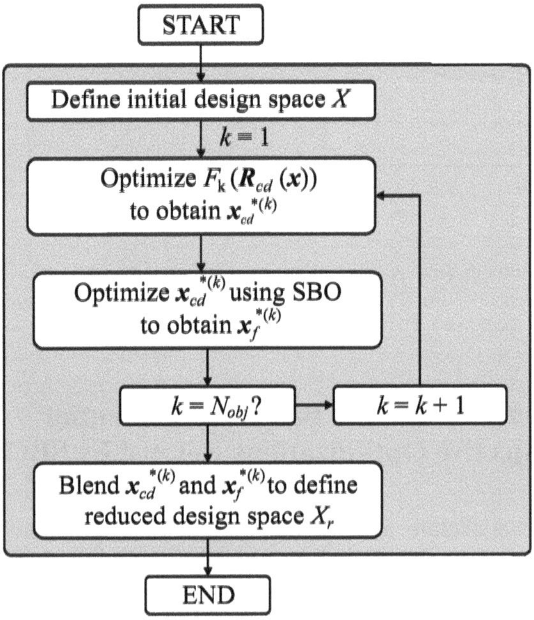

Fig. 19 Block diagram of the design space reduction algorithm [7]. An initial solution space X is refined using iterative single-objective optimizations. Subsequently, the obtained designs are corrected using the SBO algorithm. Dimensions of the final design solutions $\boldsymbol{x}_{cd}^{*(k)}$ and $\boldsymbol{x}_f^{*(k)}$ are blended to define the new frontiers of the refined design space

for a detailed algorithm flow). Let us consider $\boldsymbol{l}$ and $\boldsymbol{u}$ as lower/upper bounds of the initially defined solution space X. Then (2) may be utilized for the determination of optimal designs with respect to each objective. The frontiers defined in such a way may be inaccurate to some extent, thus, in the next step a high-fidelity $\boldsymbol{R}_f$ model representation of optimal points is obtained

$$\boldsymbol{x}_f^{*(k)} = \arg\min_{\boldsymbol{l} \leq \boldsymbol{x} \leq \boldsymbol{u}} F_k\left(\boldsymbol{R}_f(\boldsymbol{x})\right) \quad (5)$$

The designs $\boldsymbol{x}_f^{*(k)}$ are found using SBO (typically, frequency scaling combined with additive response correction is utilized [51]).

The introduced procedure allows for the determination of $2N_{obj}$ extreme points determined for the models with various fidelities. Dimensions of obtained designs may be subsequently blended to form a frontiers of the refined design space:

$$\boldsymbol{l}^* = \min\left\{\boldsymbol{x}_{cd}^{*(1)}, \ldots, \boldsymbol{x}_{cd}^{*(N_{obj})}, \boldsymbol{x}_f^{*(1)}, \ldots, \boldsymbol{x}_f^{*(N_{obj})}\right\} \quad (6)$$

and

$$\boldsymbol{u}^* = \max\left\{\boldsymbol{x}_{cd}^{*(1)}, \ldots, \boldsymbol{x}_{cd}^{*(N_{obj})}, \boldsymbol{x}_f^{*(1)}, \ldots, \boldsymbol{x}_f^{*(N_{obj})}\right\} \quad (7)$$

For typical shapes of the Pareto optimal set, the refined solution space X_r contains fronts of both the surrogate and high-fidelity models. The former is important because the RSA model created in $[\boldsymbol{l}^*, \boldsymbol{u}^*]$ is a representation of $\boldsymbol{R}_{cd}$. The latter is essential to ensure sufficient room for improving the high-fidelity designs during the refinement stage (cf. (1)). It should be noted that the utilization of high-fidelity models in the design space reduction step increases the computational cost of the design procedure. Notwithstanding, a number of extreme designs are relatively low depending on a number of design objectives (mostly 2–3) [7]. Furthermore, the refinement stage usually requires two to three evaluations of $\boldsymbol{R}_f$ model, thus the influence of SBO on the overall simulation cost during design space reduction is small. A conceptual illustration of the described procedure is shown in Fig. 20.

6.2 Design Example: Planar Monopole Antenna

Consider a planar monopole antenna described by 13 independent design variables [25]. The structure consists of a radiator formed by three stacked trapezoids (see Fig. 21). The antenna is designated to work on Taconic RF-35 dielectric substrate ($\varepsilon_r = 3.5$ $\tan\delta = 0.0018$, $h = 0.762$ mm). Design variables considered for optimization are represented by a vector $\boldsymbol{x} = [a_1\ a_2\ a_3\ a_4\ a_5\ a_6\ b_1\ b_2\ b_3\ w_2\ l\ d\ o]^T$. Variable w_1 is fixed to 1.7 to ensure 50 Ω input impedance

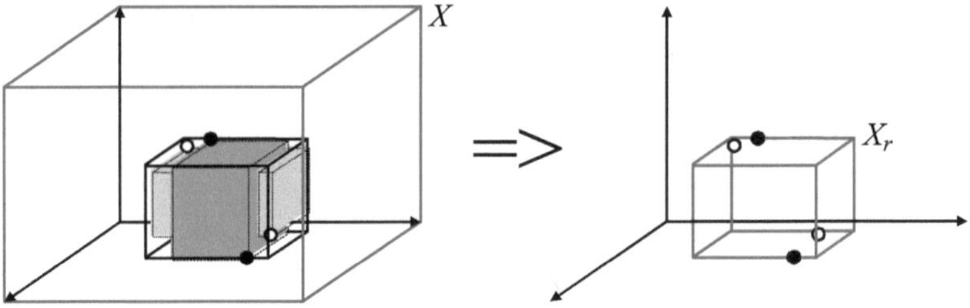

Fig. 20 Conceptual illustration of the modified design space reduction technique for $n = 3$ independent design variables and $k = 2$ design objectives. The initial design space X is reduced by means of a sequential single-objective optimizations toward each objective. The dimensions of extreme designs $\boldsymbol{x}_{cd}{}^{*(k)}$ (*filled circle*) are then refined using SBO scheme. Auxiliary designs $\boldsymbol{x}_f{}^{*(k)}$ (*open circle*) are subsequently utilized together with $\boldsymbol{x}_{cd}{}^{*(k)}$ to set boundaries of a refined solution space X_r

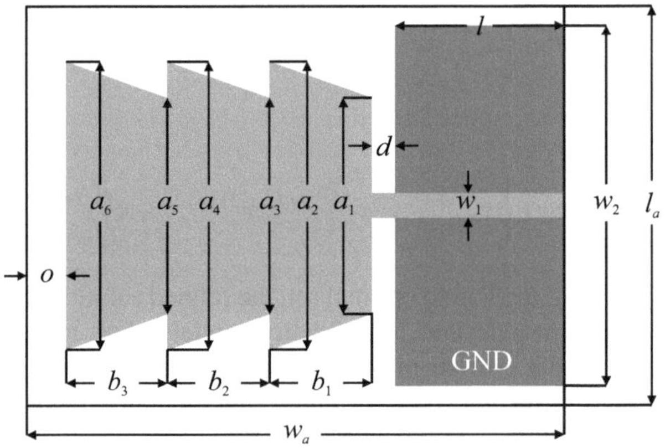

Fig. 21 Geometry of the considered planar monopole antenna with 13 independent design variables [25]

of antenna (all dimensions are in mm). The high-fidelity model $\boldsymbol{R}_f$ of the structure (~2,500,000 mesh cells, average simulation time: 10 min) and its low-fidelity counterpart $\boldsymbol{R}_{cd}$ (~33,600 mesh cells, average simulation time: 22 s) are both prepared and evaluated in CST Microwave Studio [83]. The initial design frontiers are: $\boldsymbol{l} = [5\ 5\ 5\ 5\ 5\ 5\ 1\ 1\ 1\ 0.2\ 8\ 20\ 5]^T$ and $\boldsymbol{u} = [25\ 25\ 25\ 25\ 25\ 25\ 15\ 15\ 15\ 2\ 15\ 40\ 10]^T$. The monopole is optimized with respect to the following design objectives: (1) minimization of reflection within 3.1–10.6 GHz frequency band of interest (objective F_1), and (2) reduction of the overall antenna size defined as $w_a \times l_a$ rectangle, where $w_a = l + d + b_1 + b_2 + b_3 + o$ and $l_a = w_2 + o$ (objective F_2).

The antenna is designed using the generic procedure of Sect. 2.3 and design space reduction methodology explained in Sect. 6.1. Determination of the refined frontiers required 800 evaluations of $\boldsymbol{R}_{cd}$ model during single-objective optimization (pattern search is utilized as an optimization engine) and four simulations of the $\boldsymbol{R}_f$ model

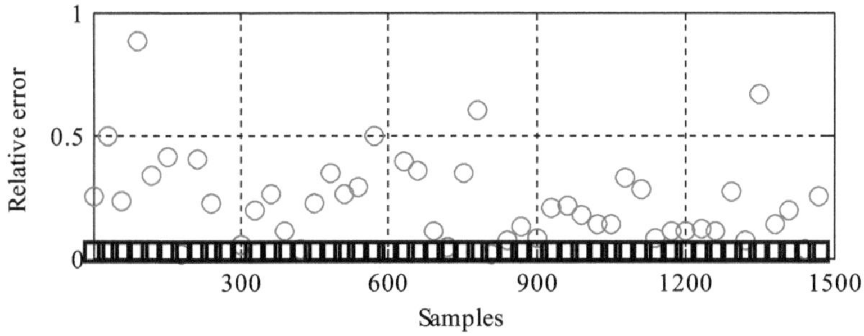

Fig. 22 Relative error of the kriging interpolation model constructed using 1,500 $\boldsymbol{R}_{cd}$ samples obtained using LHS scheme in the initial design space (*open circle*) and in the refined design space (*open square*). The errors are estimated with respect to objective F_1

during SBO driven refinement of extreme designs. Frontiers of the reduced design space X_r are as follows: $\boldsymbol{l}^* = [10.07\ 21.63\ 22.2\ 21\ 20.8\ 22.7\ 3.2\ 3.8\ 12.32\ 0.57\ 8.3\ 22.07\ 5.0]^T$, $\boldsymbol{u}^* = [11.3\ 21.96\ 24.3\ 24.15\ 21.27\ 24.6\ 3.9\ 4\ 13.08\ 0.74\ 11.2\ 39.35\ 5.75]^T$. The refined design space is 14 orders of magnitude smaller (volume-wise) than the initial one [7].

In this example, a RSA model is sequentially generated within the refined design space X_r starting from 500 $\boldsymbol{R}_{cd}$ samples allocated using the LHS scheme (average relative error estimated through cross-validation is 6 %) [7]. After three iterations the average error of final RSA model composed using 1,500 $\boldsymbol{R}_{cd}$ samples is reduced to only 3.5 %. It should be emphasized that the accuracy improvement of the final model compared to the initial one, i.e., 1.72, is much better than $3^{1/13} = 1.08$ (cf. Sect. 3.2). This is due to the "flattening" of the design space. Moreover, average error of the model generated using the same set of 1,500 samples within initial solution space is over 22 % which definitely turns it useless for the multi-objective optimization. A comparison of both RSA models errors is shown in Fig. 22.

The initial Pareto optimal set is found using MOEA. Subsequently, a set of 12 designs selected from the initial front is refined using OSM algorithm [54]. Moreover, two designs obtained during the determination of extreme points are added to the final Pareto front. The minimum antenna in-band reflection is −16.3 dB (corresponding antenna footprint is 1,526 mm^2). The smallest footprint of an antenna that still satisfies the conditions upon acceptable reflection ($|S_{11}| \leq -10$ dB) is 1,134 mm^2. Furthermore, the difference between the minimal and maximal antenna size that satisfies the requirements upon reflection is over 26 %. The comparison of Pareto representations constituted by $\boldsymbol{R}_f$ and $\boldsymbol{R}_s$ models is shown in Fig. 23, while the comparison of antenna reflection characteristics is presented in Fig. 24. Detailed dimensions of selected designs based on high-fidelity model evaluations are collected in Table 5.

The overall cost of antenna optimization including design space reduction, generation of RSA model, MOEA optimization, and design refinement using SBO engine is about 21 h. A detailed evaluation cost with respect to each model is

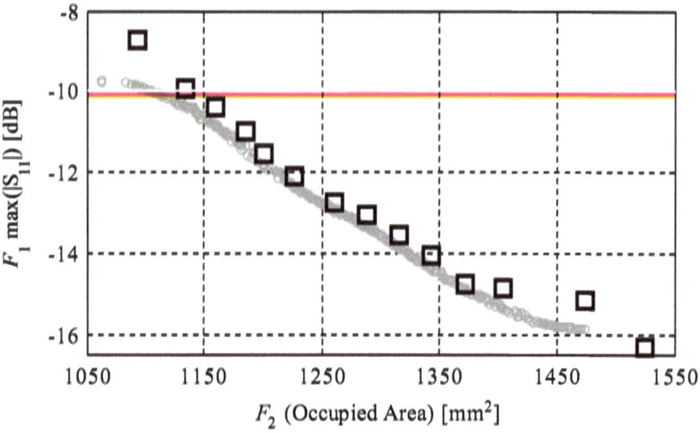

Fig. 23 Pareto optimal set of a planar UWB monopole antenna obtained for low- (*open circle*) and high-fidelity (*open square*) model

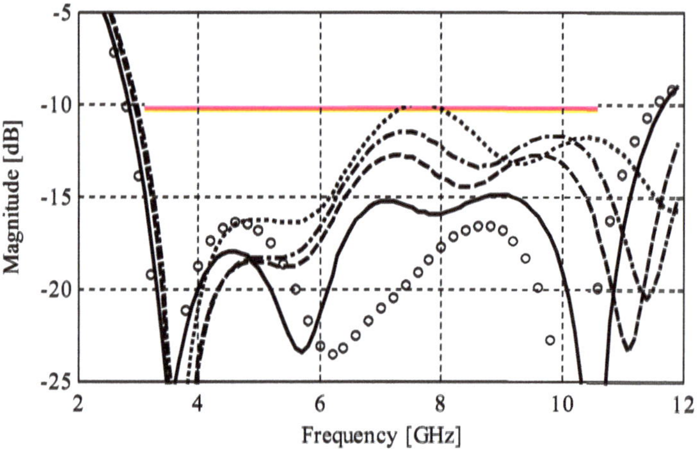

Fig. 24 Frequency responses of the selected designs from Table 5: $x_f^{(1)}$ (*open circle*), $x_f^{(3)}$ (*solid line*), $x_f^{(8)}$ (*dashed line*), $x_f^{(10)}$ (*dotted dashed line*), $x_f^{(13)}$ (*dotted line*)

presented in Table 6. The total estimated cost of direct multi-objective optimization is about 347 days, thus the utilization of introduced design procedure together with the design space reduction algorithm speeds up the multi-objective antenna design by over two orders of magnitude compared to conventional approach.

Table 5 Multi-objective optimization results of a planar UWB monopole antenna

		Selected designs						
		$x_f^{(1)}$	$x_f^{(2)}$	$x_f^{(3)}$	$x_f^{(5)}$	$x_f^{(8)}$	$x_f^{(10)}$	$x_f^{(13)}$
F_1 [mm^2]		1,526	1,475	1,405	1,342	1,261	1,202	1,134
F_2 [dB]		−16.3	−15.2	−14.9	−14.1	−12.7	−11.5	−10.0
Antenna parameters [mm]	a_1	11.3	10.9	11.0	11.1	11.1	10.9	10.1
	a_2	21.6	21.8	21.8	21.9	21.7	21.6	21.6
	a_3	22.4	22.4	22.2	22.2	22.2	22.2	22.2
	a_4	22.3	22.4	22.8	21.6	21.3	21.0	21.0
	a_5	21.3	20.9	21.0	20.9	21.0	21.0	20.8
	a_6	24.6	24.1	23.7	23.8	24.1	24.6	22.7
	b_1	3.9	3.9	3.9	3.9	3.6	3.5	3.9
	b_2	4.0	4.0	3.9	3.9	3.9	3.8	3.8
	b_3	13.0	13.1	13.0	12.7	12.4	12.3	12.3
	w_2	0.6	0.6	0.7	0.6	0.6	0.6	0.6
	l	11.0	10.6	10.6	10.7	11.0	11.0	11.1
	d	37.9	37.0	35.4	33.9	32.0	30.6	28.3
	o	5.2	5.1	5.0	5.0	5.0	5.0	5.0

Table 6 Planar UWB monopole antenna: optimization cost

	Number of model evaluations[a]	CPU time	
Algorithm component	Absolute [min]	Relative to $\boldsymbol{R}_f$	
Evaluation of $\boldsymbol{R}_s$	50,000	20	2
Evaluation of $\boldsymbol{R}_{cd}$	2,300	843	84.3
Evaluation of $\boldsymbol{R}_f$	43	430	43
Total cost[a]	N/A	1,293	**129.3**

[a]Excludes $\boldsymbol{R}_f$ and $\boldsymbol{R}_{cd}$ evaluation at the initial design

7 Conclusion

In this chapter, a technique for multi-objective optimization of computationally expensive antenna models is discussed. The method exploits population-based metaheuristic in the form of an evolutionary algorithm. Fast determination of the trade-off solutions between non-commensurable objectives is possible by the utilization of the multi-fidelity EM-simulated antenna models. The physics-based models are evaluated only during the construction and the refinement of the response surface approximation model. The latter bears the burden of multi-objective optimization. Pareto optimal solutions obtained from RSA model are refined by means of SBO. The overall cost of the method is negligible in comparison with direct multi-objective optimization of the high-fidelity antenna model.

The presented method is further extended to handle antennas with multiple independent design variables. This is realized by means of suitable design space reduction schemes. The goal of design space reduction is to identify the region that contains relevant fraction of the Pareto front. Three different schemes for solving this task based on sequential single-objective optimizations and analysis of the designs with respect to Pareto dominance criteria are discussed. Furthermore, introduced design and optimization methodology is illustrated using three exemplary planar antennas: a 6-variable UWB dipole, an 8-variable Yagi-Uda structure, and 13-variable UWB monopole that are successfully optimized in a timeframe being only a fraction of conventional multi-objective setup. Despite promising results, the proposed optimization methodology is restricted to designs with about a dozen of independent design variables. Moreover, the methods for design space reduction discussed in this chapter cannot guarantee that the entire Pareto front of interest is accounted in the refined space. Expanding the presented methods to handle highly dimensional cases will be the subject of the future research.

References

1. Chen, N.Z.N.: Wideband microstrip antennas with sandwich substrate. IET Microw. Antennas Propag. **2**, 538–546 (2008)
2. Milligan, T.A.: Modern Antenna Design, 2nd edn. Wiley-IEEE Press, Hoboken (2005)
3. Fujimoto, K.: Mobile Antenna Systems Handbook, 3rd edn. Artech House, Norwood (2008)
4. Dziunikowski, W.: Multiple-input multiple-output (MIMO) antenna systems. In: Chandran, S. (ed.) Adaptive Antenna Arrays: Trends and Applications. Signals and Communication Technology, pp. 259–273. Springer, Berlin (2004)
5. Koziel, S., Ogurtsov, S.: Rapid optimisation of omnidirectional antennas using adaptively adjusted design specifications and kriging surrogates. IET Microw. Antennas Propag. **7**, 1194–1200 (2013)
6. Guha, D., Gupta, B., Kumar, C., Antar, Y.M.M.: Segmented hemispherical DRA: new geometry characterized and investigated in multi-element composite forms for wideband antenna applications. IEEE Trans. Antennas Propag. **60**, 1605–1610 (2012)
7. Koziel, S., Bekasiewicz, A., Zieniutycz, W.: Expedited EM-driven multi-objective antenna design in highly-dimensional parameter spaces. IEEE Antennas Wirel. Propag. Lett. **13**, 631–634 (2014)
8. Bekasiewicz, A., Koziel, S.: A concept and design optimization of compact planar UWB monopole antenna. In: IEEE Int. Symp. Antennas Prop. (2014)
9. Koziel, S., Bekasiewicz, A.: Novel structure and EM-driven design of small UWB monople antenna. In: Int. Symp. Antenna Technol. Appl. Electromagn. (2014)
10. Koziel, S., Ogurtsov, S.: Multi-objective design of antennas using variable-fidelity simulations and surrogate models. IEEE Trans. Antennas Propag. **61**, 5931–5939 (2013)
11. Sharaqa, A., Dib, N.: Position-only side lobe reduction of a uniformly excited elliptical antenna array using evolutionary algorithms. IET Microw. Antennas Propag. **7**, 452–457 (2013)
12. Kuwahara, Y.: Multiobjective optimization design of Yagi-Uda antenna. IEEE Trans. Antennas Propag. **53**, 1984–1992 (2005)
13. Afshinmanesh, F., Marandi, A., Shahabadi, M.: Design of a single-feed dual-band dual-polarized printed microstrip antenna using a Boolean particle swarm optimization. IEEE Trans. Antennas Propag. **56**, 1845–1852 (2008)
14. Chamaani, S., Mirtaheri, S.A., Abrishamian, M.S.: Improvement of time and frequency domain performance of antipodal Vivaldi antenna using multi-objective particle swarm optimization. IEEE Trans. Antennas Propag. **59**, 1738–1742 (2011)
15. Cao, W., Zhang, B., Liu, A., Yu, T., Guo, D., Wei, Y.: Broadband high-gain periodic endfire antenna by using I-shaped resonator (ISR) structures. IEEE Antennas Wirel. Propag. Lett. **11**, 1470–1473 (2012)

16. Venkatarayalu, N.V., Ray, T.: Optimum design of Yagi-Uda antennas using computational intelligence. IEEE Trans. Antennas Propag. **52**, 1811–1818 (2004)
17. Deb, K.: Multi-Objective Optimization Using Evolutionary Algorithms. Wiley, Chichester (2001)
18. Talbi, E.-G.: Metaheuristics – From Design to Implementation. Wiley, Hoboken (2009)
19. Carvalho, R., Saldanha, R.R., Gomes, B.N., Lisboa, A.C., Martins, A.X.: A multi-objective evolutionary algorithm based on decomposition for optimal design of Yagi-Uda antennas. IEEE Trans. Magn. **48**, 803–806 (2012)
20. Eichfelder, G.: Adaptive scalarization methods in multiobjective optimization. SIAM J. Opt. **19**, 1694–1718 (2009)
21. Coleman, C.M., Rothwell, E.J., Ross, J.E.: Investigation of simulated annealing, ant-colony optimization, and genetic algorithms for self-structuring antennas. IEEE Trans. Antennas Propag. **52**, 1007–1014 (2004)
22. Holland, J.H.: Adaptation in Natural and Artificial Systems. MIT Press, Cambridge (1992)
23. Khodier, M.: Optimisation of antenna arrays using the cuckoo search algorithm. IET Microw. Antennas Propag. **7**, 458–464 (2013)
24. Ramos, R.M., Saldanha, R.R., Takahashi, R.H.C., Moreira, F.J.S.: The real-biased multiobjective genetic algorithm and its application to the design of wire antennas. IEEE Trans. Magn. **39**, 1329–1332 (2003)
25. Yang, X.-S., Ng, K.-T., Yeung, H.S., Man, K.F.: Jumping genes multiobjective optimization scheme for planar monopole ultrawideband antenna. IEEE Trans. Antennas Propag. **56**, 3659–3666 (2008)
26. Kuwahara, Y.: Multiobjective optimization design of Yagi-Uda antenna. IEEE Trans. Antennas Propag. **53**, 1984–1992 (2005)
27. Robinson, J., Rahmat-Samii, Y.: Particle swarm optimization in electromagnetics. IEEE Trans. Antennas Propag. **52**, 397–407 (2004)
28. Nanbo, J., Rahmat-Samii, Y.: Hybrid real-binary particle swarm optimization (HPSO) in engineering electromagnetics. IEEE Trans. Antennas Propag. **58**, 3786–3794 (2010)
29. Chamaani, S., Mirtaheri, S.A., Abrishamian, M.S.: Improvement of time and frequency domain performance of antipodal Vivaldi antenna using multi-objective particle swarm optimization. IEEE Trans. Antennas Propag. **59**, 1738–1742 (2011)
30. Nanbo, J., Rahmat-Samii, Y.: Advances in particle swarm optimization for antenna designs: real-number, binary, single-objective and multiobjective implementations. IEEE Trans. Antennas Propag. **55**, 556–567 (2007)
31. Boeringer, D.W., Werner, D.H.: Particle swarm optimization versus genetic algorithms for phased array synthesis. IEEE Trans. Antennas Propag. **52**, 771–779 (2004)
32. Hao, W., Junping, G., Ronghong, J., Jizheng, Q., Wei, L., Jing, C., Suna, L.: An improved comprehensive learning particle swarm optimization and its application to the semiautomatic design of antennas. IEEE Trans. Antennas Propag. **57**, 3018–3028 (2009)
33. Minasian, A.A., Bird, T.S.: Particle swarm optimization of microstrip antennas for wireless communication systems. IEEE Trans. Antennas Propag. **61**, 6214–6217 (2013)
34. Grimaccia, F., Mussetta, M., Zich, R.E.: Genetical swarm optimization: self-adaptive hybrid evolutionary algorithm for electromagnetics. IEEE Trans. Antennas Propag. **55**, 781–785 (2007)
35. Yeung, S.H., Man, K.F., Luk, K.M., Chan, C.H.: A trapeizform U-slot folded patch feed antenna design optimized with jumping genes evolutionary algorithm. IEEE Trans. Antennas Propag. **56**, 571–577 (2008)
36. Mythili, P., Osoba, P.E., Michielssen, E.: A multi-objective antenna placement genetic algorithm for matched array synthesis on complex platforms. In: IEEE International Conference on Communication Systems, pp. 109–112 (2010)
37. Xiaoyan, Y., Zhouyuan, L., Rodrigo, D., Mopidevi, H.S., Kaynar, O., Jofre, L., Cetiner, B.A.: A parasitic layer-based reconfigurable antenna design by multi-objective optimization. IEEE Trans. Antennas Propag. **60**, 2690–2701 (2012)
38. Hannien, I.: Optimization of a reflector antenna system. CST AG whitepaper, pp. 1–4 (2012)

39. Kolundzija, B.M., Olcan, D.I.: Multiminima heuristic methods for antenna optimization. IEEE Trans. Antennas Propag. **54**, 1405–1415 (2006)
40. John, M., Ammann, M.J.: Antenna optimization with a computationally efficient multiobjective evolutionary algorithm. IEEE Trans. Antennas Propag. **57**, 260–263 (2009)
41. Lisboa, A.C., Vieira, D.A.G., Vasconcelos, J.A., Saldanha, R.R., Takahashi, R.H.C.: Monotonically improving Yagi-Uda conflicting specifications using the dominating cone line search method. IEEE Trans. Magn. **45**, 1494–1497 (2009)
42. Bekasiewicz, A., Koziel, S., Leifsson, L.: Low-Cost EM-Simulation-Driven Multi-Objective Optimization of Antennas. In: Int. Conf. Computational Science, in Procedia Computer Science. **29**, 790–799 (2014)
43. Koziel, S., Ogurtsov, S.: Rapid design optimization of antennas using space mapping and response surface approximation models. Int. J. RF Microw. Comput. Aid. Eng. **21**, 611–621 (2011)
44. Koziel, S., Bandler, J.W.: Accurate modeling of microwave devices using kriging-corrected space mapping surrogates. Int. J. Numer. Model. **25**, 1–14 (2012)
45. Soontornpipit, P., Furse, C.M., You, C.C.: Miniaturized biocompatible microstrip antenna using genetic algorithm. IEEE Trans. Antennas Propag. **53**, 1939–1945 (2005)
46. Altshuler, E.E., O'Donnell, T.H.: An electrically small multi-frequency genetic antenna immersed in a dielectric powder. IEEE Antennas Propag. Mag. **53**, 33–40 (2011)
47. Schantz, H.: The Art and Science of Ultrawideband Antennas. Artech House, New York (2005)
48. Redhe, M., Nilsson, L.: Optimization of the new Saab 9–3 exposed to impact load using a space mapping technique. Struct. Multidiscip. Optim. **27**, 411–420 (2004)
49. Crevecoeur, G., Dupre, L., Van De Walle, R.: Space mapping optimization of the magnetic circuit of electrical machines including local material degradation. IEEE Trans. Magn. **43**, 2609–2611 (2007)
50. Pedersen, F., Weitzmann, P., Svendsen, S.: Modeling thermally active building components using space mapping. In: Proc. Symp. Building Physics in the Nordic Countries, pp. 896–903 (2005)
51. Bandler, J.W., Cheng, Q.S., Dakroury, S.A., Mohamed, A.S., Bakr, M.H., Madsen, K., Søndergaard, J.: Space mapping: the state of the art. IEEE Trans. Microw. Theory Tech. **52**, 337–361 (2004)
52. Koziel, S., Leifsson, L., Ogurtsov, S.: Reliable EM-driven microwave design optimization using manifold mapping and adjoint sensitivity. Microw. Opt. Technol. Lett. **55**, 809–813 (2013)
53. Koziel, S., Ogurtsov, S., Szczepanski, S.: Rapid antenna design optimization using shape-preserving response prediction. Bull. Pol. Acad. Sci. Tech. Sci. **60**, 143–149 (2012)
54. Koziel, S., Cheng, Q.S., Bandler, J.W.: Space mapping. IEEE Microw. Mag. **9**, 105–122 (2008)
55. Koziel, S., Ogurtsov, S.: Antenna design by simulation-driven optimization. Springer, New York (2014)
56. Koziel, S., Ogurtsov, S.: Rapid design optimization of antennas using space mapping and response surface approximation models. Int. J. RF Microw. Comput. Aid. Eng. **21**, 611–621 (2011)
57. Ouyang, J., Yang, F., Zhou, H., Nie, Z., Zhao, Z.: Conformal antenna optimization with space mapping. J. Electromagn. Waves Appl. **24**, 251–260 (2010)
58. Zhu, J., Bandler, J.W., Nikolova, N.K., Koziel, S.: Antenna optimization through space mapping. IEEE Trans. Antennas Propag. **55**, 651–658 (2007)
59. Sorokosz, L., Zieniutycz, W.: On the approximation of the UWB dipole elliptical arms with stepped-edge polygon. IEEE Antennas Wirel. Propag. Lett. **11**, 636–639 (2013)
60. Li, L., Cheung, S.W., Yuk, T.I.: Compact MIMO antenna for portable devices in UWB applications. IEEE Trans. Antennas Propag. **61**, 4257–4264 (2013)
61. Koziel. S., Leifsson, L., Ogurtsov, S.: Space mapping for electromagnetic-simulation-driven design optimization. In: Koziel, S., Leifsson, L. (eds.) Surrogate-Based Modeling and Optimization: Applications in Engineering, pp. 1–25. Springer, New York (2013)

62. Giunta, A.A., Wojtkiewicz, S.F., Eldred, M.S.: Overview of modern design of experiments methods for computational simulations. American Institute of Aeronautics and Astronautics, Paper AIAA 2003–0649 (2003)
63. Leary, S., Bhaskar, A., Keane, A.: Optimal orthogonal-array-based Latin hypercubes. J. Appl. Stat. **30**, 585–598 (2003)
64. Ye, K.Q.: Orthogonal column Latin hypercubes and their application in computer experiments. J. Am. Stat. Assoc. **93**, 1430–1439 (1998)
65. Beachkofski, B., Grandhi, R.: Improved distributed hypercube sampling. American Institute of Aeronautics and Astronautics, Paper AIAA 2002–1274 (2002)
66. Queipo, N.V., Haftka, R.T., Shyy, W., Goel, T., Vaidynathan, R., Tucker, P.K.: Surrogate-based analysis and optimization. Prog. Aerosp. Sci. **41**, 1–28 (2005)
67. Kabir, H., Wang, Y., Yu, M., Zhang, Q.J.: Neural network inverse modeling and applications to microwave filter design. IEEE Trans. Microw. Theory Tech. **56**, 867–879 (2008)
68. Murphy, E.K., Yakovlev, V.V.: Neural network optimization of complex microwave structures with a reduced number of full-wave analyses. Int. J. RF Microw. Comput. Aid. Eng. **21**, 279–287 (2010)
69. Gutiérrez-Ayala, V., Rayas-Sánchez, J.E.: Neural input space mapping optimization based on nonlinear two-layer perceptrons with optimized nonlinearity. Int. J. RF Microw. Comput. Aid. Eng. **20**, 512–526 (2010)
70. Tighilt, Y., Bouttout, F., Khellaf, A.: Modeling and design of printed antennas using neural networks. Int. J. RF Microw. Comput. Aid. Eng. **21**, 228–233 (2011)
71. Kabir, H., Yu, M., Zhang, Q.J.: Recent advances of neural network-based EM-CAD. Int. J. RF Microw. Comput. Aid. Eng. **20**, 502–511 (2010)
72. Siah, E.S., Ozdemir, T., Volakis, J.L., Papalambros, P., Wiese, R.: Fast parameter optimization using Kriging metamodeling [antenna EM modeling/simulation]. In: IEEE Antennas and Prop. Int. Symp., pp. 76–79 (2003)
73. Siah, E.S., Sasena, M., Volakis, J.L., Papalambros, P.Y., Wiese, R.W.: Fast parameter optimization of large-scale electromagnetic objects using DIRECT with kriging metamodeling. IEEE Trans. Microw. Theory Tech. **52**, 276–285 (2004)
74. Shaker, G.S.A., Bakr, M.H., Sangary, N., Safavi-Naeini, S.: Accelerated antenna design methodology exploiting parameterized Cauchy models. J. Prog. Electromagn. Res. (PIER B) **18**, 279–309 (2009)
75. Xia, L., Meng, J., Xu, R., Yan, B., Guo, Y.: Modeling of 3-D vertical interconnect using support vector machine regression. IEEE Microw. Wirel. Comp. Lett. **16**, 639–641 (2006)
76. Lophaven, S.N., Nielsen, H.B., Søndergaard, J.: DACE: A Matlab Kriging Toolbox. Technical University of Denmark, Lyngby (2002)
77. Koziel, S., Ogurtsov, S., Couckuyt, I., Dhaene, T.: Variable-fidelity electromagnetic simulations and co-kriging for accurate modeling of antennas. IEEE Trans. Antennas Propag. **61**, 1301–1308 (2013)
78. Alexandrov, N.M., Lewis, R.M.: An overview of first-order model management for engineering optimization. Optim. Eng. **2**, 413–430 (2001)
79. Qian, Y., Deal, W.R., Kaneda, N., Itoh, T.: Microstrip-fed quasi-Yagi antenna with broadband characteristics. Electron. Lett. **34**, 2194–2196 (1998)
80. Bekasiewicz, A., Koziel, S., Zieniutycz, W.: Design space reduction and variable-fidelity EM simulations for feasible pareto optimization of antennas. Int. Rev. Prog. Appl. Comp. Electromagn. (2014)
81. Koziel, S., Bekasiewicz, A., Zieniutycz, W.: Fast multi-objective antenna design through variable-fidelity EM simulations. In: Int. Symp. Antenna Technol. Appl. Electromagn. (2014)
82. Spence, T.G., Werner, D.H.: A novel miniature broadband/multiband antenna based on an end-loaded planar open-sleeve dipole. IEEE Trans. Antennas Propag. **54**, 3614–3620 (2006)
83. CST Microwave Studio, ver. 2013: CST AG, Bad Nauheimer Str. 19, D-64289 Darmstadt, Germany (2013).
84. Kolda, T.G., Lewis, R.M., Torczon, V.: Optimization by direct search: new perspectives on some classical and modern methods. SIAM Rev. **45**, 385–482 (2003)
85. Nocedal, J., Wright, S.: Numerical Optimization, 2nd edn. Springer, New York (2006)

Numerically Efficient Approach to Simulation-Driven Design of Planar Microstrip Antenna Arrays By Means of Surrogate-Based Optimization

Slawomir Koziel and Stanislav Ogurtsov

Abstract A numerically efficient technique for simulation-driven design of planar microstrip antenna arrays is discussed. It exploits the surrogate-based optimization (SBO) paradigm and variable-fidelity electromagnetic (EM) simulations. The design process includes radiation pattern optimization and matching. Two low-fidelity models are utilized: a coarse-mesh EM model of the entire array and a model of the array based on the array factor combined with the simulated radiation response of a single element. Both models, after suitable correction, guide the optimization process towards the optimum of the high-fidelity model of the antenna array. Design optimization of microstrip antenna arrays comprising 25 and 49 elements is conducted and described to demonstrate operation as well as efficiency of the proposed technique. The computational cost of optimized designs is equivalent to a few high-fidelity simulations of the entire array despite a large number of design variables.

Keywords Antenna array design • Antenna array optimization • Microstrip antenna array • Radiation pattern synthesis • Simulation-driven design • Surrogate-based optimization • Variable-fidelity simulations

1 Introduction

Optimization of planar antenna arrays can be challenging if array radiation and reflection responses are noticeably affected by coupling, finite size of the substrate, and radomes. In all of these situations, full-wave electromagnetic (EM) simulations of the entire structure are required in the design process. Such simulations, however, are computationally expensive when accurate. In addition, antenna array problems

S. Koziel (✉) • S. Ogurtsov
Engineering Optimization & Modeling Center, School of Science and Engineering, Reykjavik University, Menntavegur 1, Reykjavik IS-101, Iceland
e-mail: koziel@ru.is

S. Koziel et al. (eds.), *Solving Computationally Expensive Engineering Problems*, Springer Proceedings in Mathematics & Statistics 97,
DOI 10.1007/978-3-319-08985-0_6

normally involve a large number of adjustable parameters such as excitation amplitudes and/or phases, spacing, dimensions of elements, location of feeds, etc. [1, 2]. As a result, the design process—with conventional numerical optimization methods such as gradient-based routines [3]—involves numerous EM simulations of the array model and, therefore, might be of prohibitive computational costs. An alternative (and popular) approach is the use of simple and fast superposition model assuming ideal (isotropic) radiators. While numerically feasible, this approach is not reliable and cannot be utilized for design of real-world antenna arrays.

Some recent approaches to array design exploit metaheuristics, such as genetic algorithms [4, 5], particle swarm optimizers [6, 7], and other population-based methods [8, 9]. These techniques are useful for handling certain challenges of array pattern synthesis, e.g., search in the presence of multiple local optima. However, metaheuristics normally need hundreds and even thousands of objective function calls. Thus, they are applicable to problems where the array evaluation cost is not of concern.

In this work we utilize discrete EM models of the entire array under design [10–13]. Unfortunately, such models are computationally expensive when accurate. To speed up the design process we use two types of auxiliary models. The first model is based on the simulated radiation responses of the single element combined with the analytical array factor [1, 2, 14]. This semi-analytical model cannot reliably account for coupling and may produce inaccurate radiation responses in the directions off the main beam. On the other hand, the model is very fast so that, upon suitable correction, it can be used to optimize the array radiation pattern.

The second utilized model is a coarse-discretization model of the entire array. Although this model cannot be used directly in the design process due to its inaccuracy and high level of numerical noise, it can speed up the design process by exploiting its correlations with the original, high-fidelity EM model.

We employ both aforementioned auxiliary models in the surrogate-based optimization (SBO) framework [15–20] to reduce the computational cost of the optimization process and make it robust. The paper is organized as follows. We begin with a description of a typical problem of a planar array design where both radiation and reflection responses should be adjusted. We describe the flow of the SBO process where the corrected and fast coarse-mesh model of the entire array is used as a predictor. An example of a 5 by 5 planar array of microstrip antennas demonstrates performance and costs of this approach. Comparison with direct optimization is also provided. As a way to make the SBO procedure even faster we introduce a surrogate model configured from the simulated single element radiation response combined with the analytical array factor. An example of a 7 by 7 array demonstrates performance and costs of this modified SBO procedure.

2 Simulation-Based Antenna Array Design

2.1 Antenna Array Design Through Numerical Optimization

Antenna array design problems are challenging because of the following reasons, to list just a few major ones: (1) necessity of accurate full-wave electromagnetic (EM) simulations to evaluate radiation and reflection responses of the array of interest at different points of the design space; (2) a large number of design variables (array dimensions and/or element excitations); (3) several often conflicting design requirements imposed on the radiation and reflection response of the array so that it is essentially a multi objective design problem.

Full-wave EM simulations are probably the only versatile and generic way to reliably estimate the radiation and reflection responses of antenna array structures and account for different non-idealities (e.g., finite size of the substrate/ground, presence of the radome/housing). Unfortunately, such simulations are computationally expensive when accurate. Simplified models, e.g., models based on the single element radiation response combined with the analytical array factor [1, 2] do not produce accurate radiation responses in the directions off the main beam and fail to account for inter element coupling.

With a large number of variables and several design objectives simulation-driven antenna array design realized as a parameter sweep, which is guided by the user, turns to be either tedious or unfeasible. On the other hand, numerical optimization [3] is a systematical way to handle antenna array design tasks [4–13]. Notice however that if objective functions are supplied by a discrete EM solver, as outlined in the diagram of Fig. 1, then many optimization approaches, e.g., gradient-based [3] and metaheuristic methods [4–8], turn to be impractical with realistic antenna array models because these optimization approaches typically need hundreds and thousands simulations of the antenna array models each of which is already computationally expensive. A SBO approach described in Sect. 3 and demonstrated in Sect. 4 offers a solution to simulation-driven antenna array design problems.

Antenna array design normally comprises two major steps: adjusting of the radiation response, e.g., directivity pattern, and adjusting the reflection response. Typical antenna array design requirements can be illustrated with Fig. 2: the required side-lobe level (SLL) is shown with the horizontal lines and the broadside peak directivity (to be maintained) is shown with a circle in Fig. 2b; the maximal level of the active reflection coefficient over the operational bandwidth is shown with horizontal line in Fig. 2c. Both these radiation and reflection figures can be obtained in general only with discrete EM simulations of the antenna array models. An additional reason to perform simulation-based antenna array design is illustrated with help of Fig. 3 where the effect of an element location in the array is clearly seen.

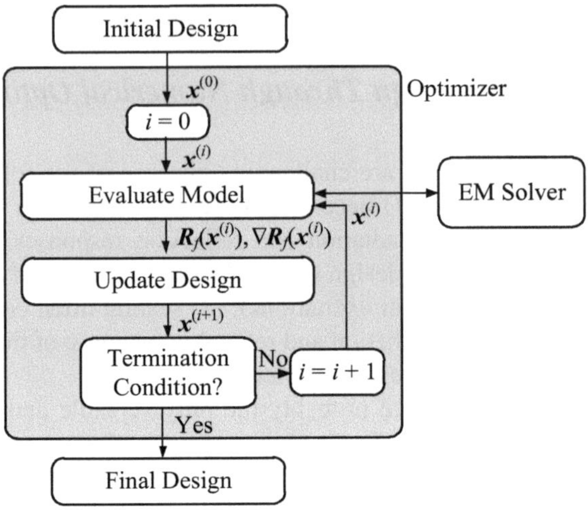

Fig. 1 Simulation-driven design by optimization: a conventional approach

2.2 *Planar Antenna Array Design Problem*

Consider an antenna array of Fig. 4. The use of discrete EM models of the entire array is necessary here to account for coupling and reliably evaluate the radiation and reflection responses. The array is required to have a linear polarization and operate at 10 GHz. Each patch is fed by a 50 ohm probe.

The design tasks are: to maintain the array peak directivity at the 20 dBi level; to have the direction of maximum radiation normal to the plane of the array; to suppress the SLL down to −20 dB; to keep returning signals lower than −10 dB, all at 10 GHz. Initial dimensions of the elements are 11 by 9 mm; a grounded 1.58 mm thick RT/duroid 5880 is the substrate; lateral extension of the substrate/metal ground is set to a half of the patch size in a particular direction. Locations of feeds at the initial design are at the center of the patch in horizontal direction and 2.9 mm off the center in the vertical direction.

The symmetry wall, shown in Fig. 4, defines the array dimensions and incident waves (amplitudes and phases) to be symmetrical with respect to the wall. With the imposed symmetry we restrict ourselves to adjusting spacing (s_1, s_2, u_1, u_2), patch size (x_1, y_1), location of probes ($d_1, \ldots, d_{15}$), amplitudes ($a_1, \ldots, a_{15}$), and phases ($b_1, \ldots, b_{15}$).

To evaluate the response of the array under design we adopt the following two EM models: (1) a high-fidelity (or fine) discrete EM model of the entire array, $\boldsymbol{R}_f$; and (2) a coarse-discretization EM model of the entire array $\boldsymbol{R}_{cd}$ which is a coarse-mesh version of $\boldsymbol{R}_f$. The use of these models in the developed SBO procedure is described in the following section.

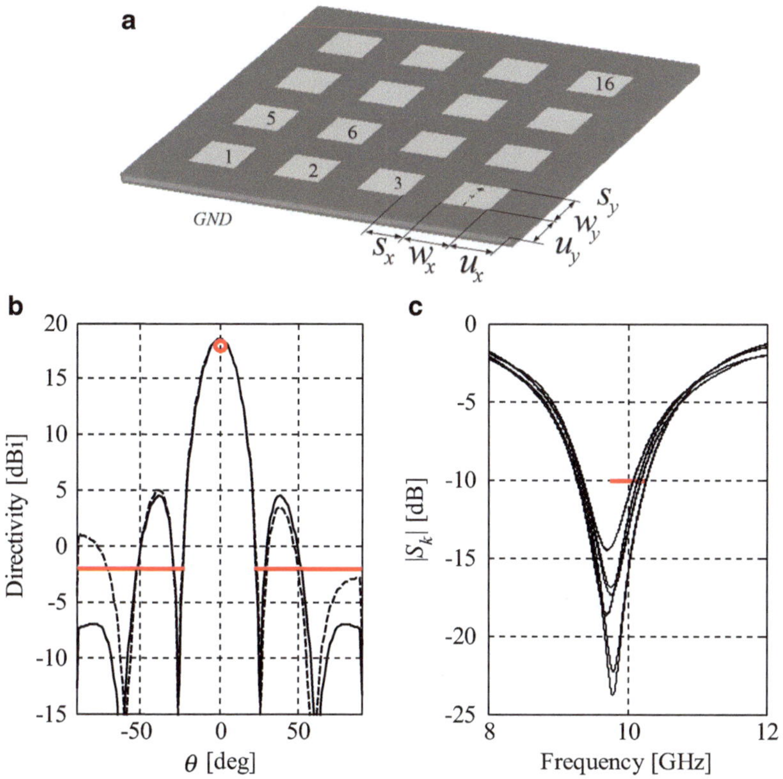

Fig. 2 A 16 element microstrip antenna array of Cartesian lattice: (**a**) view; (**b**) directivity pattern cuts in the E(*dashed line*) and H-planes (*solid line*); (**c**) active reflection coefficients at the feeds of the elements. The radiation and reflection responses are for a certain set of dimensional parameters and with the uniform excitation of the elements. Design specifications are shown with the *horizontal lines* at (**b**) and (**c**). The array is on a 1.575 mm thick RT5880 layer with finite lateral dimensions. $s_x = s_y = 8.0$ mm, $w_x = w_y = u_x = u_y = 9.15$ mm. The feeding probes are 2.9 mm off the patch center in the y-direction

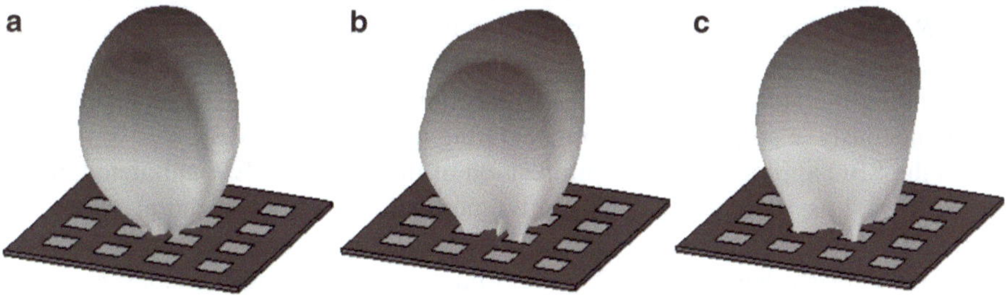

Fig. 3 Directivity pattern (linear scale) of selected elements embedded into the antenna array of Fig. 1: (**a**) element 1; (**b**) element 2; (**c**) element 10

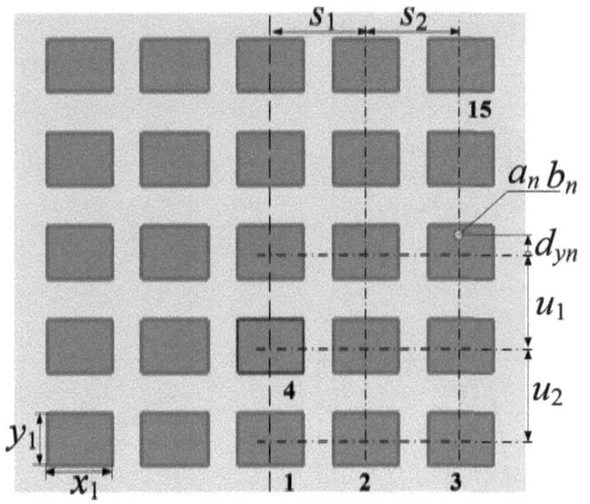

Fig. 4 Microstrip antenna array. The symmetry (magnetic) wall is shown with the *vertical dash line*

3 Design Optimization Methodology

3.1 SBO Basics

The array design is formulated here as a nonlinear minimization task, where we aim at solving the following problem

$$\mathbf{x}^* = \arg\min_{\mathbf{x}} U\left(\mathbf{R}_f\left(\mathbf{x}\right)\right) \tag{1}$$

Here, $\boldsymbol{R}_f(\boldsymbol{x}) \in R^m$ is the response vector of a high-fidelity model, representing all figures of interest, in particular, the radiation response, as well as the reflection response at all ports; U is the objective function; $\boldsymbol{x} \in R^n$ is a vector of design variables, here, representing all adjustable parameters as described in Sect. 2. The objective function is defined so that a better design corresponds to a smaller value of $U(\boldsymbol{R}_f(\boldsymbol{x}))$. Typically, minimax formulation is used, where appropriate upper/lower specification levels are imposed on both radiation response (in particular, related to minimizing the side lobes) as well as reflection response (e.g., to keep reflection simultaneously at all ports at a given operating frequency −10 dB).

In the SBO approach, direct optimization of the expensive EM-simulated model $\boldsymbol{R}_f$ in Fig. 1 is replaced by an iterative correction and optimization of its fast surrogate as shown in Fig. 5. Typically, the model $\boldsymbol{R}_f$ is only evaluated once per iteration (at every new design $\boldsymbol{x}^{(i+1)}$ after optimizing the surrogate model) to update the surrogate. The number of iterations for a well performing SBO algorithm is substantially smaller than for direct optimization methods [16].

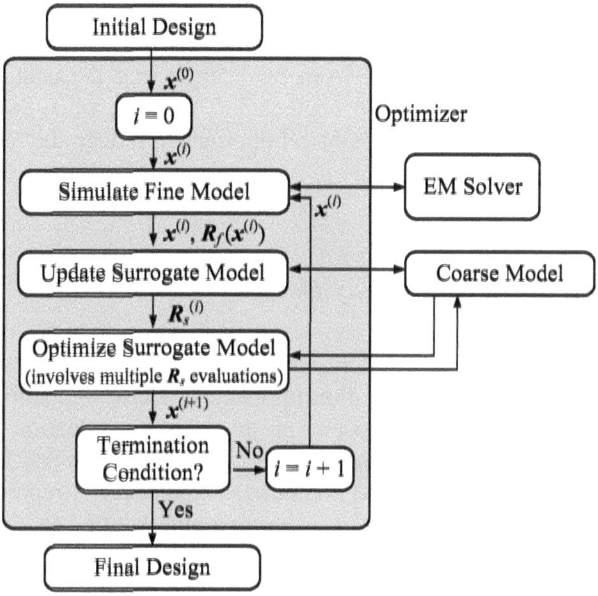

Fig. 5 Simulation-driven design by optimization: a generic SBO approach

A generic SBO scheme produces approximations $x^{(i)}$ of x^* as follows [16]:

$$\mathbf{x}^{(i+1)} = \arg\min_{\mathbf{x}} U\left(\mathbf{R}_s^{(i)}(\mathbf{x})\right) \tag{2}$$

where $R_s^{(i)}$ is the surrogate model at iteration i. In general, the surrogate model is constructed by suitable correction of the underlying low-fidelity model. In this work the low-fidelity model R_{cd} is a lossless coarse-mesh version of the original high-fidelity model R_f [19]. Often, the algorithm (2) is embedded in the trust-region framework for improving its convergence properties [21]. In any case, it is advantageous to ensure at least zero-order consistency [22] between the surrogate and the high-fidelity model, i.e., $R_s^{(i)}(x^{(i)}) = R_f(x^{(i)})$, and, whenever possible, also the first-order consistency, i.e., $J[R_s^{(i)}(x^{(i)})] = J[R_f(x^{(i)})]$, where $J[]$ stands for the Jacobian of the respective model. Satisfying the latter condition requires derivative information from both the surrogate and the high-fidelity model which is normally not available, unless adjoints sensitivities can be applied [23, 24].

The quality of model R_{cd} with a particular discretization is determined by a visual inspection of its responses simulated at and about the initial design and with respect to those of model R_f, i.e., the quality of R_{cd} is inferred from numerical experiments involving the user's judgments. A major requirement is that the responses of the low-fidelity model should capture main properties of the high-fidelity model. Optimal automatic setting of the low-fidelity model is addressed in [19].

Quality of the low-fidelity model R_{cd} turns into prediction capability of the surrogate [18]. At the same time, R_{cd} should be much faster than R_f so that the total costs of optimization and update of the surrogate are reasonably small. For coarse-discretization models of antennas the time evaluation ratio of the high- and low-fidelity models is usually from 5 to 50 so that the computational cost of the low-fidelity model cannot be neglected. Therefore, when developing SBO algorithms for antennas design, it is also important to reduce the number of low-fidelity model simulations.

3.2 SBO Procedure for Array Design

The evaluation time of the high-fidelity model $\boldsymbol{R}_f$ of the array of Fig. 4 is around 20 min. It makes its direct optimization impractical. Here, we exploit an auxiliary low-fidelity model $\boldsymbol{R}_{cd}$ but with a coarser mesh so that its evaluation time is around 1 min. Both models $\boldsymbol{R}_f$ and $\boldsymbol{R}_{cd}$ are simulated with CST MWS [25] on a 2 GHz Intel(R) Xeon(R) CPU 64 GB RAM computer. The model $\boldsymbol{R}_{cd}$ represents the array radiation pattern quite accurately but it is not particularly good for representing the reflection response.

One can split the design variable vector $\boldsymbol{x}$ into two parts: $\boldsymbol{x} = [\boldsymbol{x}_p{}^T\ \boldsymbol{x}_m{}^T]^T$, where $\boldsymbol{x}_p = [s_1\ s_2\ u_1\ u_2\ x_1\ y_1\ a_1 \ldots a_{15}]$ are variables used to optimize the array pattern, and $\boldsymbol{x}_m = [d_{y1}\ d_{y2}\ \ldots\ d_{y15}]$ are variables used to adjust the reflection. Having this in mind, the following three-step design procedure (also outlined in Fig. 6) has been developed:

Step 1: Optimize the directivity pattern of the low-fidelity model $\boldsymbol{R}_{cd}$ using $\boldsymbol{x}_p$ with fixed $\boldsymbol{x}_m = \boldsymbol{x}_{m.0}$ (the initial value); the optimized $\boldsymbol{x}_p$ will be referred to as $\boldsymbol{x}_p{}^*$. Optimization is performed using the pattern search algorithm [26] in order to overcome the problem of numerical noise present in the simulated responses of model $\boldsymbol{R}_{cd}$. Optimization of $\boldsymbol{R}_{cd}$ at this step is realized using auxiliary first-order response surface models constructed using large-step design perturbations, and the trust-region framework to ensure convergence.

Step 2: Evaluate model $\boldsymbol{R}_f$ at $\boldsymbol{x} = [(\boldsymbol{x}_p{}^*)^T\ (\boldsymbol{x}_{m.0})^T]^T$; Use $\boldsymbol{R}_{cd}$ to estimate the necessary changes in $\boldsymbol{x}_m$ to improve reflection responses; Here, it is assumed that a small change of a given $\boldsymbol{x}_m$ component will noticeably affect the reflection of the corresponding patches and not the others. It has been verified with numerical experiments that this assumption is satisfied for the structure under design for the used range of the design variables. The procedure is the following: (1) evaluate model $\boldsymbol{R}_{cd}$ at $\boldsymbol{x} = [(\boldsymbol{x}_p{}^*)^T\ (\boldsymbol{x}_{m.0})^T]^T$ and at the two perturbed designs varied by $\pm\Delta d_y$ corresponding to a reflection response that does not satisfy matching requirements (cf. Fig. 7); (2) using interpolation of the data obtained in (1), estimate the change of d_y that gives reasonable change of the response (this takes into account the fact that responses of $\boldsymbol{R}_f$ and $\boldsymbol{R}_{cd}$ are shifted both in frequency and amplitude); The modified vector $\boldsymbol{x}_m$ will be referred to as $\boldsymbol{x}_m{}^*$.

Step 3: Evaluate $\boldsymbol{R}_f$ at $\boldsymbol{x} = [(\boldsymbol{x}_p{}^*)^T\ (\boldsymbol{x}_m{}^*)^T]^T$; Adjust the global parameter y_1 (patch length) to shift the matching responses in frequency as necessary. The change of y_1 is estimatedusing evaluation of $\boldsymbol{R}_{cd}$ at $\boldsymbol{x} = [(\boldsymbol{x}_p{}^*)^T\ (\boldsymbol{x}_m{}^*)^T]^T$ and the two

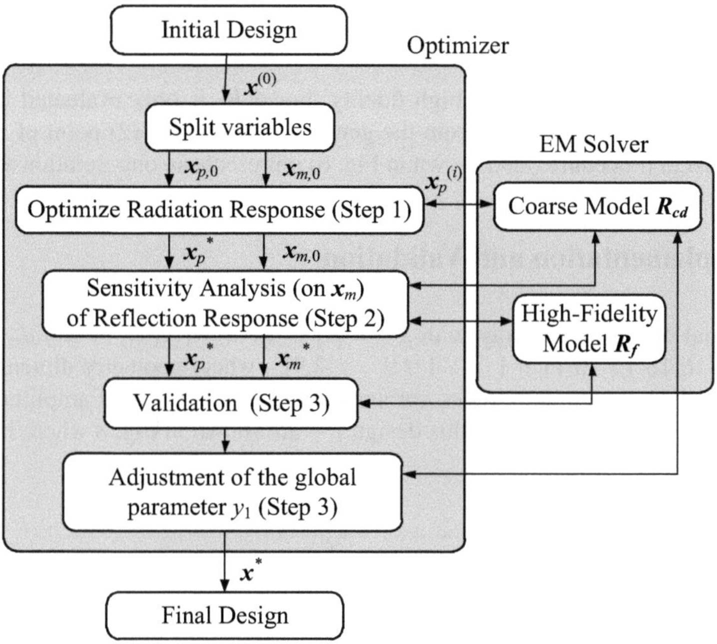

Fig. 6 SBO procedure of Sect. 3.2 with two EM models $\boldsymbol{R}_{cd}$ and $\boldsymbol{R}_f$: a block diagram

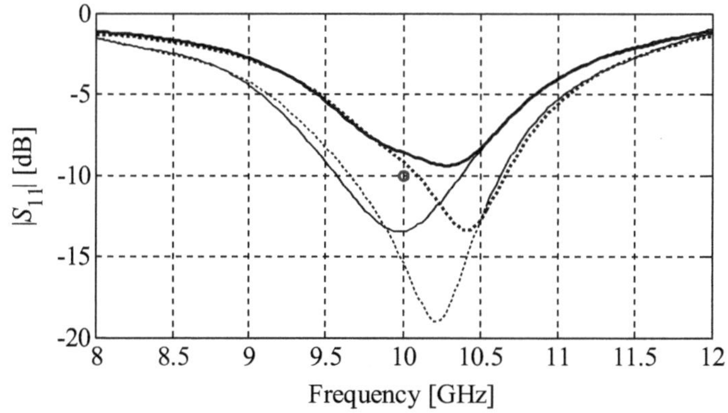

Fig. 7 SBO procedure of Sect. 3.2, reflection responses at a selected port: $\boldsymbol{R}_f$ (*thick solid line*) and $\boldsymbol{R}_{cd}$ (*thin solid line*) at $\boldsymbol{x} = [(\boldsymbol{x}_p{}^*)^T\ (\boldsymbol{x}_{m.0})^T]^T$ and at a design with variable d_{yk} corresponding to port k perturbed by certain Δd_{yk} (*thick and thin dotted lines*). Based on these responses of $\boldsymbol{R}_{cd}$ and that of $\boldsymbol{R}_f$ at $[(\boldsymbol{x}_p{}^*)^T\ (\boldsymbol{x}_{m.0})^T]^T$ a proper perturbation for d_{yk} is found as described in *Step 2*. Additional "horizontal" correction of this response may be necessary as described in *Step 3*. A *circle* denotes design specifications

perturbed designs obtained by changing y_1 and interpolating the results. The final design is obtained after this step is referred to as $\boldsymbol{x}^*$.

It should be noted that the high-fidelity model $\boldsymbol{R}_f$ is only evaluated in *Step 2* (once) and in *Step 2* (twice). From the generic SBO scheme (2) point of view, the above design procedure (also shown in Fig. 6) represents an one-iteration approach.

4 Implementation and Validation

The initial design is an array with $\boldsymbol{x}^{(0)} = [s_1\ s_2\ u_1\ u_2\ x_1\ y_1\ a_1 \ldots a_{15}\ d_{y1}\ d_{y2}\ \ldots\ d_{y15}] = [16\ 16\ 16\ 16\ 11\ 9\ 1\ \ldots\ 1\ 2.9\ \ldots\ 2.9]^T$, where geometry dimensions are in mm and excitation amplitudes are normalized to the maximal amplitude of the incident signals. Responses of this design $\boldsymbol{x}^{(0)}$ are shown in Fig. 8 where in Fig. 8b

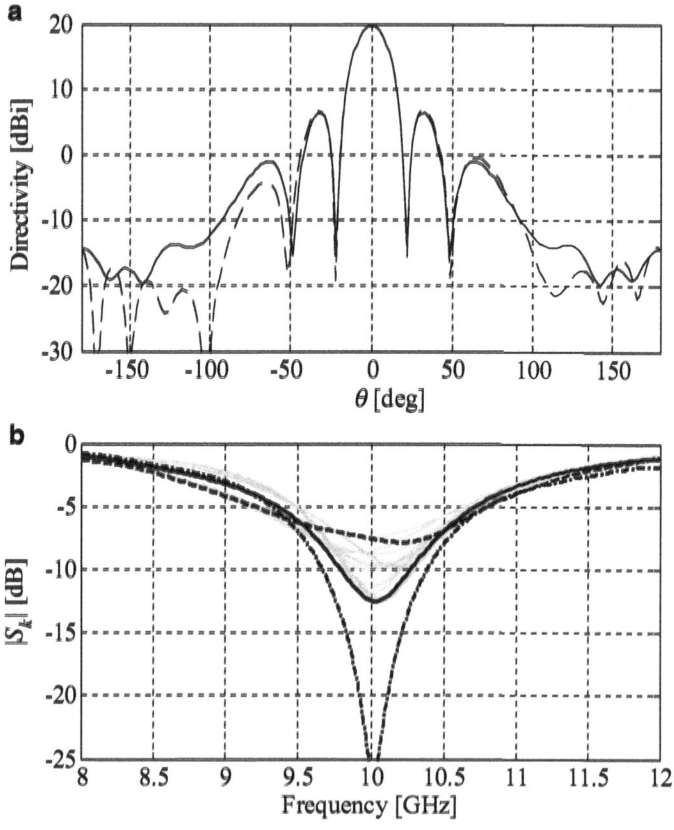

Fig. 8 High-fidelity model $\boldsymbol{R}_f$: (**a**) directivity pattern in the E (*dashed line*) and H (*solid line*) planes at the initial design at 10 GHz; (**b**) active reflection coefficient of element 3 (*solid line*) and 7 (*dashed line*) (see Fig. 4), at the initial design, and the reflection coefficient of a single element (*dotted dashed line*). The active reflection coefficients are normalized to the maximal amplitude of the incident signals. Reflection coefficients of other elements are shown with the *grey lines* in (**b**)

the reflection coefficient of the isolated single element is given for reference. Peak directivity of a single isolated element (a microstrip patch antenna) is about 7.4 dBi.

Design specifications for *Step 1* (directivity pattern optimization) are the following: minimize directivity (in the minimax sense) off the main beam of design $\boldsymbol{x}^{(0)}$, i.e., for the zenith angles off the sector [−21.5⁰, 21.5⁰]. *Step 1* (optimization of the coarse model for pattern) results in design $\boldsymbol{x}_p^{*}$ = [16.363 16.588 16.498 16.910 11.072 8.926 0.9845 0.4529 0.3718 0.9873 0.9748 0.4500 0.9970 0.9754 0.9919 0.9548 0.9369 0.5503 1.0000 0.4671 0.3621]^T. Responses of the array after *Step 1* are shown in Fig. 9. The cost of this *Step 1* is 182 evaluations of the coarse-discretization model $\boldsymbol{R}_{cd}$.

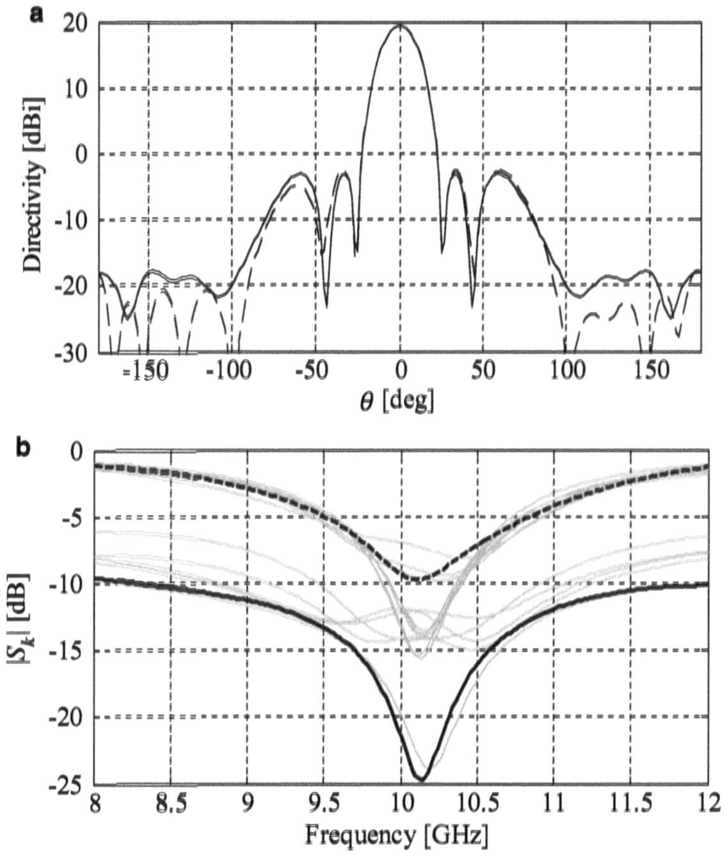

Fig. 9 High-fidelity model $\boldsymbol{R}_f$: (**a**) directivity pattern in the E (*dashed line*) and H (*solid line*) planes after *Step 1* (directivity optimization) at 10 GHz; (**b**) active reflection coefficient of elements 3 (*solid line*) and 7 (*dashed line*) (see Fig. 4). The active reflection coefficients are normalized to the maximal amplitude of the incident signals. Reflection coefficients of other elements are shown with the *grey lines* in (**b**)

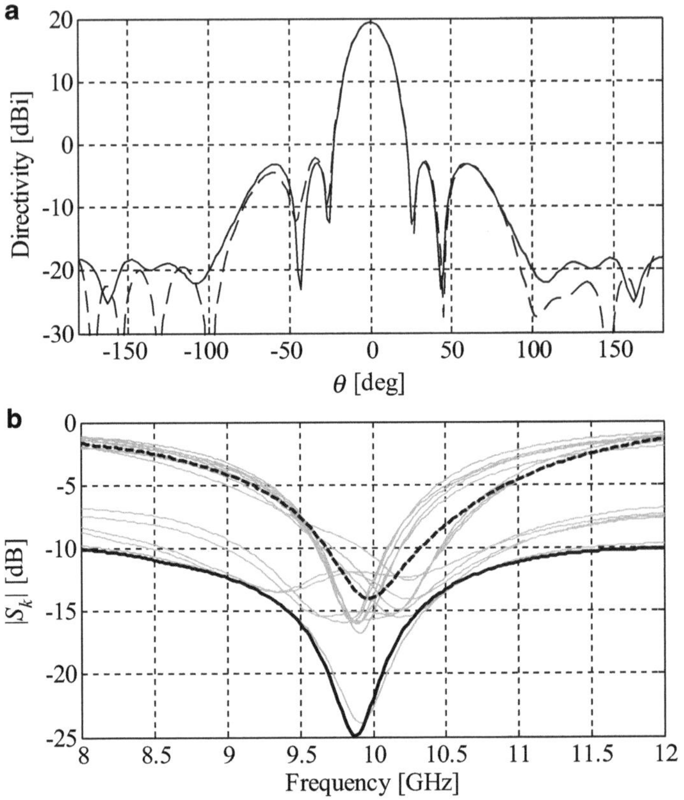

Fig. 10 High-fidelity model $\boldsymbol{R}_f$ at the final design: (**a**) directivity pattern cuts in the E (*dashed line*) and H (*solid line*) planes at 10 GHz; (**b**) active reflection coefficient of element 3 (*solid line*), and that of element 7 (*dashed line*) (see Fig. 4). The active reflection coefficients are normalized to the maximal amplitude of the incident signals

At *Step 2* (matching correction I), we change d_{yk} for ports where matching is not sufficient (i.e., $|S_k| > -10$ dB). For ports 4, 7, 8, and 10 the feed location is increased to 3.4 mm. The cost of *Step 2* is $8 \times \boldsymbol{R}_c + 1 \times \boldsymbol{R}_f$. At *Step 3*, (matching correction II) one changes the global parameter y_1 to 9.1 mm to move reflection responses to the left in frequency. This step costs $2 \times \boldsymbol{R}_c + 2 \times \boldsymbol{R}_f$.

Responses of the final design with SLL of −21 dB are shown in Fig. 10. The total cost of the design optimization process is $192 \times \boldsymbol{R}_{cd} + 3 \times \boldsymbol{R}_f = 12.5 \times \boldsymbol{R}_f$, i.e., it is equivalent in time to only 12.5 high-fidelity simulations of the entire structure.

Other cases were also considered. Responses of the final design with additional suppression (extra −10 dB) of the radiation in the sectors of [−31.5°, −21.5°] and [21.5°, 31.5°] with the incident excitation amplitudes as variables and uniform phase distribution are shown in Fig. 11. The total cost of this design is $13 \times \boldsymbol{R}_f$. It should be noted that obtaining additional suppression in the aforementioned

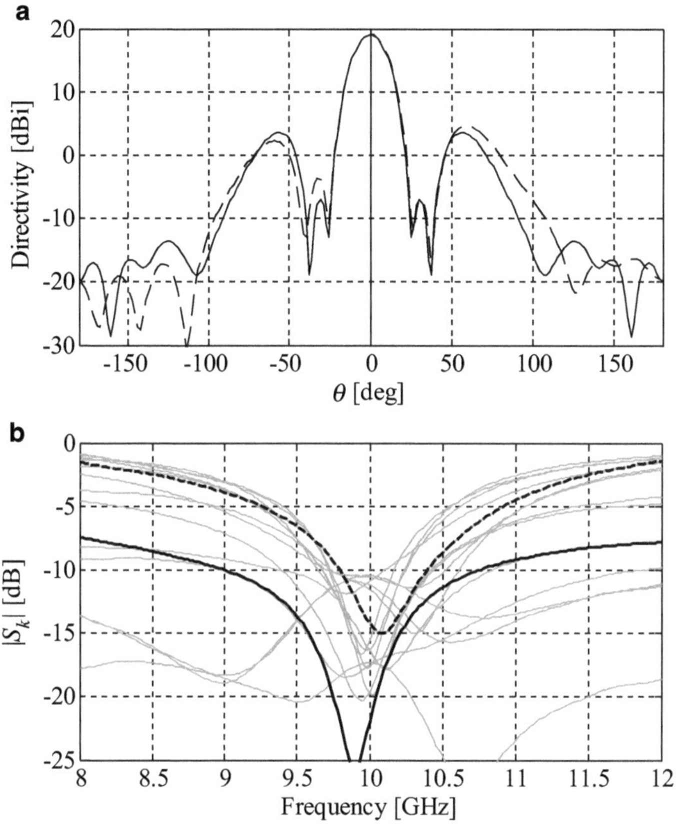

Fig. 11 Responses at the final design with additional suppression of SLL next to the main lobe, non-uniform amplitude excitation: (**a**) directivity pattern cuts in the E (*dashed line*) and H (*solid line*) planes at 10 GHz; (**b**) active reflection coefficient of element 3 (*solid line*) and that of element 7 (*dashed line*) (see Fig. 4). The active reflection coefficients are normalized to the maximal amplitude of the incident signals

sector compromises to some extent the SLL, which is now about −17 dB. It can be observed that the proposed design method is sufficiently flexible to handle various types of design specifications.

Responses of the final design with amplitudes and phases as design variables are shown in Fig. 12. In this case, the problem is much more complex from the optimization point of view (51 variables); however, the proposed approach allows for obtaining the optimized design at the total cost of only about $21 \times \boldsymbol{R}_f$. The additional degrees of freedom make it possible to further reduce the SLL to −24 dB.

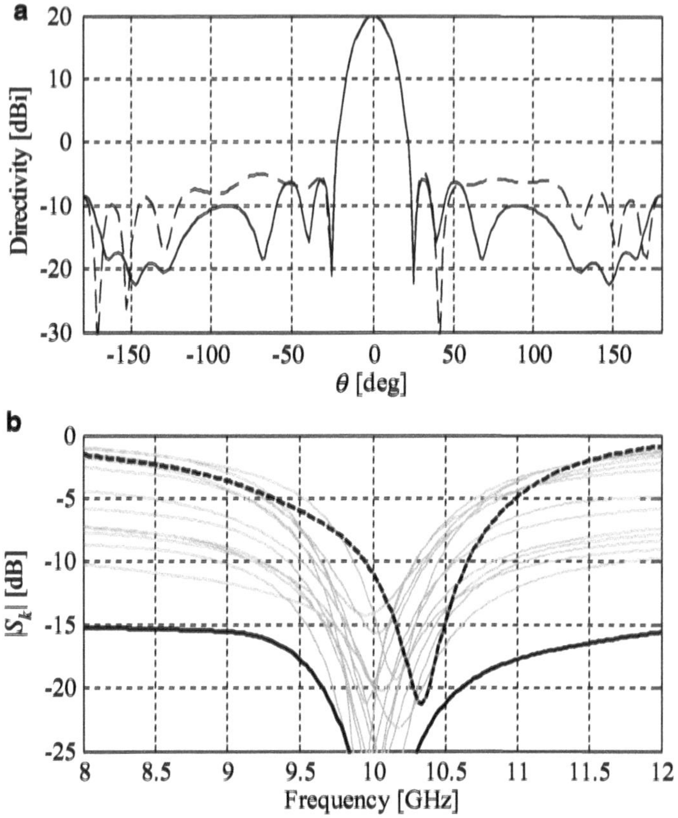

Fig. 12 Responses at the final design with non-uniform amplitude and phase excitation: (**a**) directivity pattern cuts in the E (*dashed line*) and H (*solid line*) planes at 10 GHz; (**b**) active reflection coefficient of element 3 (*solid line*) and that of element 7 (*dashed line*) (see Fig. 4). The active reflection coefficients are normalized to the maximal amplitude of the incident signals

5 Rapid Optimization of Radiation Response

5.1 Design Case: 7 by 7 Microstrip Array

Consider a 7 by 7 planar array shown in Fig. 13. The array is to operate at 10 GHz with linear polarization in the E-plane. Each patch is fed by a probe in the 50 ohm environment. Initial dimensions of elements, microstrip patches, are 11 by 9 mm; a grounded layer of 1.58 mm thick RT/duroid 5880 is the substrate. The extension of the substrate and ground, s_0 and u_0, is set to 15 mm.

The design tasks are: to have (a) SLL below −20 dB for zenith angles off the main beam with the null-to null width of 32°; (b) the peak directivity about 20 dBi; (c) the direction of the maximum radiation perpendicular to the plane of

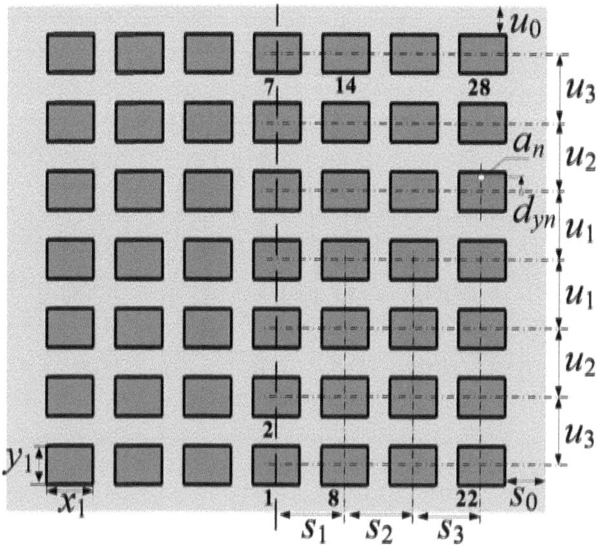

Fig. 13 Array of 49 microstrip patches: *front view*. Symmetry (magnetic) plane is shown with the *vertical dash line* at the center

the array; (d) returning signals lower than −10 dB, all at 10 GHz. A starting point for optimization is a uniform array, $\boldsymbol{x}^{(0)} = [s_1\ s_2\ s_3\ u_1\ u_2\ u_3\ x_1\ y_1\ a_1\ \ldots\ a_{28}\ d_{y1}\ \ldots\ d_{y28}]^{\mathrm{T}} = [16\ 16\ 16\ 16\ 16\ 16\ 11\ 9\ 1\ \ldots\ 1\ 2.9\ldots 2.9]^T$ where all dimensional parameters are in mm, excitation amplitudes are normalized, and phase shifts are in degrees.

Initial values of the spacings are easily found using model $\boldsymbol{R}_a$ assuming them equal to each other. The feed offsets, d_{yn} shown in Fig. 10, are 2.9 mm for all patches; it is obtained by optimizing the EM model of the single patch antenna. The SLL of this design is about −13 dB as expected, and the peak directivity is 22.7 dBi (cf. Fig. 14a).

5.2 Utilization of the Analytical Model

While the total cost of the procedure described in Sect. 3 and illustrated for the antenna array considered in Sect. 4 is quite low (corresponding to around 10–20 evaluations of the high-fidelity EM model of the entire array), the majority of this cost is related to optimization of the coarse-discretization model $\boldsymbol{R}_{cd}$. This contribution would be even larger for the 7 by 7 array considered here. This overhead can be reduced by replacing the coarse-discretization model $\boldsymbol{R}_{cd}$ by the analytical model at the stage of radiation response optimization. The resulting design methodology is a two-stage process described below.

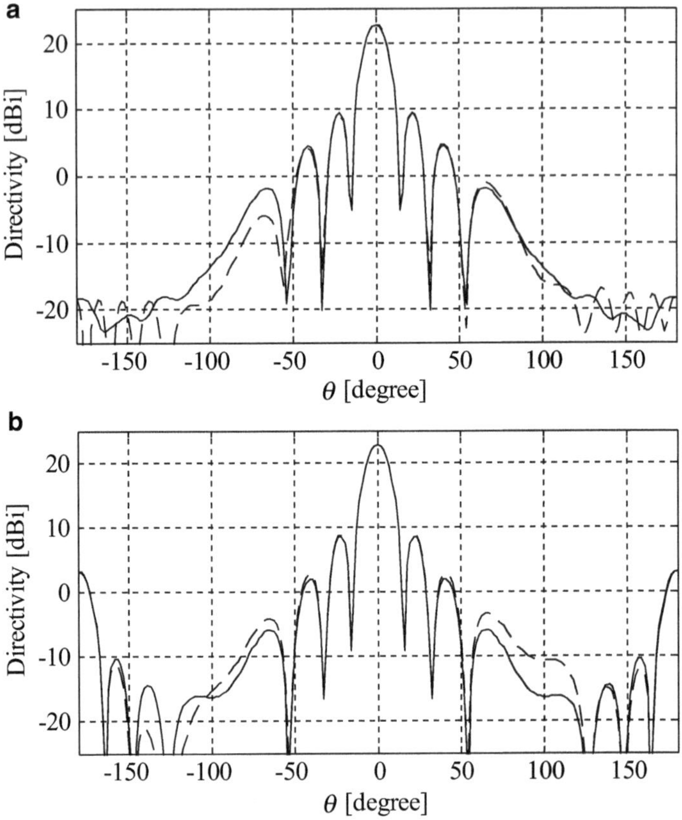

Fig. 14 Array of 49 microstrip patches, directivity pattern cuts in the E (*dashed line*) and H (*solid line*) planes at the initial design at 10 GHz: (**a**) high-fidelity model $\boldsymbol{R}_f$; (**b**) analytical model $\boldsymbol{R}_a$

Stage 1 (radiation optimization): Here, we adopt the analytical model $\boldsymbol{R}_a$ representing directivity $D_a(\theta,\phi) \sim D_e(\theta,\phi)\bullet|A(\theta,\phi)|^2$, which embeds the EM-simulated radiation response of the single microstrip patch antenna $D_e(\theta,\phi)$ and analytical array factor $A(\theta,\phi)$ [14]. Although the analytical model is extremely fast, it is not as accurate as the coarse-discretization EM one; the response of model $\boldsymbol{R}_a$ is shown in Fig. 14b.

Therefore, the radiation response is optimized iteratively, exploiting an SBO scheme shown in Fig. 15. The surrogate model is created by means of the following additive response correction (also referred to as output space mapping [15, 27])

$$\mathbf{R}_s^{(i)}(\mathbf{x}) = \mathbf{R}_a(\mathbf{x}) + \left[\mathbf{R}_f\left(\mathbf{x}^{(i)}\right) - \mathbf{R}_a\left(\mathbf{x}^{(i)}\right)\right] \tag{3}$$

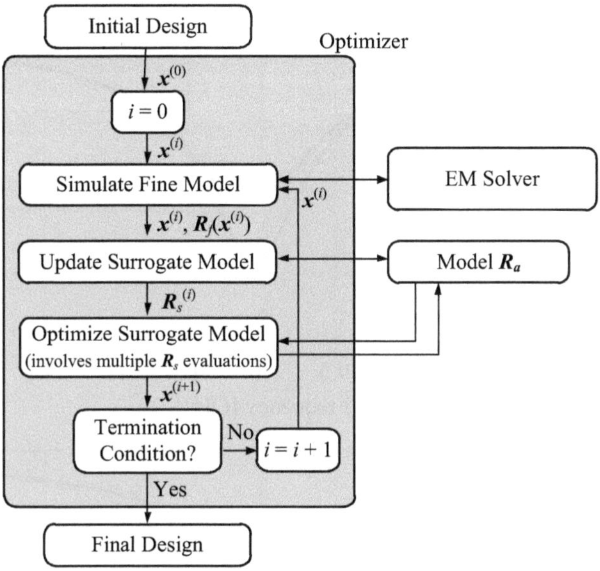

Fig. 15 SBO approach for optimization of the radiation response using the analytical model $\boldsymbol{R}_a$ as the low-fidelity model. Surrogate model $\boldsymbol{R}_s$ is constructed according to (3)

where $\boldsymbol{x}^{(i)}$ is the current design. This kind of correction ensures zero-order consistency [22] between the surrogate and the high-fidelity model at $\boldsymbol{x}^{(i)}$, i.e., $\boldsymbol{R}_s^{(i)}(\boldsymbol{x}^{(i)}) = \boldsymbol{R}_f\,(\boldsymbol{x}^{(i)})$.

It should be emphasized that the additive response correction is well suited for constructing the surrogate model in our case because the major discrepancy between the analytical and the high-fidelity EM radiation model is vertical difference as indicated in Fig. 14a, b. Usually, 2–3 iterations of the SBO algorithm (2) with the surrogate model (3) are necessary to yield a satisfactory design in terms of the radiation pattern. One iteration requires only one evaluation of $\boldsymbol{R}_f$.

Stage 2 (reflection/coupling adjustment): the coarse-discretization model $\boldsymbol{R}_{cd}$ is used to correct reflection. After completion of Stage 1, the reflection responses are shifted in frequency so that the minima of returning signals $|S_k|$ are not exactly at the required frequency. Responses $|S_k|$ can be shifted in frequency individually by adjusting the feed offsets d_n, and collectively by adjusting the patch size, y_1. The amounts of adjustments are estimated using $\boldsymbol{R}_{cd}$, for which, the dependencies of $|S_k|$ w.r.t. design variables are similar as for $\boldsymbol{R}_f$.

Both models are evaluated using the same EM solver so that the responses are well correlated despite their misalignment (both in frequency and level as illustrated in Fig. 16). The computational cost of reflection adjustment is only one evaluation of $\boldsymbol{R}_f$ and a few evaluations of $\boldsymbol{R}_{cd}$ (depending on how many reflection coefficients $|S_k|$ are to be adjusted).

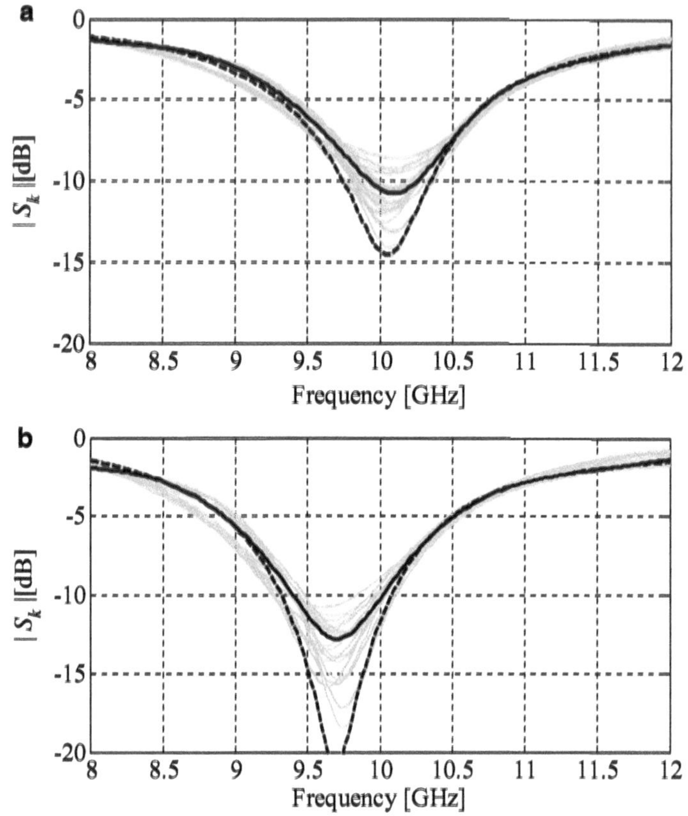

Fig. 16 Array of Fig. 10, active reflection coefficient of element 4 (*solid line*) which is at the center, that of element 22 (*dashed line*) which is at the corner: (**a**) high-fidelity model $\boldsymbol{R}_f$; (**b**) coarse model $\boldsymbol{R}_{cd}$. One can see a frequency shift as well as a vertical misalignment

5.3 Results I: Optimization with Non-uniform Amplitude Excitation

Design has been carried out with incident excitation amplitudes as design variables. Maximum allowed array spacings were restricted to 20 mm. The cost of Stage 1, directivity pattern optimization, is only three evaluation of $\boldsymbol{R}_f$ (the cost of optimizing the analytical $\boldsymbol{R}_a$ can be neglected).

At Stage 2, we change the y-size of the patches, global parameter y_1 to 9.14 mm in order to move reflection responses to the left in frequency y_1. Offsets d_n of the elements still violating the specification have been adjusted individually.

The cost of this step is $5 \times \boldsymbol{R}_{cd} + 1 \times \boldsymbol{R}_f$. The final design is found at $\boldsymbol{x}^* = [s_1\ s_2\ s_3\ u_1\ u_2\ u_3\ x_1\ y_1\ a_1\ \ldots\ a_{28}]^T = $ [15.97 17.35 20.00 14.38 17.98 19.99 11.00 9.14 0.922 0.787 1.000 0.835 0.953 0.779 0.770 0.958 0.966 1.000 0.810 0.963

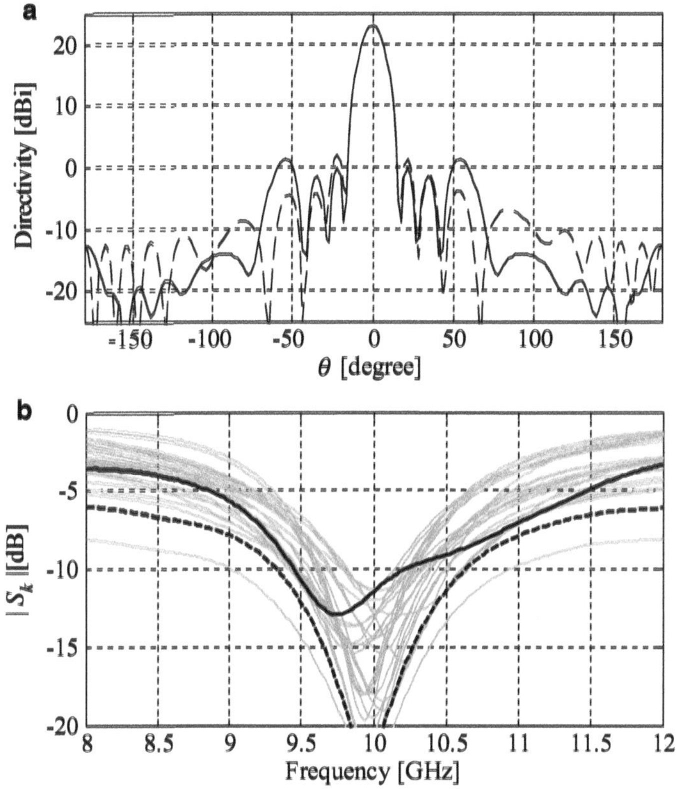

Fig. 17 Array of 49 microstrip patch antennas optimized with non-uniform amplitude excitation and spacings constrained to 20 mm: (**a**) directivity pattern cuts in the E (*dashed line*) and H (*solid line*) planes at 10 GHz; (**b**) active reflection coefficient of element 4 (*solid line*) which is at the center, that of element 22 (*dashed line*) which is at the corner. The active reflection coefficients are normalized to the maximal amplitude of the incident signals

0.989 0.925 0.452 0.620 0.832 0.842 0.814 0.631 0.576 0.072 0.752 0.697 0.872 0.821 0.703 0.037]^T where the dimensional parameters are in mm and excitation amplitudes are normalized. Most probe offsets d_n have been left of the initial design value, 2.9 mm, except four adjusted to $d_4 = d_{11} = d_{18} = 3.9$ mm and $d_{10} = 3.4$ mm.

The radiation response and reflection response of the final design are shown in Fig. 17. The SLL of this design $\boldsymbol{x}^{\#}$ is under −20 dB and the peak directivity of $\boldsymbol{x}^{\#}$ is 22.9 dBi. The total cost of optimization is about $5 \times \boldsymbol{R}_f$.

5.4 Results II: Optimization with Non-uniform Phase Excitation

Another case has been considered with the excitation phase shifts as design variables and spacings restricted to 20 mm. The final design is at $\boldsymbol{x}^* = [s_1\ s_2\ s_3\ u_1\ u_2\ u_3\ x_1\ y_1\ b_1\ \ldots\ b_{28}]^T = [15.00\ 15.00\ 20.00\ 15.15\ 5.46\ 19.95\ 11.00\ 9.10\ 0\ 8.6\ -6.3\ 1.1\ 4.3\ 2.6\ 3.1\ 33.3\ 0.3\ 11.0\ -4.9\ 5.3\ -14.6\ 45.7\ -60.7\ 17.4\ 5.8\ 29.6\ -7.0\ 39.4\ -48.9\ -17.7\ 46.5\ -13.8\ 22.5\ -1.65\ 47.9\ -38.9]^T$ where the dimensional parameters are in mm, phase shifts are in degrees and given relatively the first element. Its responses are shown in Fig. 18.

The SLL of this design $\boldsymbol{x}^*$ is under -17 dB; the peak directivity of $\boldsymbol{x}^{(0)}$ is 22.2 dBi; return signals $|S_k|$ are higher than in the previous case, their suppression should be addressed with design of the feed network. The total cost of $5 \times \boldsymbol{R}_f$ is similar to that of the previous example.

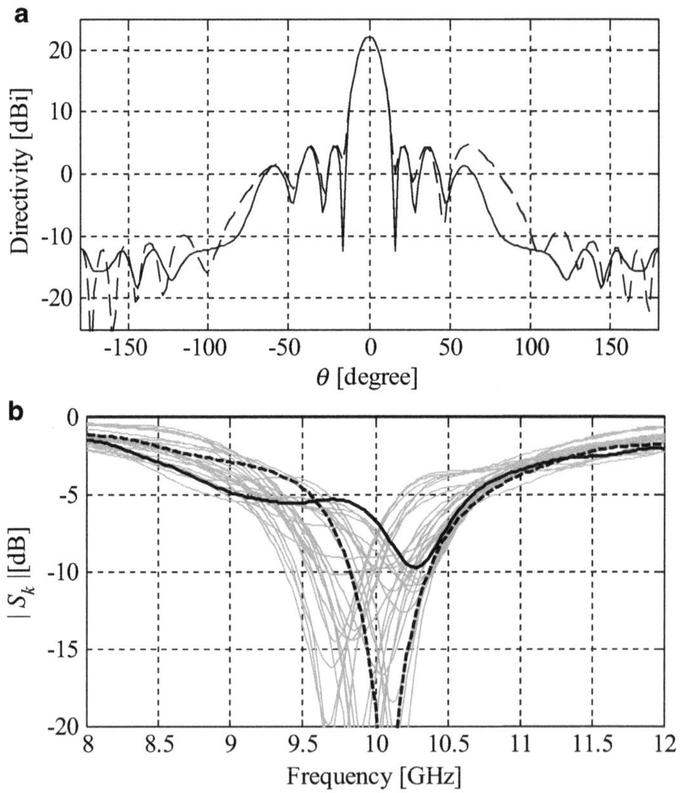

Fig. 18 Array of 49 microstrip patch antennas optimized with non-uniform phase excitation and spacings constrained by 20 mm: (**a**) directivity pattern cuts in the E (*dashed line*) and H (*solid line*) planes at 10 GHz; (**b**) active reflection coefficient of elements 4 (*solid line*) and 22 (*dashed line*)

6 Conclusions

Low-cost SBO approach for simulation-driven design of planar arrays of microstrip patch antennas has been discussed. By utilizing variable-fidelity simulations and auxiliary analytical array models with suitable correction schemes, the design goals can be met at the cost of only a few high-fidelity EM simulations of the entire array. Optimized simulation-driven designs for cases of non-uniform amplitude and non-uniform phase have been obtained. The final designs have shown, in overall, similar performance in terms of radiation and reflection.

It seems that combination of coarse-discretization simulations with SBO techniques (in particular, response correction) is a promising way to conduct EM-driven design of realistic antenna array models in a computationally feasible manner. As demonstrated, the use of quasi-analytical models offers further design speed up; however, at the expense of somehow limited control of the reflection response. An extension of this work will address design optimization of phased antenna arrays.

References

1. Volakis, J.L. (ed.): Antenna Engineering Handbook, 4th edn. McGraw-Hill, New York (2007)
2. Mailloux, R.J.: Phased Array Handbook, 2nd edn. Artech House, Norwood (2005)
3. Nocedal, J., Wright, S.J.: Numerical Optimization. Springer, New York (2006)
4. Ares-Pena, F.J., Rodriguez-Gonzales, A., Villanueva-Lopez, E., Rengarajan, S.R.: Genetic algorithms in the design and optimization of antenna array patterns. IEEE Trans. Antennas Propag. **47**, 506–510 (1999)
5. Boeringer, D.W., Werner, D.H., Machuga, D.W.: A simultaneous parameter adaptation scheme for genetic algorithms with application to phased array synthesis. IEEE Trans. Antennas Propag. **53**, 356–371 (2005)
6. Khodier, M.M., Christodoulou, C.G.: Linear array geometry synthesis with minimum sidelobe level and null control using particle swarm optimization. IEEE Trans. Antennas Propag. **53**, 2674–2679 (2005)
7. Jin, N., Rahmat-Samii, Y.: A novel design methodology for aperiodic arrays using particle swarm optimization. Nat. Radio Sci. Meeting Dig., Boulder, CO, 69–69 (2006)
8. Gies, D., Rahmat-Samii, Y.: Particle swarm optimization for reconfigurable phased-differentiated array design. Microw. Opt. Technol. Lett. **38**, 168–175 (2003)
9. Weng, W.C., Yang, F., Elsherbeni, A.Z.: Linear antenna array synthesis using Taguchi's method: a novel optimization technique in electromagnetics. IEEE Trans. Antennas Propag. **55**, 723–730 (2007)
10. Karmakar, N.C., Bialkowski, M.E., Padhi, S.K.: Microstrip circular phased array design and development using microwave antenna CAD tools. IEEE Trans. Antennas Propag. **50**, 944–952 (2002)
11. Werner, D.H., Gregory, M.D., Namin, F., Petko, J., Spence, T.G.: Ultra-wideband antenna arrays. In: Gross, F.B. (ed.) Frontiers in Antennas: Next Generation Design & Engineering. McGaw-Hill, New York (2011)
12. Vouvakis, M.N., Schaubert, D.H.: Vivaldi antenna arrays. In: Gross, F.B. (ed.) Frontiers in Antennas: Next Generation Design & Engineering. McGaw-Hill, New York (2011)
13. Weinmann, F.: Design, optimization, and validation of a planar nine-element quasi-Yagi antenna array for X-band applications. IEEE Antennas Propag. Mag. **50**, 141–148 (2008)

14. Balanis, C.A.: Antenna Theory, 3rd edn. Wiley-Interscince, Hoboken, New Jersey (2005)
15. Bandler, J.W., Cheng, Q.S., Dakroury, S.A., Mohamed, A.S., Bakr, M.H., Madsen, K., Søndergaard, J.: Space mapping: the state of the art. IEEE Trans. Microw. Theory Tech. **52**, 337–361 (2004)
16. Koziel, S., Echeverría-Ciaurri, D., Leifsson, L.: Surrogate-based methods. In: Koziel, S., Yang, X.S. (eds.) Computational Optimization, Methods and Algorithms. Series: Studies in Computational Intelligence, pp. 33–60. Springer (2011)
17. Forrester, A.I.J., Keane, A.J.: Recent advances in surrogate-based optimization. Prog. Aerosp. Sci. **45**, 50–79 (2009)
18. Koziel, S., Ogurtsov, S.: Simulation-driven design in microwave engineering: methods. In: Koziel, S., Yang, X.S. (eds.) Computational Optimization, Methods and Algorithms. Studies in Computational Intelligence. Springer, New York (2011)
19. Koziel, S., Ogurtsov, S.: Model management for cost-efficient surrogate-based optimization of antennas using variable-fidelity electromagnetic simulations. IET Microw. Antennas Propag. **6**, 1643–1650 (2012)
20. Queipo, N.V., Haftka, R.T., Shyy, W., Goel, T., Vaidynathan, R., Tucker, P.K.: Surrogate based analysis and optimization. Prog. Aerosp. Sci. **41**, 1–28 (2005)
21. Conn, A.R., et al.: Trust region methods, MPS-SIAM Series on Optimization (2000)
22. Alexandrov, N.M., Lewis, R.M.: An overview of first-order model management for engineering optimization. Optim. Eng. **2**, 413–430 (2001)
23. Nikolova, N.K., et al.: Sensitivity analysis of network parameters with electromagnetic frequency-domain simulators. IEEE Trans. Microw. Theory Tech. **54**, 670–681 (2006)
24. Li, D., et al.: Electromagnetic optimization using sensitivity analysis in the frequency domain. IET Microw. Antennas Propag. **1**, 852–859 (2007)
25. CST Microwave Studio: CST AG. Bad Nauheimer Str. 19, D-64289 Darmstadt, Germany (2011)
26. Kolda, T.G., Lewis, R.M., Torczon, V.: Optimization by direct search: new perspectives on some classical and modern methods. SIAM Rev. **45**, 385–482 (2003)
27. Bandler, J.W., Koziel, S., Madsen, K.: Space mapping for engineering optimization. SIAG/Optimization Views-and-News Special Issue on Surrogate/Derivative-free Optimization **17**, 19–26 (2006)

Optimal Design of Computationally Expensive EM-Based Systems: A Surrogate-Based Approach

Abdel-Karim S.O. Hassan, Hany L. Abdel-Malek, and Ahmed S.A. Mohamed

Abstract It is quite a challenge to find the optimal design of computationally expensive engineering systems in different areas such as electrical engineering, structural mechanics, fluid dynamics, and electromagnetic-based (EM-based) systems. The optimal design of such systems requires solving huge optimization problems involving a lot of expensive function evaluations. For example, in microwave circuit design, a function evaluation requires running a full-wave electromagnetic simulator which may exhaust hours of CPU time. The total computational overhead makes the optimization of these engineering systems practically prohibitive. Computationally cheap surrogates (Response Surfaces, Space Mapping, Kriging models, Neural Networks, etc.) offer a good solution of such problems. Throughout the optimization process, iteratively updated surrogates are employed to replace the computationally expensive function evaluations.

In this chapter, surrogate-based approaches that can be applied for optimal design of EM-based systems are presented. The first one is a novel surrogate-based trust region optimization approach. The proposed approach relies on building and successively updating quadratic surrogate models to be optimized instead of the objective function over the trust regions. The approach is applied to find the optimal design of RF cavity linear accelerator (LINAC).

In addition, a novel surrogate-based geometrical design centering technique for microwave circuits is introduced. The technique integrates generalized space mapping (GSM) surrogates with the normed distances concept. The normed distances from a point to the feasible region boundaries are evaluated using norms related to the probability distribution of the circuit parameters. The technique is applied to obtain the design center point of a microwave filter.

Keywords Computationally expensive engineering systems • Design centering • EM-based systems • Microwave circuit design • Normed distances • Optimal design • RF cavity • Space mapping surrogates

---

A.-K.S.O. Hassan (✉) • H.L. Abdel-Malek • A.S.A. Mohamed
Faculty of Engineering, Engineering Mathematics and Physics Department, Cairo University, Giza 12613, Egypt
e-mail: asho_hassan@yahoo.com; hanymalek@aucegypt.edu; aashiry@ieee.org

S. Koziel et al. (eds.), *Solving Computationally Expensive Engineering Problems*, Springer Proceedings in Mathematics & Statistics 97,
DOI 10.1007/978-3-319-08985-0_7

1 Introduction

Engineering systems, in general, are required to meet some performance measure constraints through the adjustment of a set of designable parameters. The desired performance of a system (design specifications) is set by the designer and usually described by specifying bounds on the performance measures of the system. The conventional system design aims at finding values of the system designable parameters that merely satisfy the design specifications. In general, there will be a multitude of acceptable designs. However, for contemporary engineering design, other criterion (objective function) can be chosen for selecting the best acceptable design. In this respect, optimal system design is usually accompanied with an optimization problem. Naturally, the performance measures and the objective functions of an engineering system depend on the parameter values and are evaluated through numerical simulations. Hence, the design process is heavily based on system simulations, and for computationally expensive engineering systems, the high expense of the required simulations may obstruct the optimization process. To overcome this, computationally cheap surrogates (Response Surfaces, Space Mapping, Kriging Models, Neural Networks, etc.) offer a good solution of such problems. Throughout the optimization process, iteratively updated surrogates are employed to replace the computationally expensive function evaluations.

The surrogate model is a mathematical or physical model which can take the place of the computationally expensive fine model (simulations) throughout the optimization process. Computationally expensive simulations may arise from numerical solution of large systems of, e.g., integral or differential equations describing a physical system, or it could be an actual physical system. Surrogate models may be obtained through physical simplification or by employing a less accurate numerical approximation. Often the computationally cheap surrogate model is less accurate than the expensive high fidelity model.

The surrogate model is used in all heavy computations of the optimization process, whereas the fine model is evaluated only at a limited number of points, e.g., the sequentially generated optimal points. These fine model evaluations can be used to construct or update the surrogate model and also to validate the optimization results. Surrogate models are classified into two main categories: functional and physical models [48, 49]. Functional models are constructed without any knowledge about the underlying physics of the expensive model they represent. Physical models carry some information about the system they are representing; hence, they can predict the behavior of the system with much less computations.

In practice, the optimal design process and the corresponding optimization problem have some permanent special difficulties. The high cost of the frequent evaluation of computationally expensive functions is one of these difficulties. Hence, robust optimization methods that utilize the fewest possible number of function evaluations are greatly required [1, 2]. Another difficulty is the absence of any gradient information as the required simulation cost in evaluating the gradient information is prohibitive in practice [3]. Attempting to approximate the function gradients using the finite difference approach requires much more function evaluations, which highly increase the computational cost. Another defect in estimating the gradients by finite differencing is that the estimated function values are usually affected by some numerical noise due to estimation uncertainty. Hence, small perturbations do not reflect the local behavior of the function value itself but rather that of the noise.

For such objective functions, only derivative-free optimization (DFO) is feasible [1, 2, 4–6]. Further, the derivative-free trust region methods usually handle such problems more efficiently as the trust region framework constitutes one of the most important globally convergent optimization methods, which has the ability to converge to a solution starting from any initial point [7]. In addition, these methods employ computationally inexpensive surrogates that can be constructed by using function evaluations at some selected points. These surrogate models may be response surfaces or radial basis functions.

This chapter presents new surrogate-based approaches that can be applied for optimal design of EM-based systems. In the first part of this chapter, a new derivative-free trust region optimization approach is introduced that neither requires nor approximates the gradients of the objective function. It implements a non-derivative optimization method that combines a trust region framework with quadratic fitting surrogates for the objective function [4, 5]. The approach relies on building, successively updating and optimizing quadratic surrogate models of the objective function over trust regions. The quadratic surrogate model reasonably reflects the local behavior of the objective function in a trust region around the current iterate. The efficiency of the new approach is investigated by applying it to a numerical example and comparisons with a recent optimization technique are also included. The effective shunt impedance per unit length is maximized to find the optimal design of the structure of RF cavity. The RF cavity is the major part of particle linear accelerators (LINACs).

The second part of this chapter treats the problem of design centering of computationally expensive microwave circuits. Design centering is an optimal design problem which attempts to find the nominal values of designable circuit parameters that maximize the probability of satisfying the design specifications (yield function). These design specifications are functions of circuit parameter values. Practically, the values of circuit parameters are subjected to known but unavoidable fluctuations due to model uncertainty, manufacturing process, or environmental changes. To simulate these fluctuations, the circuit parameters are considered to be random variables with certain probability distribution. The fluctuations in the circuit parameters may lead to violation of the design specification. Design centering seeks for the best nominal values of circuit parameters which make the design more robust against circuit fluctuations.

Generally, there are two main approaches for design centering. The first approach is based upon statistical analysis techniques, e.g., Monte Carlo method, for the estimation of the yield function values during the optimization process [1, 2, 8–14]. The second approach is classified as a geometric approach. The approach maximizes the yield function implicitly [15–29, 46–47]. In this approach, the feasible region (a region in the parameter space where all design specifications are satisfied) is approximated using a convex body, e.g., a hyperellipsoid. The center of this approximating body is considered as a design center [30].

The main obstacle in the design centering process is the computational effort required in evaluating the circuit performance measures, especially in case of microwave circuits [31]. These evaluations depend on circuit simulations. Generally, circuit simulations need running a full-wave electromagnetic simulator which may exhaust hours of CPU time. The total computational overhead makes the design centering of these microwave systems practically prohibitive. To overcome this problem, space mapping surrogates are employed instead of the computationally expensive high fidelity fine model. The surrogate model is used to make an approximation to the feasible region. Hence, centering process is performed on that approximation and then validated by the fine model. The process is repeated till the final center is obtained [31].

In the second part of this chapter, a novel surrogate-based geometrical design centering technique for microwave circuits is introduced. The technique integrates generalized space mapping (GSM) surrogates [32] with the normed distances concept [25]. The normed distances from a point to the feasible region boundaries are evaluated using norms related the probability distribution of the circuit parameters. The technique is applied to obtain the design center point of a six-section H-plane waveguide filter.

2 The New Surrogate-Based Trust Region Optimization Approach

The majority of the existing derivative-free trust region techniques have the following features:

- They require a relatively large number of function evaluations, $O(n^2)$ (where n is the number of system variables) to construct the initial quadratic model.
- The quadratic surrogate models are constructed via interpolating the objective function at a constant number of points; when a point is obtained, a previous point is dropped. In addition, these algorithms usually ignore the valuable information contained in all previously evaluated expensive function values.

In this section, a novel surrogate-based trust region optimization approach is presented. The proposed approach relies on building and successively updating quadratic surrogate models for the objective function. These surrogate models are optimized instead of the objective function over trust regions. Truncated conjugate gradients [33] are used to find the optimal point within each trust region. The approach constructs the initial quadratic surrogate model using few data points of order $O(n)$, where n is the number of design variables. In each iteration of the proposed approach, the surrogate model is updated using a weighted least squares fitting. The weights are assigned to give more emphasis to points close to the current center point. The accuracy and efficiency of the proposed approach are demonstrated by applying it on a set of classical benchmark test problems. Also, the approach is employed to find the optimal design of RF cavity LINAC. A comparison analysis with a recent optimization technique is also included.

The proposed algorithm, which belongs to the trust region DFO class, will be introduced. The computationally expensive objective function is locally approximated around a current iterate $\mathbf{x}_k$ by a computationally cheaper quadratic surrogate model $M(\mathbf{x})$ which can be given by:

$$M(\mathbf{x}) = a + \mathbf{b}^T(\mathbf{x} - \mathbf{x}_k) + \frac{1}{2}(\mathbf{x} - \mathbf{x}_k)^T \mathbf{B}(\mathbf{x} - \mathbf{x}_k), \tag{1}$$

where, $a \in \mathbb{R}$, the vector $\mathbf{b} \in \mathbb{R}^n$, and the symmetric matrix $\mathbf{B} \in \mathbb{R}^{n \times n}$ are the unknown parameters of $M(\mathbf{x})$. The total number of the model parameters is $q = (n+1)(n+2)/2$. These parameters can be evaluated by interpolating the objective function at q points.

2.1 Initial Model

Let $\mathbf{x}_0$ be the initial point that is provided by the user. Initially, assuming that $\mathbf{B}$ is a diagonal matrix, then the number of points required to construct the initial model is $m = 2n + 1$ [34]. The initial m points $\mathbf{x}_i, i = 1, 2, \ldots, m$, can be chosen as follows [6, 35]:

$$\mathbf{x}_1 = \mathbf{x}_0 \; and \; \begin{cases} \mathbf{x}_{i+1} = \mathbf{x}_0 + \Delta_1 \mathbf{e}_i, \; i = 1, 2, \ldots, n \\ \mathbf{x}_{i+n+1} = \mathbf{x}_0 - \Delta_1 \mathbf{e}_i, \; i = 1, 2, \ldots, m - n - 1 \end{cases} \tag{2}$$

where Δ_1 is the initial trust region radius that is provided by the user, and $\mathbf{e}_i$ is the ith coordinate vector in $\mathbb{R}^n$.

The initial quadratic model $M^{(1)}(\mathbf{x})$ will have the parameters $a^{(1)}$, the vector $\mathbf{b}^{(1)}$, and the n diagonal elements of the model Hessian matrix $\mathbf{B}^{(1)}$. These parameters are computed by requiring that the initial model interpolates the objective function $f(\mathbf{x})$ at the initial m points given in (2). Therefore, the initial model parameters are obtained by satisfying the matching conditions:

$$M^{(1)}(\mathbf{x}_i) = f(\mathbf{x}_i), \quad i = 1, 2, \ldots, m. \tag{3}$$

2.2 Model Optimization

At the kth iteration, assume that $\mathbf{x}_k$ is the current solution point. The model $M^{(k)}(\mathbf{x})$ is then minimized, in place of the objective function, over the current trust region and a new point is produced by solving the trust region sub-problem:

$$\min_s M^{(k)}(\mathbf{s}), \quad \text{subject to } \|\mathbf{s}\| \leq \Delta_k, \tag{4}$$

where $\mathbf{s} = \mathbf{x} - \mathbf{x}_k$, Δ_k is the current trust region radius, and $\| \cdot \|$ throughout is the l_2-norm. This problem is solved by the method of truncated conjugate gradient which is proposed in [33]. It is identical to the standard conjugate gradient method as long as iterates are inside the trust region. If the conjugate gradient method terminates at a point within the trust region, this point is a global minimizer of the objective function. If the new iterate is outside the trust region, a truncated step which is on the region boundary is considered. Also, the method treats the case where the minimum is in the opposite direction of the conjugate direction, which is due to the non-convexity of the model [33]. One good property of this method is that the solution computed has a sufficient reduction property which was proved in [36].

Let $\mathbf{s}*$ denote the solution of (4), and then a new point $\mathbf{x}_n = \mathbf{x}_k + \mathbf{s}*$ is obtained. The achieved actual reduction in the objective function is compared to that predicted reduction using the model by computing the reduction ratio which is given by:

$$r_k = \frac{\text{actual reduction}}{\text{predicted reduction}} = \frac{f(\mathbf{x}_k) - f(\mathbf{x}_n)}{M^{(k)}(\mathbf{x}_k) - M^{(k)}(\mathbf{x}_n)}. \tag{5}$$

This ratio reflects how much the surrogate model agrees with the objective function within the trust region. The trust region radius and the current iterate will be updated such that, if r_k is sufficiently high, i.e., $r_k \geq 0.7$, there is a good agreement between the model and the objective function over this step. Hence, it is beneficial to expand the trust region for the next iteration, and to use $\mathbf{x}_n$ as the new center of the trust region. If r_k is positive but not close to 1, i.e., $0.1 \leq r_k < 0.7$, the trust region radius is not altered. On the other hand, if r_k is smaller than a certain threshold, $r_k < 0.1$, the trust region radius is reduced. The updating formula used for updating Δ_k and $\mathbf{x}_k$ can be expressed as follows:

$$r_k \begin{cases} r_k < 0.1: & \Delta_{\text{k}+1} = \frac{1}{2}\Delta_\text{k} \\ 0.1 \leq r_k \leq 0.7: & \Delta_{\text{k}+1} = \Delta_\text{k} \\ r_k \geq 0.7 \begin{cases} \|\mathbf{s}^*\| < \Delta_\text{k}: \Delta_{\text{k}+1} = \Delta_\text{k} \\ \|\mathbf{s}^*\| \geq \Delta_\text{k}: \Delta_{\text{k}+1} = 1.5 * \Delta_\text{k} \end{cases} \end{cases}, \tag{6}$$

$$\mathbf{x}_{k+1} = \begin{cases} \mathbf{x}_k + \mathbf{s}^*, & \text{if } r_k > 0 \\ \mathbf{x}_k, & \text{otherwise} \end{cases}. \tag{7}$$

It is to be mentioned that the current center is the point of least function value achieved so far.

2.3 Model Update

When a new point is available, the current quadratic model $M^{(k)}(\mathbf{x})$ is updated so that the point of lowest objective function value $\mathbf{x}_k$ is now the center of the kth trust region. The model will take the form:

$$M^{(k)}(\mathbf{s}) = a^{(k)} + \mathbf{s}^T\mathbf{b}^{(k)} + \frac{1}{2}\mathbf{s}^T\mathbf{B}^{(k)}\mathbf{s}; \quad \mathbf{s} = \mathbf{x} - \mathbf{x}_k \text{ and } \mathbf{s} \in \mathbb{R}^n. \tag{8}$$

The parameters $a^{(k)}$, $\mathbf{b}^{(k)}$, and $\mathbf{B}^{(k)}$ are evaluated employing the parameter values of the previous model $M^{(k-1)}(\mathbf{x})$ in addition to all available function values. The constant $a^{(k)}$ is assigned the value of $f(\mathbf{x}_k)$, i.e., $a^{(k)} = f(\mathbf{x}_k)$. The model will be updated in two steps. First, the vector $\mathbf{b}^{(k)}$ is updated then the Hessian matrix $\mathbf{B}^{(k)}$ is updated as follows:

Step1: Updating the vector $\mathbf{b}^{(k)}$

The vector $\mathbf{b}^{(k)}$ can be obtained using only n points. However, using the n recent points may result in ill-conditioned system of linear equations. In order to avoid this, it is proposed to use the least squares approximation with the most recent $2n$ points. So, the vector $\mathbf{b}^{(k)}$ is evaluated such that the model $M^{(k)}(\mathbf{x})$ fits the last $2n$ points obtained, $\mathbf{x}_i$, $i = 1, 2, \ldots, 2n$, i.e., the following condition should be satisfied:

$$M^{(k)}(\mathbf{s}_i) = f(\mathbf{s}_i), \quad \text{where } \mathbf{s}_i = \mathbf{x}_i - \mathbf{x}_k \quad \text{and} \quad i = 1, 2, \ldots, 2n. \tag{9}$$

When computing the vector $\mathbf{b}^{(k)}$, the matrix $\mathbf{B}^{(k)}$ is assigned temporarily the value of the previous model Hessian matrix, $\mathbf{B}^{(k-1)}$, hence the vector $\mathbf{b}^{(k)}$ is obtained by solving the following system of linear equations:

$$\mathbf{A}\mathbf{b}^{(k)} = \mathbf{v}, \tag{10}$$

where

$$\mathbf{A} = \begin{bmatrix} \mathbf{s}_1^T \\ \mathbf{s}_2^T \\ \vdots \\ \mathbf{s}_{2n}^T \end{bmatrix} \text{ and } \mathbf{v} = \begin{bmatrix} f(\mathbf{s}_1) - a^{(k)} - \frac{1}{2}\mathbf{s}_1^T\mathbf{B}^{(k-1)}\mathbf{s}_1 \\ f(\mathbf{s}_2) - a^{(k)} - \frac{1}{2}\mathbf{s}_2^T\mathbf{B}^{(k-1)}\mathbf{s}_2 \\ \vdots \\ f(\mathbf{s}_{2n}) - a^{(k)} - \frac{1}{2}\mathbf{s}_{2n}^T\mathbf{B}^{(k-1)}\mathbf{s}_{2n} \end{bmatrix}. \tag{11}$$

The previous system is an over-determined system. The least squares approximation for $\mathbf{b}^{(k)}$ is

$$\mathbf{b}^{(k)} = \left(\mathbf{A}^T\mathbf{A}\right)^{-1}\mathbf{A}^T\mathbf{v}. \tag{12}$$

Step2: Updating the matrix $\mathbf{B}^{(k)}$

The model Hessian matrix $\mathbf{B}^{(k)}$ is evaluated using the following updating formula:

$$\mathbf{B}^{(k)} = c\,\mathbf{B}^{(k-1)} + \mathbf{q}\mathbf{p}^T, \tag{13}$$

where c is a positive constant, $0.5 < c < 1$, and the vector $\mathbf{p} \in \mathbb{R}^n$,

$$\mathbf{q} = \left[\text{sign}\left(\text{diag}\left(\mathbf{B}^{(k-1)}\right)\right)\right] * \sqrt{(1-c) * \left|\text{diag}\left(\mathbf{B}^{(k-1)}\right)\right|}. \tag{14}$$

This choice of $\mathbf{q}$ ensures that changes in $\mathbf{B}^{(k)}$ occur gradually. The vector $\mathbf{p}$ is evaluated such that the model $M^{(k)}(\mathbf{x})$ tries to fit all the available m points obtained so far, $\mathbf{x}_i$, $i = 1, 2, \ldots, m$, i.e., the following condition should be satisfied

$$M^{(k)}(\mathbf{s}_i) = f(\mathbf{s}_i), \text{where} \quad \mathbf{s}_i = \mathbf{x}_i - \mathbf{x}_k \quad \text{and} \quad i = 1, 2, \ldots, m, \tag{15}$$

i.e., the vector $\mathbf{p}$ is obtained by solving the weighted system of linear equations

$$\mathbf{A}\mathbf{p} = \mathbf{v}, \tag{16}$$

where

$$\mathbf{A} = \begin{bmatrix} \frac{1}{2}\mathbf{s}_1^T\mathbf{q}\mathbf{s}_1^T w_1 \\ \frac{1}{2}\mathbf{s}_2^T\mathbf{q}\mathbf{s}_2^T w_2 \\ \vdots \\ \frac{1}{2}\mathbf{s}_m^T\mathbf{q}\mathbf{s}_m^T w_m \end{bmatrix}, \quad \mathbf{v} = \begin{bmatrix} w_1 * \left(f(\mathbf{s}_1) - a^{(k)} - \mathbf{s}_1^T\mathbf{b}^{(k)} - \frac{1}{2}\mathbf{s}_1^T c\mathbf{B}^{(k-1)}\mathbf{s}_1\right) \\ w_2 * \left(f(\mathbf{s}_2) - a^{(k)} - \mathbf{s}_2^T\mathbf{b}^{(k)} - \frac{1}{2}\mathbf{s}_2^T c\mathbf{B}^{(k-1)}\mathbf{s}_2\right) \\ \vdots \\ w_m * \left(f(\mathbf{s}_m) - a^{(k)} - \mathbf{s}_m^T\mathbf{b}^{(k)} - \frac{1}{2}\mathbf{s}_m^T c\mathbf{B}^{(k-1)}\mathbf{s}_m\right) \end{bmatrix}. \tag{17}$$

To obtain more accurate model in the neighborhood of the current center, the available points are assigned different weights w_i, $i = 1, 2, \ldots, m$ according to their distances from the trust region center. In the proposed approach the weight w_i associated with each equation takes the form:

$$w_i = \begin{cases} 1 & \text{if } \|\mathbf{s}_i\| \le c_1\Delta \\ \frac{c_1\Delta}{\|\mathbf{s}_i\|} & \text{if } \|\mathbf{s}_i\| > c_1\Delta \end{cases}, \quad i = 1, 2, \ldots, m, \tag{18}$$

where c_1 is a positive constant, $c_1 \ge 1$.

The previous system in (16) is an over-determined system ($m > n$). The least squares approximation for $\mathbf{p}$ is

$$\mathbf{p} = \left(\mathbf{A}^T\mathbf{A}\right)^{-1}\mathbf{A}^T\mathbf{v}. \tag{19}$$

After getting the vector $\mathbf{p}$, the term $\mathbf{qp}^T$ is calculated and the matrix is made symmetric by resetting the off-diagonal elements to their average values, i.e., $b_{ij} = b_{ji} \leftarrow (b_{ij} + b_{ji})/2$, then the new Hessian matrix $\mathbf{B}^{(k)}$ is updated according to Eq. (13). It is to be mentioned that making the matrix $\mathbf{B}^{(k)}$ symmetric does not affect the model neither in value nor in gradient.

2.4 Comment

If $r_k < 0$, for two consecutive iterations, then $\mathbf{x}_k$ is kept unchanged and the trust region radius is not altered. A procedure aiming to improve the quality of the model is employed. The model can be improved by generating a new point $\mathbf{s}_{new} = \mathbf{x}_{new} - \mathbf{x}_k$, which is chosen to be on the boundary of the trust region so that it improves the distribution of points around the center of the trust region.

2.5 Numerical Example

The effectiveness of the proposed algorithm is demonstrated through the 6D Watson benchmark example. The function was proposed in [37]. All results are compared with those obtained by NEWUOA (NEW Unconstrained Optimization Algorithm) [6]. The performance is measured by the number of function evaluations N required to reach the optimal solution.

$$f(\mathbf{x}) = x_1^2 + \left(x_2 - x_1^2 - 1\right)^2 + \sum_{i=2}^{30} \left\{ \sum_{j=2}^{6} (j-1)\, x_j \left(\frac{i-1}{29}\right)^{j-2} - \left\{ \sum_{j=1}^{6} x_j \left(\frac{i-1}{29}\right)^{j-1} \right\}^2 - 1 \right\}^2. \tag{20}$$

This minimization problem is ill-conditioned, and rather difficult to solve. This function has a minimum of 2.2877×10^{-3} at $(-0.0157\ 1.0124 - 0.233\ 1.2604 - 1.5137\ 0.993)^T$. The initial values used for $\mathbf{x}_0$ and Δ_1 are $(0\ 0\ 0\ 0\ 0\ 0)^T$ and 0.25, respectively. The optimal value obtained using the proposed technique and NEWUOA for different N function evaluations are shown in Table 1 and Fig. 1.

In this numerical example, it is to be noticed that at the beginning of the optimization process, the proposed algorithm is much faster than NEWUOA. However, as the optimization gets close to the optimum, the methods based on interpolation will be more accurate as expected. This explains why the proposed algorithm is well suited for objective functions that have some uncertainty in their values or subject to statistical variations.

Table 1 The 6D Watson function: proposed approach versus NEWUOA

N	Proposed algorithm	NEWUOA
23	1.0399	1.5134
37	0.6976	0.2405
51	0.3369	0.1854
65	0.2358	0.0535
80	0.0780	0.0216

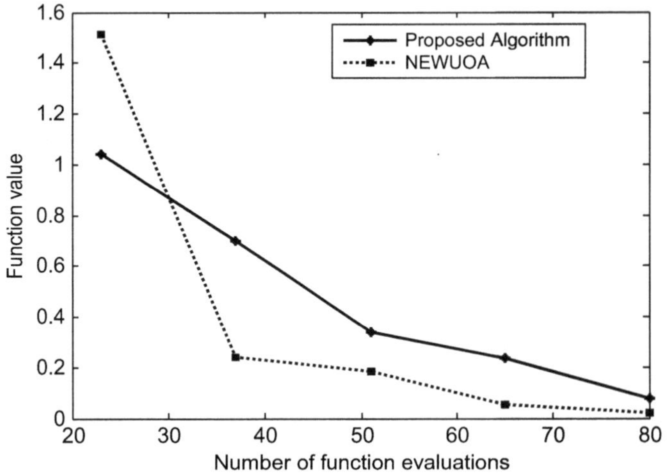

Fig. 1 The 6D Watson example

2.6 *Optimal Design of RF Cavity*

The RF cavity is a major component of LINACs [38, 39]. The structure of RF cavity must efficiently transfer the electromagnetic energy to the beam.

The most useful figure of merit for high field concentration along the beam axis and low ohmic power loss in the cavity walls is the effective shunt impedance per unit length ZT^2 where T is the transient-time factor (a measure of the energy gain reduction caused by the sinusoidal time variation of the field in the cavity, [40]).

The technique is applied to an RF cavity with resonance frequency 9.4 GHz, shown in Fig. 2. The objective is to maximize effective shunt impedance per unit length. In order to do that, we optimize the axial z positions of ten points that describe the cavity curvature through a spline curve. The axial positions $z = (z_1, z_2, \ldots, z_{10})^T$ in the z-direction are taken as the design parameters. The radial positions of these points are chosen on a logarithmic scale along r-direction. It is to be noted that during the variation of the curvature, the resonance frequency is always kept at 9.4 GHz. The initial values used for the ten radial positions z_0 are all set to 0.6 cm and Δ_1 is set to 0.02 cm.

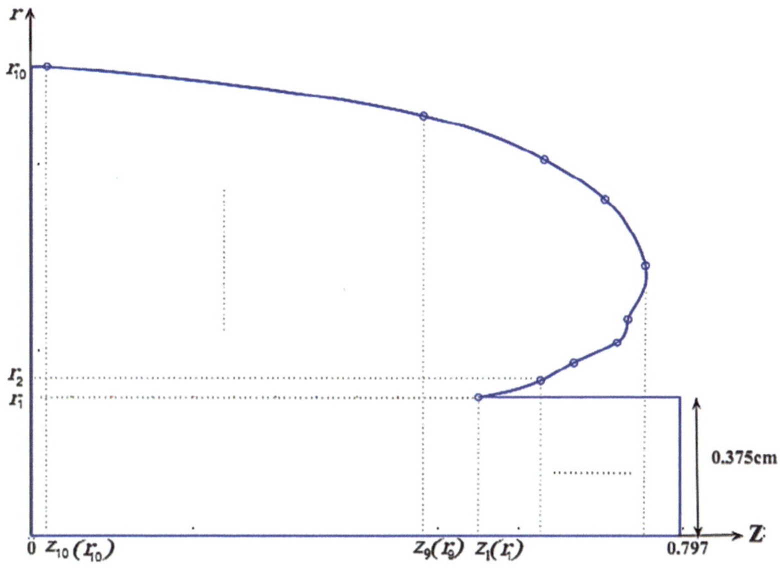

Fig. 2 Structure of the RF cavity

Table 2 RF cavity design results: proposed approach versus NEWUOA

N	Proposed algorithm	NEWUOA
50	111.771	112.587
75	115.207	116.833
90	117.183	119.316
120	119.01	120.511
160	120.5	120.910
200	121.01	121.211
260	121.301	121.521

The results of the effective shunt impedance per unit length for RF cavity in mega ohm per meter after N function evaluations for both the proposed algorithm and NEWUOA are shown in Table 2.

It is to be mentioned that starting from the same initial point, the convergence of the proposed algorithm is nearly the same as NEWUOA algorithm. However, the advantage of the proposed algorithm is its easy implementation and accessibility for update and modification.

The figure of optimal cavity using the proposed algorithm and the NEWUOA are shown in Fig. 3.

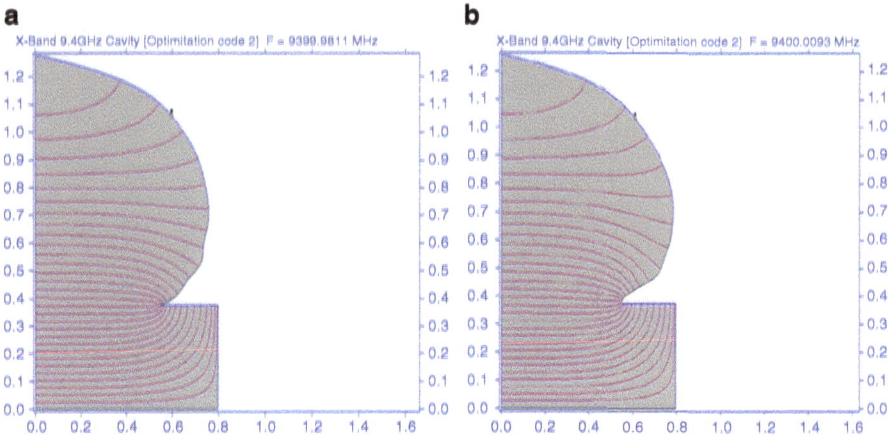

Fig. 3 The optimal cavity: (**a**) using the proposed algorithm with effective shunt impedance per unit length = 121.301 M ohm/m. (**b**) using NEWUOA with effective shunt impedance per unit length = 121.521 M ohm/m

3 Microwave Circuit Design Centering Approach Exploiting Normed Distances and Space Mapping Surrogates

In this section, GSM surrogates are integrated with the normed distances concept to develop a novel surrogate-based geometrical design centering technique for microwave circuit applications. The design centering problem is formulated as a max–min optimization problem using normed distances from a point to the feasible region boundaries. The norm used in evaluating the distances is related to the probability distribution of the circuit parameters. The normed distance is evaluated by solving a nonlinear optimization problem. A convergent iterative boundary search technique is used to solve the nonlinear optimization problem concerning the normed distance. In the new approach of microwave design centering, a GSM surrogate is initially constructed based on the coarse model and then updated through space mapping (SM). In each SM iteration, a current SM feasible region approximation is available and the centering process using normed distances is implemented with this region approximation leading to a better design center. The new center point is validated by the fine model and is used to update the next GSM surrogate. The process is repeated to obtain the next center point. Practical circuit examples are given to show the effectiveness of the new design centering method.

3.1 Generalized Space Mapping (GSM)

A GSM [32] with input and output mappings is used where a matching in response and gradient between surrogate and fine models is performed. The matching will be made at every center point $\mathbf{x}^k$. The response of the surrogate at any point is given by:

$$\mathbf{R}_s^k(\mathbf{x}) = \mathbf{A}^k \cdot \mathbf{R}_c\left(\mathbf{B}^k\mathbf{x} + \mathbf{c}^k\right) + \mathbf{d}^k + \mathbf{E}^k\left(\mathbf{x} - \mathbf{x}^k\right), \tag{21}$$

where $\mathbf{R}_c$ is the coarse model response vector, $\mathbf{A}^k \in M_{m \times m}$ is a diagonal matrix, $\mathbf{B}^k \in \mathrm{M}_{n \times n}$, $\mathbf{c}^k \in M_{n \times 1}$, and $\mathbf{d}^k \in M_{m \times 1}$ (n is the number of design variables and m is the number of constraints) is given by:

$$\mathbf{d}^k = \mathbf{R}_f\left(\mathbf{x}^k\right) - \mathbf{A}^k \cdot \mathbf{R}_c\left(\mathbf{B}^k\mathbf{x}^k + \mathbf{c}^k\right), \tag{22}$$

where $\mathbf{R}_f$ is the fine model response vector, and $\mathbf{E}^k \in \mathrm{M}_{m \times n}$ is given by:

$$\mathbf{E}^k = \mathbf{J}_f\left(\mathbf{x}^k\right) - \mathbf{A}^k \cdot \mathbf{J}_c\left(\mathbf{B}^k\mathbf{x}^k + \mathbf{c}^k\right)\mathbf{B}^k, \tag{23}$$

where $\mathbf{J}_f$ and $\mathbf{J}_c$ are the Jacobian matrices of the fine and coarse model, respectively. The mapping parameters $\mathbf{A}$, $\mathbf{B}$, and $\mathbf{c}$ are obtained through an optimization process called parameter extraction:

$$\left(\mathbf{A}^k, \mathbf{B}^k, \mathbf{c}^k\right) = \arg \min_{\mathbf{A},\mathbf{B},\mathbf{c}} \mathbf{e}^k(\mathbf{A}, \mathbf{B}, \mathbf{c}), \tag{24}$$

where $\mathbf{e}^k$ represents the response deviation residual of the surrogate from the fine model and is given by:

$$\begin{aligned}\mathbf{e}^k(\mathbf{A}, \mathbf{B}, \mathbf{c}) = \sum_{l=0}^{k} w_l \left\|\mathbf{R}_f\left(\mathbf{x}^l\right) - \mathbf{A}\mathbf{R}_c\left(\mathbf{B}\mathbf{x}^l + \mathbf{c}\right)\right\| + \\ \sum_{l=0}^{k} v_l \left\|\mathbf{J}_f\left(\mathbf{x}^l\right) - \mathbf{A}\mathbf{J}_c\left(\mathbf{B}\mathbf{x}^l + \mathbf{c}\right)\mathbf{B}\right\|,\end{aligned} \tag{25}$$

where the coefficients w_l and v_l are weights chosen according to the nature of the problem.

3.2 Feasible Region and Yield Function

In general, design specifications define a region in the parameter space called the feasible region F which can be defined as:

$$F = \left\{\mathbf{x} \in \mathbb{R}^n \middle| \mathbf{f}\left(\mathbf{R}_f(\mathbf{x})\right) \leq 0\right\}, \tag{26}$$

where $\mathbf{x} \in \mathbb{R}^n$ is a vector of the design parameters, n is the number of design parameters, $\mathbf{R}_f : \mathbb{R}^n \to \mathbb{R}^m$, $\mathbf{R}_f(\mathbf{x})$ is a fine model response vector, m is the number of the constraints, and $\mathbf{f}: \mathbb{R}^m \to \mathbb{R}^m$ is the constraint vector function. As the response of the fine model is very expensive, it will be replaced by a GSM surrogate model response $\mathbf{R}_s$ $(\mathbf{x})$ given by (21) which is computationally cheap. Hence, we have a feasible region approximation F_s which will replace the actual feasible region and is defined as:

$$F_s = \left\{ \mathbf{x} \in \mathbb{R}^n \middle| \mathbf{f}(\mathbf{R}_s\ (\mathbf{x})) \leq 0 \right\}. \tag{27}$$

The design parameters are assumed to be random variables distributed with joint probability density function (pdf) $\varnothing(\mathbf{x}, \mathbf{x}^0)$, where $\mathbf{x}^0 \in \mathbb{R}^n$ the nominal parameter vector. Accordingly, the yield function is defined as:

$$Y\left(\mathbf{x}^0\right) = \text{Prob}\ \{\mathbf{x} \in F_s\} = \int_{F_S} \varnothing\left(\mathbf{x}, \mathbf{x}^0\right) d\mathbf{x}. \tag{28}$$

Hence, the design centering problem can be formulated as:

$$\max_{\mathbf{x}^0} \left[Y\left(\mathbf{x}^0\right) = \int_{F_S} \varnothing\left(\mathbf{x}, \mathbf{x}^0\right) \mathrm{d}\mathbf{x} \right]. \tag{29}$$

According to practical requirements of the system design, the system parameters are assumed to be normally distributed with pdf given by:

$$\varnothing\left(\mathbf{x}, \mathbf{x}^0\right) = \frac{1}{(2\pi)^{n/2}\sqrt{|\mathbf{C}|}} e^{-\frac{1}{2}\left(\mathbf{x}-\mathbf{x}^0\right)^T \mathbf{C}^{-1}\left(\mathbf{x}-\mathbf{x}^0\right)}, \tag{30}$$

where $\mathbf{C}$ is $n \times n$ covariance matrix which is symmetric positive definite. Other distributions like the unimodal are commonly approximated by normal pdf's [20].

3.3 Normed Distances

There exists a correspondence between the level contours of a given probability density function and a particular norm [41]. For example, the level contours of the normal pdf given by (30) can be described using the l_2-norm such that:

$$\|\mathbf{x}\| = \sqrt{\mathbf{x}^T \mathbf{C}^{-1} \mathbf{x}}, \tag{31}$$

where $\mathbf{C}$ is the covariance matrix and $\mathbf{x} \in \mathbb{R}^n$.

According to the assumption that the circuit parameters are normally distributed, by using the pdf norm given by (30), the distance between the two points $\mathbf{x}_1, \mathbf{x}_2 \in \mathbb{R}^n$ will be given by:

$$d\left(\mathbf{x}_1, \mathbf{x}_2\right) = \|\mathbf{x}_1 - \mathbf{x}_2\| = \sqrt{\left(\mathbf{x}_1 - \mathbf{x}_2\right)^T \mathbf{C}^{-1}\left(\mathbf{x}_1 - \mathbf{x}_2\right)}. \quad (32)$$

Let $\mathbf{x}^0 \in \mathbb{R}^n$ be a feasible point, i.e., $\mathbf{x}^0 \in F_s$ where F_s is the feasible region which is given by (27). Using the norm defined by the normal pdf, the distance between the point $\mathbf{x}^0$ and the feasible region boundary ($f_i(\mathbf{R}_s(\mathbf{x})) = 0$, $i = 1, 2, \ldots, m$) is given by [20, 25]:

$$\beta_i = min_{\mathbf{x}} \left\{ d\left(\mathbf{x}^0, \mathbf{x}\right) \middle| f_i\left(\mathbf{R}_s\left(\mathbf{x}\right)\right) = 0 \right\}. \quad (33)$$

If the point $\mathbf{x}^0$ violates the ith constraint, β_i could be defined as:

$$\beta_i = -min_x \left\{ d\left(\mathbf{x}^0, \mathbf{x}\right) \middle| f_i\left(\mathbf{R}_s\left(\mathbf{x}\right)\right) = 0 \right\}. \quad (34)$$

The normed distance β_i is denoted in [20] as a worst-case distance as it is the minimum distance from $\mathbf{x}^0$ to violate the constraint boundary $f_i(\mathbf{R}_s(\mathbf{x})) = 0$ in the norm defined by the pdf. The normed distance β_i between the point $\mathbf{x}^0$ and the feasible region boundary $f_i(\mathbf{R}_s(\mathbf{x})) = 0$ defines a normed body (ellipsoid) as shown in Fig. 4.

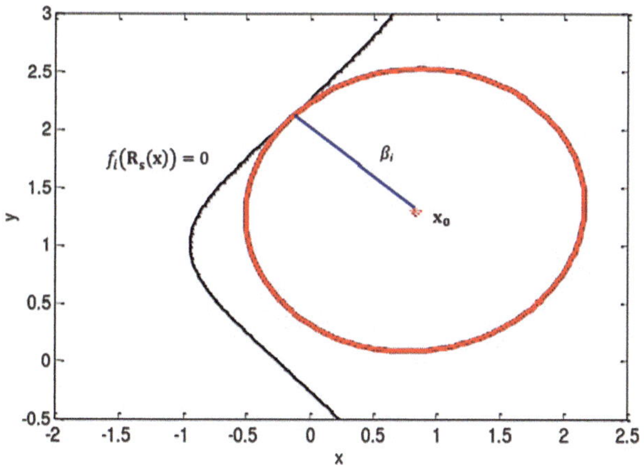

Fig. 4 Normed distance between the point $\mathbf{x}^0$ and the feasible region boundary

For simplicity, drop the suffix i, then the normed distance between $\mathbf{x}^0$ and the hypersurface $f(\mathbf{R}_s(\mathbf{x})) = 0$ will be given by:

$$\beta = min_{\mathbf{x}} \sqrt{(\mathbf{x} - \mathbf{x}^0)^T \mathbf{C}^{-1} (\mathbf{x} - \mathbf{x}^0)} \quad \text{such that } f(\mathbf{R}_s(\mathbf{x})) = 0. \tag{35}$$

Hence, finding the normed distance β requires solving the nonlinear optimization problem (35). The solution x^* of this problem [25] is given by:

$$\mathbf{x}^* = \mathbf{x}^0 + \beta \frac{\mathbf{C}\,\mathbf{g}}{\sqrt{\mathbf{g}^T\,\mathbf{C}\,\mathbf{g}}}, \tag{36}$$

where, $\mathbf{g} = \nabla f(\mathbf{R}_s(\mathbf{x}^*))$ and $\beta = \left| \frac{\mathbf{g}^T(\mathbf{x}^* - \mathbf{x}^0)}{\sqrt{\mathbf{g}^T\,\mathbf{C}\,\mathbf{g}}} \right|$.

It is to be noticed that the solution point $\mathbf{x}^*$ of (35) is a boundary point, i.e., $f(\mathbf{R}_s(\mathbf{x}^*)) = 0$. In [25] a convergent search technique to locate the solution point $\mathbf{x}^*$ is proposed as follows:

1. Locate a boundary point $\mathbf{x}_1$ on the constraint boundary $f(\mathbf{R}_s(\mathbf{x})) = 0$, by performing a line search in the $\mathbf{C}\,\mathbf{g}$ direction starting from $\mathbf{x}^0$. This search takes small steps in the $\mathbf{C}\,\mathbf{g}$ direction till it reaches the boundary point $\mathbf{x}_1$ ($f(\mathbf{R}_s(\mathbf{x}_1)) = 0$). The gradient of the constraint function is updated during the search. This process will permit the rotation of the search to locate a good boundary point on the constraint boundary $f(\mathbf{R}_s(\mathbf{x})) = 0$ [23, 42].
2. Starting from the boundary point $\mathbf{x}_1$, a boundary search technique on the boundary $f(\mathbf{R}_s(\mathbf{x})) = 0$ for the location of solution boundary point $\mathbf{x}^*$ is performed as follows:

   Starting from the boundary point $\mathbf{x}_1$, a point $\mathbf{x}_{c1} \in \mathbb{R}^n$ is obtained which is given by:

$$\mathbf{x}_{c1} = \mathbf{x}_1 + \gamma \frac{\mathbf{C}\,\mathbf{g}_1}{\sqrt{\mathbf{g}_1{}^T\,\mathbf{C}\,\mathbf{g}_1}}, \tag{37}$$

where,

$$\left.\begin{aligned} \mathbf{g}_1 &= \nabla f(\mathbf{R}_s(\mathbf{x}_1)) \\ \gamma &= \theta\beta_1,\ \theta \in (0,1) \\ \beta_1 &= \frac{\mathbf{g}_1^T(\mathbf{x}_1 - \mathbf{x}^0)}{\sqrt{\mathbf{g}_1{}^T\,\mathbf{C}\,\mathbf{g}_1}} \end{aligned}\right\}. \tag{38}$$

After obtaining $\mathbf{x}_{c1}$, a line search starting from $\mathbf{x}_{c1}$ along the direction $(\mathbf{x}^0 - \mathbf{x}_1)$ is performed to obtain a boundary point $\mathbf{x}_2 \in \mathbb{R}^n$ such that $f(\mathbf{R}_s(\mathbf{x}_2)) = 0$, where

$$\mathbf{x}_2 = \mathbf{x}_{c1} + \mu_1 \left(\mathbf{x}^0 - \mathbf{x}_1\right), \tag{39}$$

where μ_1 is the step of the line search starting from $\mathbf{x}_{c1}$ in the $(\mathbf{x}^0 - \mathbf{x}_1)$ direction.

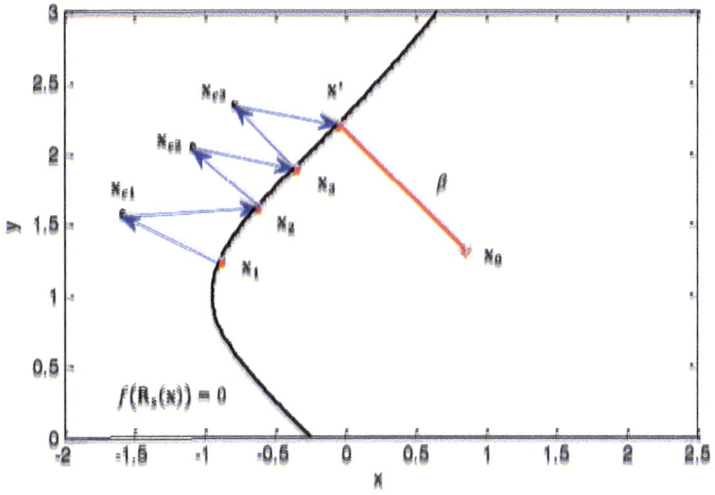

Fig. 5 Boundary search technique

This process is repeated till the convergence occurs and the solution boundary point $\mathbf{x}^*$ is located. See Fig. 5.

3.4 *Design Centering using Normed Distances*

The design centering problem can be formulated using normed distances as follows [20]:

$$\max_{\mathbf{x}^0} \left(\min_{i=1,2,\ldots,m} \beta_i \right), \tag{40}$$

where β_i , $i = 1, 2 \ldots, m$ is the normed distance from $\mathbf{x}^0$ to the constraint boundary $f_i(\mathbf{R}_s(\mathbf{x})) = 0$ in the norm defined by the normal probability distribution.

The above max–min problem can be transformed into a nonlinear programming problem by using an additional variable z, the problem will be:

$$\begin{aligned} &\max_{\mathbf{x}^0,\, z} z \\ \text{Such that} \quad &z - \beta_i \le 0,\ i = 1, 2 \ldots \ \ldots \ ., m. \end{aligned} \tag{41}$$

It is to be noticed the final center obtained from the algorithm with the minimum β_i defines the largest ellipsoid (with certain orientation given by the covariance matrix **C**) that can be inscribed in the feasible region.

3.5 Illustrative Examples

Consider the following two-dimensional nonlinearly convex feasible region given by the following constraints:

$$\begin{aligned} &\left((x_2-1)^2+1\right) * \exp(1-x_1) \le 7 \\ &\exp(x_1 - 2x_2 + 1) \le 7 \\ &x_1^2 + x_2^2 - 1 \le 7, \end{aligned} \tag{42}$$

Case1 By starting from an initial infeasible point $\mathbf{x}^0 = [5\ 5]^T$ and applying the algorithm with covariance matrix $\mathbf{C} = \begin{bmatrix} 0.25 & 0 \\ 0 & 0.25 \end{bmatrix}$ (independent parameters case) with initial yield = 0 % , the final point is reached at $[0.8455,\ 1.3410]^T$ with final yield = 97.2 % (see Fig. 6).

Case2 Starting from an initial infeasible point $\mathbf{x}^0 = [5\ 5]^T$ and applying the algorithm with covariance matrix $\mathbf{C} = \begin{bmatrix} 0.25 & 0.2 \\ 0.2 & 0.25 \end{bmatrix}$ (correlated parameters case) with initial yield = 0 % and final point = $[0.907,\ 0.967]^T$ with final yield = 96.6 %. (See Fig. 7).

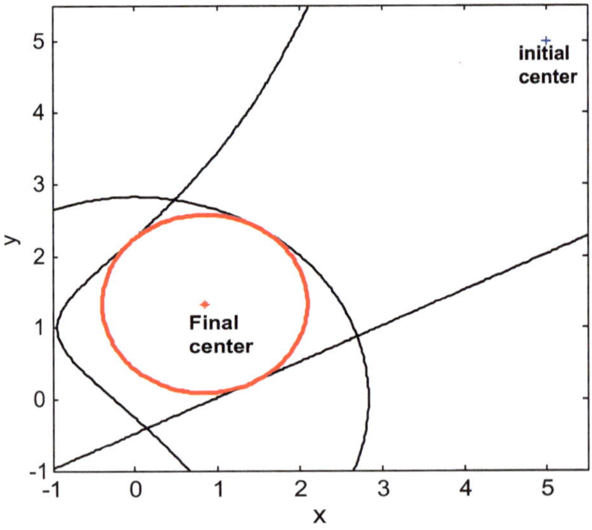

Fig. 6 The maximum volume ellipsoid inscribed in the feasible region of *case1*

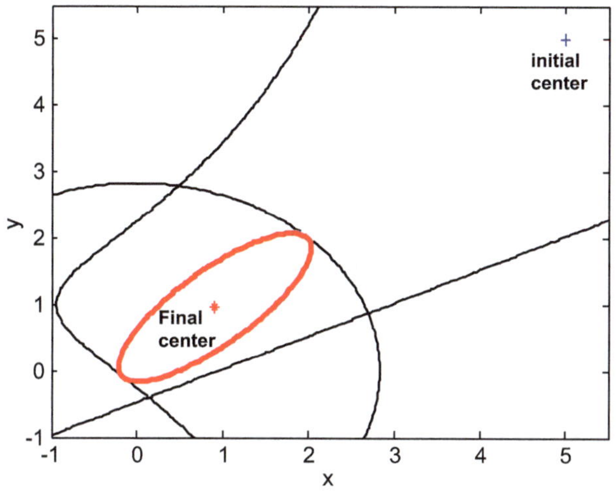

Fig. 7 The maximum volume ellipsoid inscribed in the feasible region of *case2*

3.6 Microwave Design Centering Algorithm

The proposed algorithm to solve design centering problem can be performed through application of two sub-algorithms. The first algorithm uses GSM to perform feasible region approximation by extracting the parameters needed to match the response of fine and surrogate models while the second algorithm performs a design centering process using normed distance to obtain a new design center. The new center point is validated by the fine model and is used to update the next GSM surrogate. The process is repeated to obtain the next center point.

The algorithm can be summarized by the following steps:

Step 1: Define the fine model $\mathbf{R}_f$, its associative coarse model $\mathbf{R}_c$, starting point $\mathbf{x}^0$, stopping criterion ϵ_1, and initialize $k = 0$

Step 2: Build the surrogate model $\mathbf{R}_s^k$ and the corresponding feasible region approximation F_s^k using the GSM technique at the current design center $\mathbf{x}^k$

Step 3: Apply the normed distance design centering approach on the feasible region approximation F_s^k to obtain a new center $\mathbf{x}^{k+1}$

Step 4: Increment k

Step 5: If $\frac{\|\mathbf{x}^k - \mathbf{x}^{k-1}\|}{\|\mathbf{x}^k\|} > \epsilon_1$ go to step 2,
else the solution is obtained and it is $\mathbf{x}^k$.

3.7 *Practical Example: Six-Section H-Plane Waveguide Filter*

The six-section H-plane filter [43] is a waveguide with a width 3.485 cm. The propagation mode is TE10 with a cutoff frequency of 4.3 GHz. The six-waveguide sections are separated by seven H-plane septa (as shown in Fig. 8a) which have a finite thickness of 0.6223 mm [44]. In this problem, there exist seven design parameters: three waveguide-section lengths L_1, L_2, and L_3 and four septa widths W_1, W_2, W_3, and W_4. The feasible region is constrained by the magnitude of the reflection coefficients at 44 frequencies {5.2, 5.3, ..., 9.5 GHz}. These magnitudes have to satisfy the upper and lower design specifications given by:

$$f_l\left(\mathbf{R}_f(\mathbf{x})\right) = \begin{cases} |S_{11}(\mathbf{x},\omega_l)| \geq 0.85 & \omega_l \leq 5.2\text{ GHz} \\ |S_{11}(\mathbf{x},\omega_l)| \leq 0.16 & 5.4\text{ GHz} \leq \omega_l \leq 9.0\text{ GHz} \\ |S_{11}(\mathbf{x},\omega_l)| \geq 0.5 & \omega_l \geq 9.5\text{ GHz} \end{cases}.$$

The coarse model [45] consists of lumped inductances and dispersive transmission line sections (as shown in Fig. 8b) while the fine model is performed using High Frequency Structure Simulator (HFSS).

The normed distance centering technique with GSM is applied using at most two SM iterations. The yield values are calculated by performing Monte Carlo method with 100 sample points. The starting point is [0.0161614, 0.0161899, 0.0166975, 0.0133376, 0.0120823, 0.0117456, 0.0115212] which is infeasible point. The results of independent parameter case with parameter spread $\sigma = 10^{-4}[0.7629, 0.7665, 0.7977, 0.6326, 0.5850, 0.5588, 0.5516]$ are shown in Table 3, while the results of correlated case are shown in Table 4 using covariance matrix $\mathbf{C}_1$ and $\mathbf{C}_2$ where

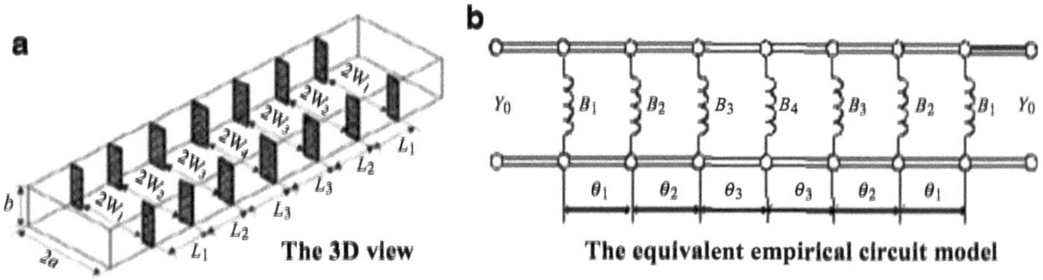

Fig. 8 The six-section H-plane waveguide filter

Table 3 Yield results for independent parameters case

Parameter spread	Initial yield (%)	Final point	Final yield (%)
σ	19	[0.016405823, 0.0161035450, 0.016587334, 0.013201395, 0.012225853, 0.011711792, 0.011557065]	76
$\sigma/2$	27	[0.016347863, 0.0161306, 0.016589344, 0.013204043, 0.012248974, 0.011707544, 0.011561786]	100

Table 4 Yield results for correlated parameters case

Parameter spread	Initial yield (%)	Final point	Final yield (%)
$\mathbf{C}_1$	23	[0.015961198 0.01617798 0.016556444 0.013416461 0.012223545 0.011697941 0.011578781]	100
$\mathbf{C}_2$	15	[0.016117571, 0.016208453, 0.016561121, 0.013355854, 0.012194497, 0.011702117, 0.011561046]	94

$$\mathbf{C}_1 = \frac{10^{-9}}{16}\begin{bmatrix} 352.483 & -128.704 & 27.484 & -8.318 & 4.011 & -1.382 & 6.665 \\ -128.704 & 188.716 & -86.261 & 7.295 & -18.624 & -1.568 & 9.168 \\ 27.484 & -86.261 & 65.081 & -13.953 & 9.197 & -3.139 & -9.842 \\ -8.318 & 7.295 & -13.953 & 56.679 & -2.301 & 8.995 & 3.324 \\ 4.011 & -18.624 & 9.197 & -2.301 & 18.288 & -8.795 & 10.197 \\ -1.382 & -1.568 & -3.139 & 8.995 & -8.795 & 18.971 & -14.011 \\ 6.665 & 9.168 & -9.842 & 3.324 & 10.197 & -14.011 & 23.016 \end{bmatrix}$$

$$\mathbf{C}_2 = \frac{10^{-9}}{49}\begin{bmatrix} 352.483 & -128.704 & 27.484 & -8.318 & 4.011 & -1.382 & 6.665 \\ -128.704 & 188.716 & -86.261 & 7.295 & -18.624 & -1.568 & 9.168 \\ 27.484 & -86.261 & 65.081 & -13.953 & 9.197 & -3.139 & -9.842 \\ -8.318 & 7.295 & -13.953 & 56.679 & -2.301 & 8.995 & 3.324 \\ 4.011 & -18.624 & 9.197 & -2.301 & 18.288 & -8.795 & 10.197 \\ -1.382 & -1.568 & -3.139 & 8.995 & -8.795 & 18.971 & -14.011 \\ 6.665 & 9.168 & -9.842 & 3.324 & 10.197 & -14.011 & 23.016 \end{bmatrix}$$

Acknowledgments Authors would like to thank Prof. Slawomir Koziel, School of Science and Engineering, Reykjavik University, for his invitation to contribute to this book. Authors also would like to acknowledge the contributions to the original work by Dr. Tamer Abuelfadl, Eng. Ahmed El-Qenawy, and Eng Ahmed Etman, Faculty of Engineering, Cairo University, which has been reviewed in this chapter.

References

1. Hassan, A.S.O., Abdel-Malek, H.L., Rabie, A.A.: Non-derivative design centering algorithm using trust region optimization and variance reduction. Eng. Optim. **38**, 37–51 (2006)
2. Hassan, A.S.O., Mohamed, A.S.A., El-Sharabasy, A.Y.: Statistical microwave circuit optimization via a non-derivative trust region approach and space mapping surrogates. Presented at IEEE MTT-S International Microwave Symposium Digest, Baltimore, pp. 1—4 (2011)
3. Graeb, H.: Analog Design Centering and Sizing. Springer, Berlin (2007)
4. Conn, A.R., Toint, P.L.: An algorithm using quadratic interpolation for unconstrained derivative free optimization. In: Di Pillo, G., Giannes, F. (eds.) Nonlinear Optimization and Applications, pp. 27–47. Plenum Publishing, New York (1996)
5. Powell, M.J.D.: UOBYQA: unconstrained optimization by quadratic approximation. Math. Program. **92**, 555–582 (2002)
6. Powell, M.J.D.: The NEWUOA software for unconstrained optimization without derivatives. In: Di Pillo, G., Roma, M. (eds.) Large-Scale Nonlinear Optimization, pp. 225–297. Springer, New York (2006)
7. Nocedal, J., Wright, S.J.: Numerical Optimization. Springer, New York (1999)
8. Singhal, K., Pinel, J.: Statistical design centering and tolerancing using parametric sampling. IEEE Trans. Circuits Syst. **28**, 692–702 (1981)
9. Hocevar, D.E., Lightner, M.R., Trick, T.N.: An extrapolated yield approximation for use in yield maximization. IEEE Trans. Comput. Aid. Des. **3**, 279–287 (1984)
10. Styblinski, M.A., Oplaski, L.J.: Algorithms and software tools for IC yield optimization based on fundamental fabrication parameters. IEEE Trans. Comput. Aid. Des. **5**, 79–89 (1986)
11. Yu, T., Kang, S.M., Hajj, I.N., Trick, T.N.: Statistical performance modeling and parametric yield estimation of MOS VLSI. IEEE Trans. Comput. Aid. Des. **6**, 1013–1022 (1987)
12. Elias, N.J.: Acceptance sampling: an efficient accurate method for estimating and optimizing parametric yield. IEEE J. Solid State Circuits **29**, 323–327 (1994)
13. Zaabab, A.H., Zhang, Q.I., Nakhla, M.: A neural network modeling approach to circuit optimization and statistical design. IEEE Trans. Microw. Theory Tech. **43**, 1349–1358 (1995)
14. Keramat, M., Kielbasa, R.: A study of stratified sampling in variance reduction techniques for parametric yield estimations. IEEE Trans. Circuits Syst. II Analog Digit. Signal Process. **45**(5), 575–583 (1998)
15. Bandler, J.W., Abdel-Malek, H.L.: Optimal centering, tolerancing and yield determination via updated approximation and cuts. IEEE Trans. Circuits Syst. **25**, 853–871 (1978)
16. Director, S.W., Hachtel, G.D., Vidigal, L.M.: Computationally efficient yield estimation procedures based on simplicial approximation. IEEE Trans. Circuits Syst. **25**, 121–130 (1978)
17. Abdel-Malek, H.L., Bandler, J.W.: Yield optimization for arbitrary statistical distributions, part I: theory. IEEE Trans. Circuits Syst. **CAS-27**, 245–253 (1980)
18. Abdel-Malek, H.L., Hassan, A.S.O.: The ellipsoidal technique for design centering and region approximation. IEEE Trans. Comput. Aid. Des. **10**, 1006–1014 (1991)
19. Wojciechowski, J.M., Vlach, J.: Ellipsoidal method for design centering and yield estimation. IEEE Trans. Comput. Aid. Des. Integr. Circuits Syst. **12**, 1570–1579 (1993)
20. Antreich, K.J., Graeb, H.E., Wieser, C.U.: Circuit analysis and optimization driven by worst-case distances. IEEE Trans. Comput. Aid. Des. **13**(1), 57–71 (1994)
21. Sapatnekar, S.S., Vaidya, P.M., Kang, S.: Convexity-based algorithms for design centering. IEEE Trans. Comput. Aid. Des. **13**(12), 1536–1549 (1994)
22. Abdel-Malek, H.L., Hassan, A.S.O., Bakr, M.H.: Statistical circuit design with the use of a modified ellipsoidal technique. Int. J. Microw. Millimet. Wave Comput. Aid. Eng. **7**, 117–128 (1997)
23. Abdel-Malek, H.L., Hassan, A.S.O., Bakr, M.H.: A boundary gradient search technique and its applications in design centering. IEEE Trans. Comput. Aid. Des. Integr. Circuits Syst. **18**, 1654–1661 (1999)

24. Hassan, A.S.O., Rabie, A.A.: Design centering using parallel-cuts ellipsoidal technique. J. Eng. Appl. Sci. **47**, 221–239 (2000)
25. Hassan, A.S.O.: Normed distances and their applications in optimal circuit design. J. Optim. Eng. **4**, 197–213 (2003)
26. Hassan, A.S.O., Abdel-Malek, H.L., Rabie, A.A.: Design centering and polyhedral region approximation via parallel-cuts ellipsoidal technique. Eng. Optim. **36**, 37–49 (2004)
27. Wojciechowski, J., Opalski, L., Zantyniski, K.: Design centering using an approximation to the constraint region. IEEE Trans. Circuits Syst. **51**(3), 598–607 (2004)
28. Abdel-Malek, H.L., Hassan, A.S.O., Soliman, E.A., Dakroury, S.A.: The ellipsoidal technique for design centering of microwave circuits exploiting space-mapping interpolating surrogates. IEEE Trans. Microw. Theory Tech. **54**(10), 3731–3738 (2006)
29. Hassan, A.S.O., Abdel-Naby, A.: Design centering and region approximation using semidefinite programming. International Journal of Research and Reviews in Applied Sciences, **6**(3), 366–375 (2011).
30. Dakroury, S.A.: Optimization of computationally expensive engineering systems exploiting space mapping surrogates, Ph.D. Thesis, Faculty of Engineering, Cairo University, (2008)
31. Hassan, A.S.O., Mohamed, A.S.A.: Surrogate-based circuit design centering. In: Koziel, S., Leifsson, L. (eds.) Surrogate-Based Modeling and Optimization, pp. 27–49. Springer, New York (2013)
32. Koziel, S., Bandler, J.W., Madsen, K.: Space-mapping based interpolation for engineering optimization. IEEE Trans. Microw. Theory Tech. **54**, 2410–2420 (2006)
33. Steihaug, T.: The conjugate gradient method and trust regions in large scale optimization. SIAM J. Numer. Anal. **20**, 626–637 (1983)
34. Powell, M.J.D.: A view of algorithms for optimization without derivatives. Technical Report NA 2007/03, Department of Applied Mathematics and Theoretical Physics, University of Cambridge, Cambridge (2007)
35. Bandler, J.W., Abdel-Malek, H.L.: Optimal centering, tolerancing and yield determination using multidimensional approximation. In: Proc. IEEE International Symposium on Circuits and Systems, Phoenix, pp. 219—222 (1977)
36. Yuan, Y.: On the truncated conjugate gradient method. Technical Report, ICM99-003, ICMSEC, Chinese Academy of Sciences, Beijing (1999)
37. Box, M.J.: A comparison of several current optimization methods, and the use of transformations in constrained problems. Comput. J. **9**(1), 67–77 (1966)
38. Nimje, V.T., Bhattacharjee, D., Dixit, K., Jayaprakash, D., Mittal, K.C., Ray A.K.: Design and development of 30 MeV 3 KW, RF electron linac. In: Asian Particle Accelerator Conference, APAC 2007, Indore, pp. 491—493 (2007)
39. Pichoff, N. *Introduction to RF linear accelerators.* In CAS - CERN Accelerator School: Intermediate Course on Accelerator Physics, Zeuthen, Germany, 15-26 Sep 2003, pp.105—128
40. Wangler, T.: Principles of RF Linear Accelerators. Wiley, New York (1998)
41. Brayton, R.K., Diretor, S.W., Hachtel, G.D.: Yield maximization and worst case design with arbitrary statistical distributions. IEEE Trans. Circuits Syst. **27**, 756–764 (1980)
42. Hassan, A.S.O., Abdel-Malek, H.L.: A geometric technique for design centering and a polytope approximation of the feasible region. J. Eng. Appl. Sci. **43**, 871–885 (1996)
43. Matthaei, G.L., Young, L., Jones, E.M.T.: Microwave Filters, Impedance-Matching Network and Coupling Structures, 1st edn. McGraw-Hill, New York (1964)
44. Bandler, J.W., Cheng, Q.S., Dakroury, S.A., Hailu, D.M., Madsen, K., Mohamed, A.S., Pedersen, F.: Space mapping interpolating surrogates for highly optimized EM-based design of microwave devices. In: IEEE MTT-S Int. Microw. Symp. Dig., Fort Worth, vol. 3, pp. 1565–1568 (2004)
45. Koziel, S., Bandler, J.W.: Space-mapping optimization with adaptive surrogate model. IEEE Trans. Microw. Theory Tech. **55**, 541–547 (2007)
46. Hassan, A.S.O., Abdel-Naby, A.: A new hybrid method for optimal circuit design using semidefinite programming. Eng. Optim. **44**(6), 725–740 (2012)

47. Seifi A, Ponnambalam K. and Vlach J., 1999. A unified approach to statistical design centering of integrated circuits with correlated parameters. IEEE Trans. Circuits And Systems, 46: 190–196
48. Serafini, D.B. A framework for managing models in nonlinear optimization of computationally expensive functions. Ph.D. Thesis, Department of Computational and Applied Mathematics, Rice University, Houston, Texas, USA, 1999
49. J. Søndergaard, *Optimization Using Surrogate Models–by the Space Mapping Technique*, Ph.D. Thesis, Informatics and Mathematical Modelling (IMM), Technical University of Denmark (DTU), Lyngby, Denmark, 2003

Atomistic Surrogate-Based Optimization for Simulation-Driven Design of Computationally Expensive Microwave Circuits with Compact Footprints

Piotr Kurgan and Adrian Bekasiewicz

Abstract A robust simulation-driven design methodology for computationally expensive microwave circuits with compact footprints has been presented. The general method introduced in this chapter is suitable for a wide class of N-port unconventional microwave circuits constructed as a deviation from classic design solutions. Conventional electromagnetic (EM) simulation-driven design routines are generally prohibitive when applied to numerically demanding microwave circuits with highly miniaturized and complex topologies. The key idea of the approach proposed here lies in an iterative redesign of a conventional circuit by a sequential modification and optimization of its atomic building blocks. The speed and accuracy of the presented method has been acquired by solving a number of simple optimization problems through surrogate-based optimization (SBO) techniques. Two exemplary designs have been supplied to verify the proposed method. An abbreviated wideband quarter-wave impedance matching transformer (MT) and a miniaturized hybrid branch-line coupler (BLC) have been developed. Diminished dimensions of the constructed circuits have been achieved by means of compact microstrip resonant cells (CMRCs). In the given examples, an implicit space mapping (ISM) technique has been utilized as a SBO engine. In general, the proposed method is compatible with other SBO routines as well. The final results have been acquired in only a fraction of time that is necessary for a direct EM optimization to generate competitive results. Numerical results have been validated experimentally.

Keywords Computationally expensive design • Computer-aided design (CAD) • Microwave engineering • Simulation-driven optimization • Electromagnetic (EM) simulation • Surrogate-based optimization (SBO) • Implicit space mapping (ISM) • Surrogate model • High-fidelity model • Coarse model

---

P. Kurgan (✉) • A. Bekasiewicz
Faculty of Electronics, Telecommunications and Informatics, Gdansk University of Technology, Narutowicza 11/12, 80-233 Gdansk, Poland
e-mail: piotr.kurgan@eti.pg.gda.pl

S. Koziel et al. (eds.), *Solving Computationally Expensive Engineering Problems*, Springer Proceedings in Mathematics & Statistics 97,
DOI 10.1007/978-3-319-08985-0_8

1 Introduction

Reliable design of compact microwave circuits is the focus of ongoing research in the field of microwave and antenna engineering. It has recently gained an increased attention with the rapid development and expansion of industrial, commercial, and military markets aimed at low-cost, small-size, and multi-functional microwave devices [1–4]. The diversity of their applications includes, but is not limited to mobile phones, wireless personal digital assistants, portable vector network analyzers, smart meters, defense electronics equipment, wireless routers, repeaters, and many others [5–10]. The fundamental issue here is the excessive physical dimensions of traditional microwave components that are comparable to the guided wavelength and require a certain (and fixed) relation with respect to it in order to maintain a proper operation of the entire component. This becomes particularly troublesome for the ultra-high frequency (UHF) band, where the wavelength varies from 1 m down to 10 cm, and conventional microwave passives such as baluns, matching transformers, phase shifters, filters, power dividers, couplers, etc. [11] exceed practical limitations of miniaturization-oriented designs. For this reason, the development of reliable design methodologies for compact microwave circuits is of utmost importance for the progress in applied microwave technologies.

The overly large substrate area consumption of conventional microwave circuits is due to their modular architecture primarily based on distributed-element uniform transmission lines (UTLs), whose electrical lengths and characteristic impedances are strictly defined. One of the representative examples is a rat-race coupler that occupies 89×89 mm estate area, when designed for 1-GHz operating frequency on Taconic RF-35 substrate having the thickness of 0.508 mm [12]. The problem becomes even more pronounced when microwave components are designed to meet the requirements of wideband or ultra-wideband (UWB) applications. In such cases, the most common approach is to use a multi-section topology, which inevitably results in much larger substrate area consumption [13].

The aforementioned issues can be conveniently mitigated by exploiting a slow-wave phenomenon, which increases the electrical to physical length ratio without altering the operation of a given component [14]. To date, four main size reduction methods have been suggested in the scientific literature to capitalize on this concept: (1) application of lumped or lumped-distributed elements instead of conventional TLs [15–17], (2) utilization of high-permittivity dielectric substrates [18–20], (3) exploitation of artificially engineered substrates composed of periodic inclusions or cavities [21–23], and (4) spatially separated storage of electric and magnetic energy, realized by using short (usually smaller than half-quarter wavelength) UTL sections of low and high characteristic impedances, respectively [24–28]. The first technique offers substantial miniaturization capabilities (even threefold length diminution compared to conventional TLs [29, 30]), but also suffers from common unavailability of suitable lumped elements with high-quality factors, narrowband operation, and a hindered assembly of lumped capacitors in the microstrip technology requiring the use of via-holes [30, 31]. The second one enables the achievement of a linear reduction in physical dimensions of the circuit, proportional to the square root of the relative dielectric coefficient [32], but at the same time it poses serious obstacles, such as high material cost, difficulty in the realization of high characteristic impedance UTLs, and high sensitivity to small variations in physical dimensions [18]. The third approach creates a viable chance of obtaining a compact microwave component, which can be attributed to unusual EM properties of an artificially engineered transmission medium [33]. These are achieved by placing structural perturbations with a half-wavelength period, which makes the accommodation of their physical size a challenging task itself [34]. Neither of the above methods is suitable for cost-efficient miniaturization, mostly due to the increased complexity of the fabrication process or the cost of materials. On the other hand, the fourth technique is devoid of the aforementioned limitations and, most importantly, it is fully compatible with a standard printed circuit board fabrication process. In this approach, traditional microwave components are redesigned so that their atomic building blocks, i.e., conventional TLs, are systematically replaced with slow-wave structures constructed, in general, from high and low impedance TL segments that are usually tightly assembled to achieve a high scale of compactness. Although it is nowadays the most frequently used technique dedicated to low-cost miniaturization of microwave circuits [35–40], there are still serious methodological issues that remain to be solved.

A reliable design of compact microwave circuits requires an accurate analysis of the entire structure under development. This can be done by means of high-fidelity full-wave EM simulations, however the task proves to be difficult, because miniaturized microwave components, due to the high complexity of their layouts, are numerically demanding and their evaluation is not only extremely time-consuming, but also involves the use of massive CPU resources. For example, a single-frequency simulation of a miniaturized hybrid ring coupler of [41] takes $\sim$75 min and requires $\sim$3 GB of RAM memory. In case of designs that comprise a larger number of components, e.g., a planar Butler matrix composed of hybrid couplers, phase shifters, crossovers, and UTLs, the simulation may be incomparably more expensive or even unattainable on regular PC machines. Moreover, highly complex layouts of typical compact structures are parameterized by many variables, which have to be simultaneously adjusted to meet given specifications, both geometry- and performance-wise. As opposed to conventional microwave circuits with only two designable parameters per unique element, the operation of their miniaturized counterparts is often counter-intuitive in terms of parameter setups and requires not only some sort of preliminary studies (e.g., based on principal component analysis [42] or lumped-element equivalent circuit [40]), but also an excessively large and multi-dimensional designable space to make sure that the target solution (or optimum design) can be found within the prescribed lower and upper bounds. For these reasons, standard simulation-driven design methodologies that perform an accurate EM analysis hundreds or thousands of times in the course of a single design routine, e.g., repetitive parameter sweeps or gradient-based optimization, are normally prohibitive. Figure 1 illustrates exemplary miniaturized microwave structures, indicating high complexity of contemporary microwave engineering design problems.

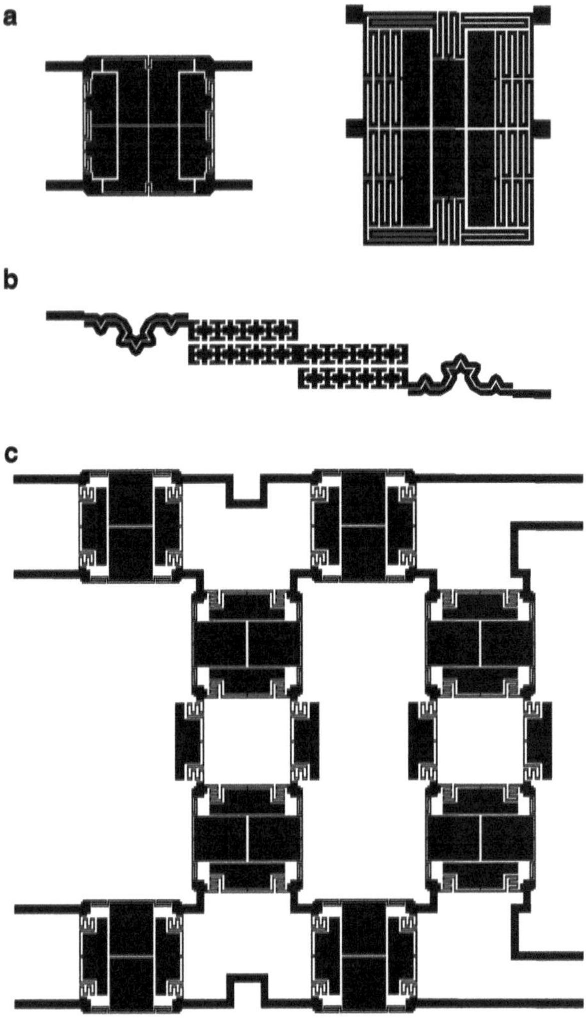

Fig. 1 Examples of miniaturized microwave structures: (**a**) branch-line coupler (*left*) and rat-race coupler (*right*) [43, 44]; (**b**) band-pass filter [45]; (**c**) 4 × 4 Butler matrix [46]

Although the straightforward use of high-fidelity, but CPU-intensive EM simulations is necessary for the reliability of the results, in most design methods available in the literature, e.g., [44, 47–57], it is exploited only in the design closure and preceded with theoretical analysis that rests on the simplified TL approach. For these instances, basic T- [3, 44, 47–52, 58] or π-shaped [3, 28, 37, 51, 53, 54, 59, 60] topologies are most commonly chosen for the slow-wave physical realization. Also, their various modifications are possible, e.g., where UTLs are substituted by stepped-impedance sections or quasi-periodic structures of alternating impedances [27, 48, 49, 57, 61, 62]. These rather plain networks can be easily analyzed by using simplified analytical formulas of the TL theory and ABCD matrix calculus [11], which offer a relatively good approximation to the solution of the corresponding EM problem, assuming the lack of cross-coupling effects and a negligible influence of TL discontinuities on the performance of the entire microwave component. This, however, is acceptable only for conventional circuits [63]. When dealing with highly miniaturized passives, characterized by complex and densely arranged layouts, the exploitation of simplified theoretical models is useful only to provide initial design solutions that require further EM fine-tuning [44, 47–57]. On the other hand, for more sophisticated slow-wave structures that are poorly describable by theoretical models, it is preferable to apply a high-fidelity EM analysis from the early stages of the design to yield reliable results [25, 26]. In other words, EM simulation-driven design optimization of the entire miniaturized structure is a design step that cannot be avoided, but which is too expensive to be commonly used in engineering practice, particularly when supplemented by conventional numerical optimization routines.

Design difficulties related to high computational cost of accurate EM simulation can be alleviated to some extent by using surrogate-based optimization (SBO) techniques [64–67]. The most popular SBO approach in microwave engineering is undoubtedly space mapping (SM) [68, 69]. Unfortunately, straightforward utilization of an algorithm such as SM is problematic in case of miniaturized structures for several reasons. Conventional SBO methods exploit the low-fidelity model (e.g., equivalent circuit) of the entire structure [68], which is of limited accuracy because it does not account for EM couplings between tightly allocated atomic building blocks of the structure. On the other hand, large number of design variables makes the extraction of the surrogate model parameters (as well as subsequent surrogate model optimization) numerically complex with issues such as non-uniqueness of the extraction process and poor generalization capability of the surrogate [66]. Also, EM simulation of the entire structure has to be performed from the very first iteration of the algorithm, which greatly affects the overall cost of the design process.

In summary, there is an urgent need for the development of computationally efficient, yet reliable methods for EM-simulation-driven design of compact microwave structures. Availability of such techniques would be of great importance for simplifying and shortening the design cycles for compact structures and, consequently, lowering their manufacturing costs in numerous application areas as elaborated at the beginning of this section.

In this chapter, a new approach to the design of computationally expensive microwave circuits with compact footprints has been presented and showcased. The novelty of the proposed method lies in the reduction of the overall design cost by replacing a single complex optimization problem by a number of simple optimization problems that are solved sequentially to reach a satisfactory approximation of the optimal solution. Furthermore, the method presented here exploits a SBO concept to capitalize on its high speed and accuracy. In the initial steps of the demonstrated design scheme, a logical decomposition of a conventional circuit into its atomic building blocks is performed. Next, each elementary constitutive element of the circuit under development is rebuilt in a sequential manner and undergoes a SBO. For illustration purposes, the proposed method has been used to design two exemplary microwave circuits with compact, yet complex layouts at a low computational cost, realized in microstrip technology. The substantial accuracy of the circuit optimization with only a handful of EM simulations has been achieved by means of an implicit space mapping (ISM) technique. The diminished dimensions of the circuits have been acquired by using slow-wave compact microstrip resonant cells (CMRCs) as a substitution for initial fundamental building blocks. However, it is noteworthy that the proposed method is also compatible with other SBO techniques (e.g., explicit SM, aggressive SM, neural SM, fuzzy SM, tuning SM [70–74], etc.) as well as with alternative means of circuit refinement (e.g., composite right/left-handed TLs, defected ground structures [75, 76], etc.), if only adequate computationally cheap surrogate models are available.

The chapter is organized as follows. Section 2 provides a general description of the proposed method—an overall design flow, surrogate model update, and SBO. Section 3 demonstrates the operation of the proposed design technique. A comprehensive comparison with benchmark optimization methods (conventional optimization schemes as well as traditional ISM) is also included. Section 4 provides experimental results, whereas Sect. 5 concludes the chapter.

2 Sequential Space Mapping: Methodology

Miniaturized microwave circuits, due to their novelty and considerable geometrical complexity, lack accurate analytical models of good generalization capabilities. For this reason, conventional design methods, i.e., theory-based and EM-simulation-driven, are typically prohibitive—the former one, due to its considerable inaccuracy, and the latter one, due to the excessive computational cost associated with high-fidelity, but extremely CPU-intensive and time-consuming EM simulations. Typically, computationally expensive design problems can be effectively solved by means of SBO methodologies. However, the aforementioned obstacle also limits the usefulness of traditional SBO techniques—in particular various types of SM—by causing convergence problems of the process. In order to address this inconvenience, the following design flow has been proposed.

2.1 General Design Flow

The new approach introduced in this chapter and illustrated in Fig. 2 offers a robust simulation-driven design methodology enabling the achievement of an accurate design solution in relatively short time. The initial step of the diagram of Fig. 2 is a definition of design specifications, e.g., substrate parameters, frequency-dependent performance of the circuit (e.g., desired S_{11} and S_{21} over a given frequency band), physical dimension requirements, etc. Subsequently, a conventional circuit is constructed from UTL segments and various discontinuities. This can be done without major obstacles as conventional microwave circuits are supplied with good theoretical models ready to be used instantly after a few iterations of fine-tuning [78].

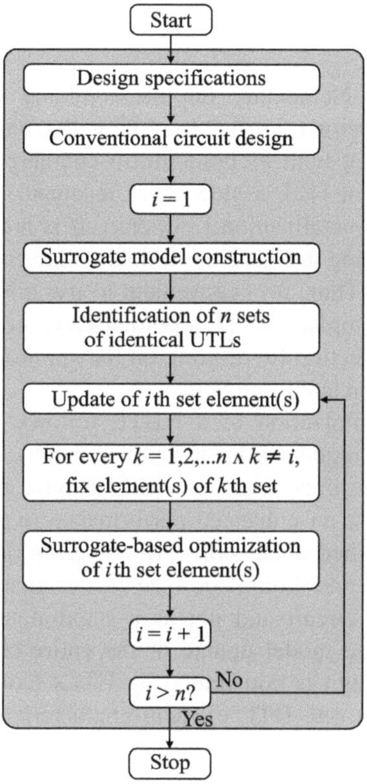

Fig. 2 Design diagram of microwave circuits with compact footprints [80]

Next, a surrogate model of a conventional circuit is constructed (or re-used from the previous step). Afterwards, atomic building blocks of a conventional circuit are identified. This is done by performing a logical decomposition of the initial circuit, i.e., dividing it into n sets of identical UTLs and a $(n + 1)$th set of all required discontinuities. For example, a logical decomposition of a rat-race coupler would result in three sets ($n = 2$)—the first one containing three UTLs of 90° electrical length and Z characteristic impedance, the second one containing a single UTL of 270° electrical length and Z characteristic impedance, and the third one containing four T-junctions. The following steps of the design flow are performed iteratively (from $i = 1$ to n). In each iteration, the circuit under consideration is updated, i.e., the ith set of identical UTL surrogate models is replaced with a set of identical nonuniform transmission line (NUTL) surrogate models, after which a new circuit is composed from fixed and updated building blocks and undergoes a SBO aimed at the satisfaction of the desired design specifications. In each iteration of the main algorithm, only designable parameters corresponding to NUTL surrogate models of the ith set become optimization variables used during a SBO, while all other design parameters remain fixed. The main algorithm ends after a successful optimization of n updated circuits.

2.2 Surrogate Model Update

A UTL segment is an elementary (atomic) building block of a conventional microwave circuit. In order to construct a circuit with compact footprint, one should substitute a UTL segment with its nonuniform counterpart (e.g., a discontinuous transmission line segment [12], a slow-wave resonant structure [40], a TL with perturbed ground plane metallization [35], etc.). It is noteworthy that, in general, the process of constructing a NUTL is difficult, time-consuming, and guided by engineering experience. Thus, it is convenient to use a NUTL library of [79] that contains a number of exemplary NUTL topologies and provides theoretical tools for their comparison and selection (depending on the application), as well as practical guidelines for their design and improvement.

Subsequently, an optimization of a NUTL follows. In practice, a NUTL is optimized to match the frequency-dependent parameters of a UTL (e.g., scattering parameters, characteristic impedance, electrical length, etc.) in a given frequency range and to demonstrate an enhanced performance (e.g., out-of-band characteristics) as well as diminished dimensions. Moreover, in order to omit the final EM fine-tuning of the circuit, one should optimize the designable parameters of a NUTL as a part of the whole circuit and not as a stand-alone component. A general illustration of a surrogate model update of the entire circuit has been presented in Fig. 3. An initial design is constituted by UTLs (collected in n sets, each of them containing at least one UTL element characterized by a certain electrical length and a characteristic impedance) and various discontinuities (gathered in the $(n + 1)$th set). In the first iteration, the first set containing UTL_1 elements is replaced by a set containing $NUTL_1$ elements, while all the other sets remain unchanged. In the ith iteration, the ith set is updated in a similar fashion and the circuit is

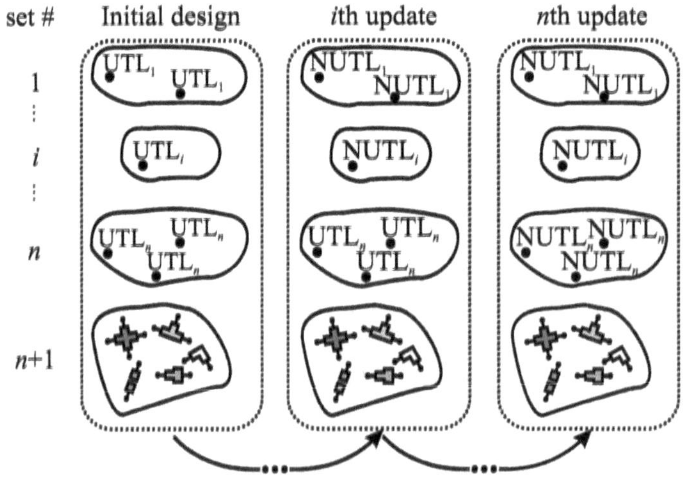

Fig. 3 General scheme for surrogate model update of the entire microwave component [80]

composed of NUTL_1, NUTL_2, ..., NUTL_i, UTL_{i+1}, UTL_{i+2}, ..., UTL_n elements and discontinuities collected in the $(n+1)$th set. After the nth update, no UTL segments remain and the whole circuit can be considered as completely refined.

2.3 Surrogate-Based Optimization

In order to design an unconventional microwave circuit with a complex topology accordingly to the prescribed specifications, a nonlinear minimization problem of the following form should be solved:

$$\mathbf{x}^* = \arg\min_{\mathbf{x}} U(\mathbf{R}_f(\mathbf{x})) \quad (1)$$

where U denotes an objective function formulated on the basis of design specifications, $\boldsymbol{R}_f$ stands for a high-fidelity model evaluation (a fine model response), whereas $\boldsymbol{x}$ represents a vector of designable variables. The optimal design solution vector is denoted by $\boldsymbol{x}^*$. The optimization problem from (1), when solved directly, is usually extremely CPU-intensive and time consuming and can be found grossly impractical in the case of numerically complex structures. The SBO approach addresses this issue by using a computationally cheap surrogate model evaluation $\boldsymbol{R}_s$ and an iterative formulation that follows [78, 80]:

$$\mathbf{x}^{(j+1)} = \arg\min_{\mathbf{x}} U(\mathbf{R}_s^{(j)}(\mathbf{x})) \quad (2)$$

where $\boldsymbol{x}^{(j+1)}$ represents the optimized solution of the jth iteration surrogate model $\boldsymbol{R}_s^{(j)}$, which is assumed to represent the fine model $\boldsymbol{R}_f$ in a relatively accurate manner [78]. Within these theoretical constraints, the algorithm formulated in (2) is aimed at approaching a quasi-optimal solution located in the vicinity of the global optimum $\boldsymbol{x}^*$. A SBO flowchart illustrating a general implementation of the theory described above is presented in Fig. 4. For detailed description of the SBO concept see, e.g., [66].

3 Design Examples

The general method described in Sect. 2 and schematically presented in Fig. 2 has been applied to design two exemplary microwave circuits with compact footprints. A design flow of an exemplary 2-port device, i.e., an abbreviated impedance matching transformer has been described in Sect. 3.1. A step-by-step design procedure of an exemplary 4-port device, i.e., miniaturized branch-line coupler (BLC), has been discussed in Sect. 3. In both design examples, ISM [81] has been used as a SBO engine due to the simplicity of its implementation [82]. A detailed flowchart of the ISM algorithm exploited in this work is depicted in Fig. 5. The EM fine model evaluation has been performed by ADS Momentum EM solver [83].

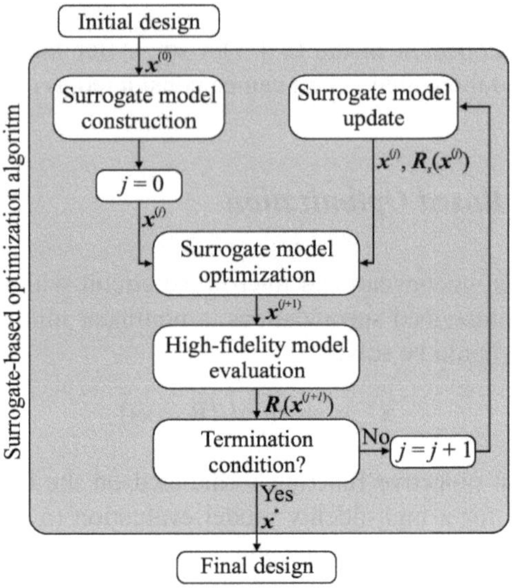

Fig. 4 Surrogate-based optimization flowchart [80]. The computational burden of a conventional design process is shifted from the iterative evaluation of a numerically demanding EM model to a computationally cheap surrogate model exploited in an iterative prediction–correction loop. High-fidelity model is evaluated only once per iteration for verification purposes [68]

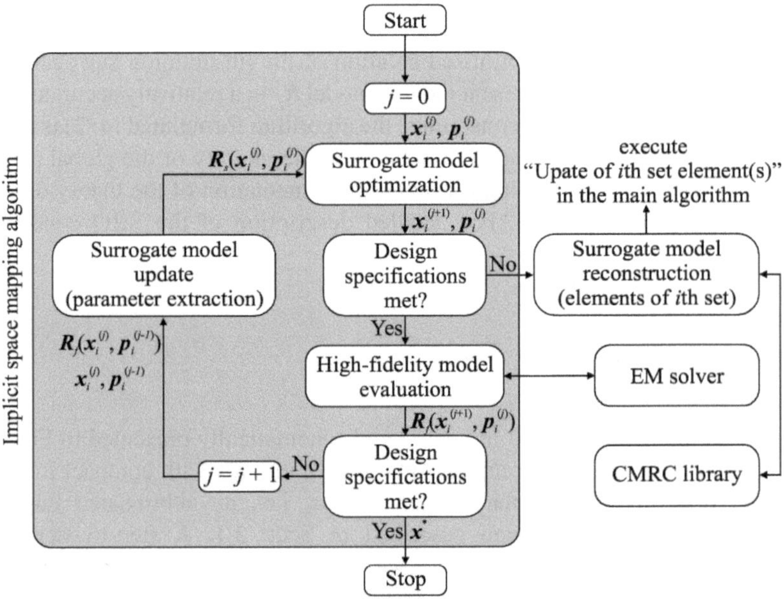

Fig. 5 Block diagram of the ISM algorithm used in this work [80]. The algorithm has been specifically tailored for the design process of compact microwave circuits with complex topologies. CMRC library can be built based on Ref. [65]

It has been assumed that the circuit improvement in terms of its diminution (example 1) and its miniaturization (example 2) is to be obtained by means of metallization perforations, i.e., intentional defects implemented in the signal line metallization plane (termed here CMRCs [77, 79]). Surrogate models of CMRCs used in this work are supplied with a geometric and preassigned parameter description of the following general form: $\boldsymbol{x}_i^{(j)} = [L_1\ H_1\ L_2\ H_2\ \ldots]^T$ and $\boldsymbol{p}_i^{(j)} = [h_1\ \varepsilon_1\ h_2\ \varepsilon_2\ \ldots]^T$, respectively ($i$ being the iteration index of the main algorithm and j denotes the iteration index of the ISM algorithm). The former parameters undergo the ISM optimization, while the latter auxiliary parameters (substrate height and relative permittivity) are used in the extraction process (see Fig. 5). Interested reader can find detailed information on ISM and extraction of preassigned parameters in the literature (e.g., [66] or [81]).

In general, the ISM algorithm should successfully finish after the jth iteration, when the EM circuit surrogate model evaluation satisfies the initially defined design specifications. In such a case, the iteration counter i of the main algorithm is incremented and a novel refined circuit with the ith UTL replaced by a NUTL is developed in a similar fashion. Conversely, when no convergence of the ISM algorithm can be found or the design specification is not met after the circuit optimization, surrogate model(s) under optimization should be replaced or improved (e.g., following the procedure of [79]) and the ith iteration of the main algorithm should be repeated with a new NUTL surrogate model.

3.1 Abbreviated Matching Transformer

The design specifications have been defined as follows: (a) Rogers RO3210 substrate ($\varepsilon_r = 10.2$, $h = 0.635$ mm, $\tan\delta = 0.0027$); (b) circuit functionality ($|S_{11}| \leq -20$ dB over 1–2.5 GHz frequency band), $Z_{\text{source}} = 50\,\Omega$, $Z_{\text{load}} = 6\,\Omega$; (c) physical requirements (minimal length reduction equals 30 %). Following the above-stated design specifications, a four-section conventional microstrip MT has been designed and fine-tuned using ADS software [83]. Subsequently, a simple MT surrogate model composed of four UTL components and several elements representing microstrip step discontinuities has been constructed in ADS circuit simulator. Next, four ($n = 4$) single-element sets of identical quarter-wavelength UTLs have been identified (46.8-Ω UTL_1, 33.6-Ω UTL_2, 17.3-Ω UTL_3, and 8.9-Ω UTL_4). Afterwards, as presented in detail in Fig. 6, an iterative part of the main algorithm follows. In the first three iterations of the main algorithm, successive UTL surrogate models have been successfully substituted with a T-shaped CMRC surrogate model (termed here CMRC_A). However, in the fourth iteration of the main algorithm, no convergence of the ISM algorithm has been achieved due to a very low impedance of the UTL_4 section. Therefore, a different CMRC model (named here CMRC_B) has been introduced for the circuit to meet the design specifications. Both CMRC models have been presented in Fig. 7. The initial design variable vector used in each iteration is $x_i^{(0)} = [1\ 1\ 1\ 1\ \ldots]^T$ mm, whereas

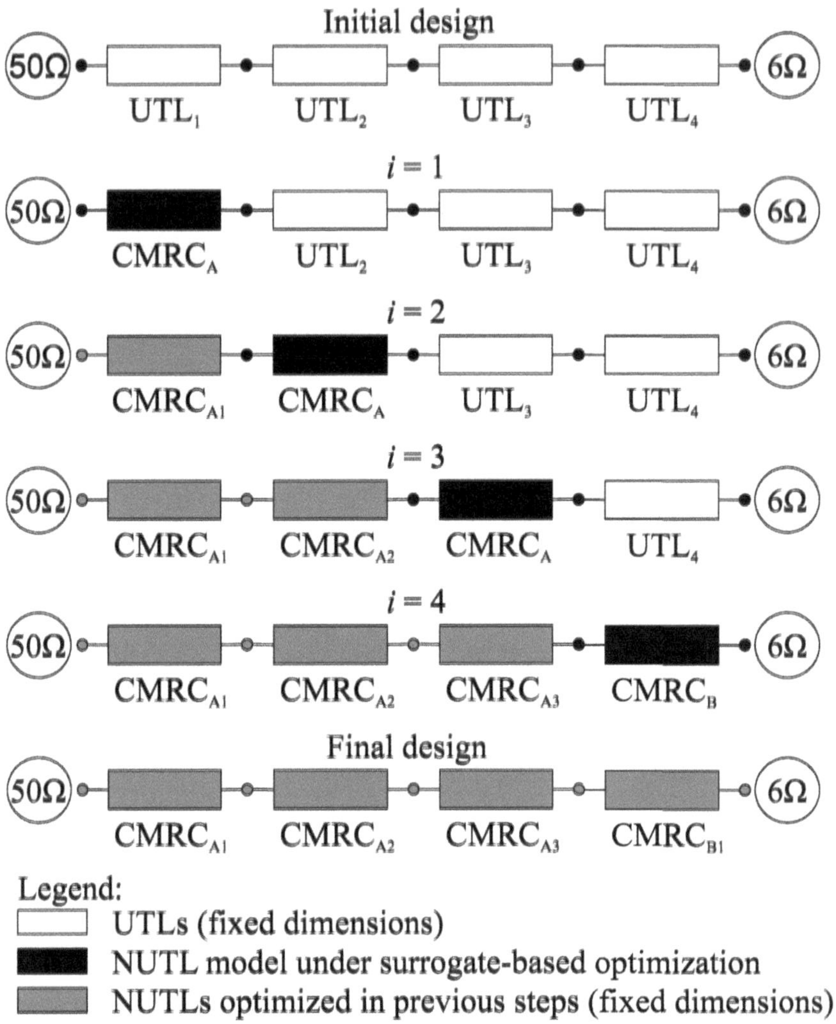

Fig. 6 A schematic description of the MT under iterative construction [80]

the initial preassigned parameter vector is $p_i^{(0)} = [0.635\ 10.2\ 0.635\ 10.2\ \ldots]^T$, in which odd elements are in mm and even elements are unitless. After four iterations of the main algorithm, a completely refined MT design has been obtained, revealing a satisfactory performance and a considerable 34 % length reduction. Particular CMRC design solutions are labeled as $CMRC_{A1}$, $CMRC_{A2}$, $CMRC_{A3}$, and $CMRC_{B1}$. A convergence plot for the ISM algorithm executed in each iteration of the main algorithm has been illustrated in Fig. 8. Error functions from Fig. 8 have been calculated at frequencies from 1 to 2.5 GHz with a 0.4 GHz step. The ISM algorithm has been set to terminate when the error function is less than 10^{-1}. Final

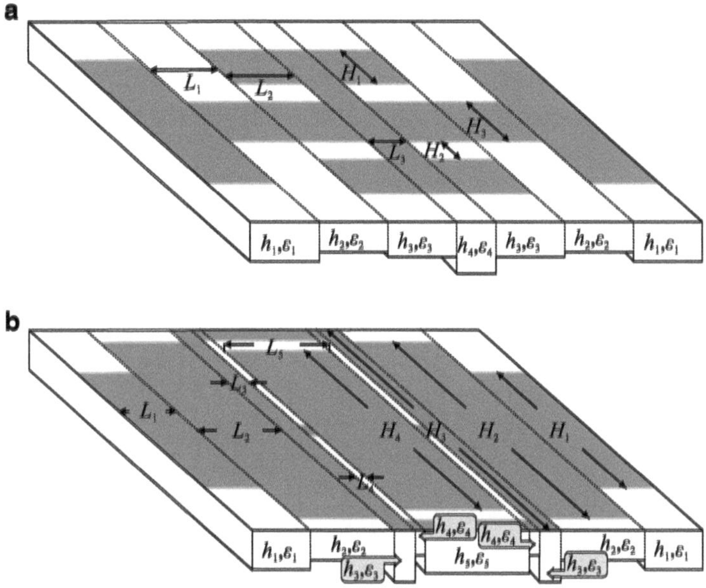

Fig. 7 Layout representations of surrogate models: (**a**) $CMRC_A$. (**b**) $CMRC_B$ [80]. Models include geometrical parameterization as well as preassigned parameter description

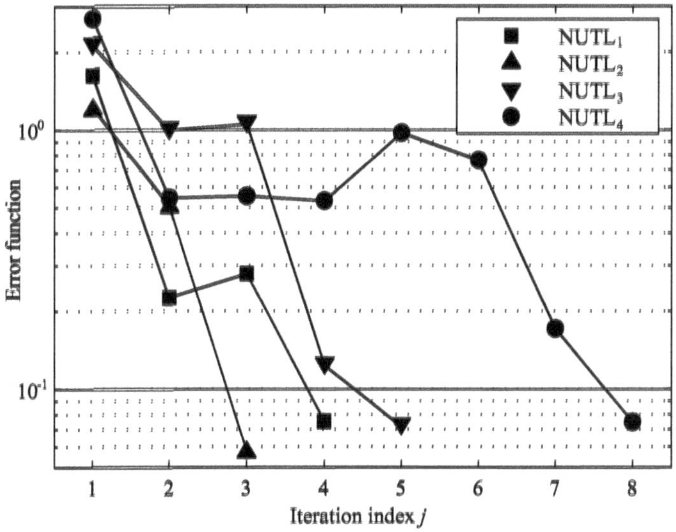

Fig. 8 Convergence plot for the ISM algorithm executed for MT design example. Iteration index corresponds to the number of high-fidelity model evaluations [80]

Table 1 Final designable geometric parameters

		$CMRC_A$			$CMRC_B$
Final design dimensions		$NUTL_1$	$NUTL_2$	$NUTL_3$	$NUTL_4$
x^* [mm]	L_1	0.4	0.8	0.4	1.05
	H_1	0.35	1.05	2.45	7.5
	L_2	2.4	1.4	1.4	1.6
	H_2	0.25	0.4	0.2	11.4
	L_3	0.25	0.4	0.2	0.4
	H_3	0.25	0.4	0.2	13
	L_4	–	–	–	0.2
	H_4	–	–	–	11.8

Table 2 Abbreviated MT: total design cost

	Optimization method			
	Holistic		Atomistic	
Total design cost	Direct	SBO	Direct	SBO
Convergence	Yes	No	Yes	Yes
Total number of high-fidelity model evaluations	2,120	N/A	746	20

geometric parameters of the abbreviated MT are listed in Table 1. The total design cost of the proposed method is presented in Table 2. Additionally, the iterative method based upon multiple SBOs applied to atomic CMRC building blocks (this work) has been compared against several other approaches, i.e., an iterative method based on multiple atomistic EM optimizations, a method based on a holistic SBO (traditional SBO where the entire design optimization problem is solved within a single optimization routine), and a method based on a holistic EM optimization (direct EM optimization of the entire circuit). It should be concluded upon data collected in Table 2 that a multiple atomistic optimization approach requires less high-fidelity model evaluations than a direct holistic optimization. Moreover, a combination of a multiple atomistic optimization approach and a SBO technique results in a method that outclasses other competitive design methodologies included in this comparison. The atomistic SBO method introduced in this work presents a design cost of 20 EM simulations for the first example, which proves its considerable computational efficiency in comparison to a classic holistic optimization method offering 2,120 EM simulations for the same example. Average CPU time of a single 64-point EM model evaluation is approximately 51 s for the atomistic SBO approach and 61 s for the holistic EM optimization approach (both methods used i7 2,600 k 8 GB RAM PC). Respective CPU times differ as the complexity of the circuit iteratively increases in case of the atomistic SBO approach, reaching the greatest complexity in the last iteration, whereas the holistic EM optimization approach utilizes the most complex model in every iteration. For these reasons, the total time of EM model evaluations is 17 min for the atomistic SBO method and 36 h for the holistic EM optimization approach.

3.2 Miniaturized BLC

The design specifications have been formulated as follows: (a) FR4 substrate ($\varepsilon_r = 4.4$, $h = 0.508$ mm, $\tan\delta = 0.02$); (b) circuit performance ($|S_{11}|$ and $|S_{41}| \leq -20$ dB over a 10 % bandwidth with a 2.2 GHz operating frequency, $|S_{11}| \leq -40$ dB at 2.2 frequency, and $|S_{21}| = |S_{31}|$ at 2.2 frequency); (c) physical requirements (minimal circuit size reduction equals 30 %). Using ADS simulation environment [83], a conventional BLC has been designed and fine-tuned to meet the above defined design specifications. Next, a simple BLC surrogate model comprising two pairs of UTL components and four T-junctions has been built in ADS circuit simulator. Subsequently, two ($n = 2$) double-element sets of identical UTLs have been identified (35.35-Ω $UTLs_1$ and 50-Ω $UTLs_2$). Then, two iterations of the main algorithm have been performed in order to construct a completely refined BLC. A schematic illustration of the iterative part of the main algorithm is presented in Fig. 9.

In each iteration i, the ith pair of UTL surrogate models has been replaced with the ith pair of NUTL surrogate models. Each NUTL surrogate model has been constituted by a cascade of two CMRCs (termed here $CMRCs_C$). A surrogate model layout representation described by geometric and preassigned parameters is shown in Fig. 10. The initial design variable vector used in each iteration is $\boldsymbol{x}_i^{(0)} = [1\ 1\ 1\ 1\ \ldots]^T$ mm, whereas the initial preassigned parameter vector is $\boldsymbol{p}_i^{(0)} = [0.508\ 4.4\ 0.508\ 4.4\ \ldots]^T$ (odd elements are in mm, while even elements are unitless). The final design demonstrates an acceptable performance and a notable 36 % circuit area miniaturization. Particular CMRC optimization results are denoted as $CMRC_{C1}$ and $CMRC_{C2}$. Error functions plotted against ISM algorithm iterations (see Fig. 11) have been calculated at frequencies from 2.09 to 2.31 GHz with a 20 MHz step and at the 2.2 GHz operating frequency alone. Final geometric parameters of the miniaturized BLC are listed in Table 3. Table 4 demonstrates total design cost of the method implemented in this work in comparison to other competitive approaches. The ISM algorithm implemented in the atomistic SBO method has converged in five iterations total—two iterations for UTLs1 ($i = 1$) and three iterations for UTLs2 ($i = 2$)—which is more than eight times more efficient than a classic holistic EM optimization method applied to the same example. Average CPU time of a single

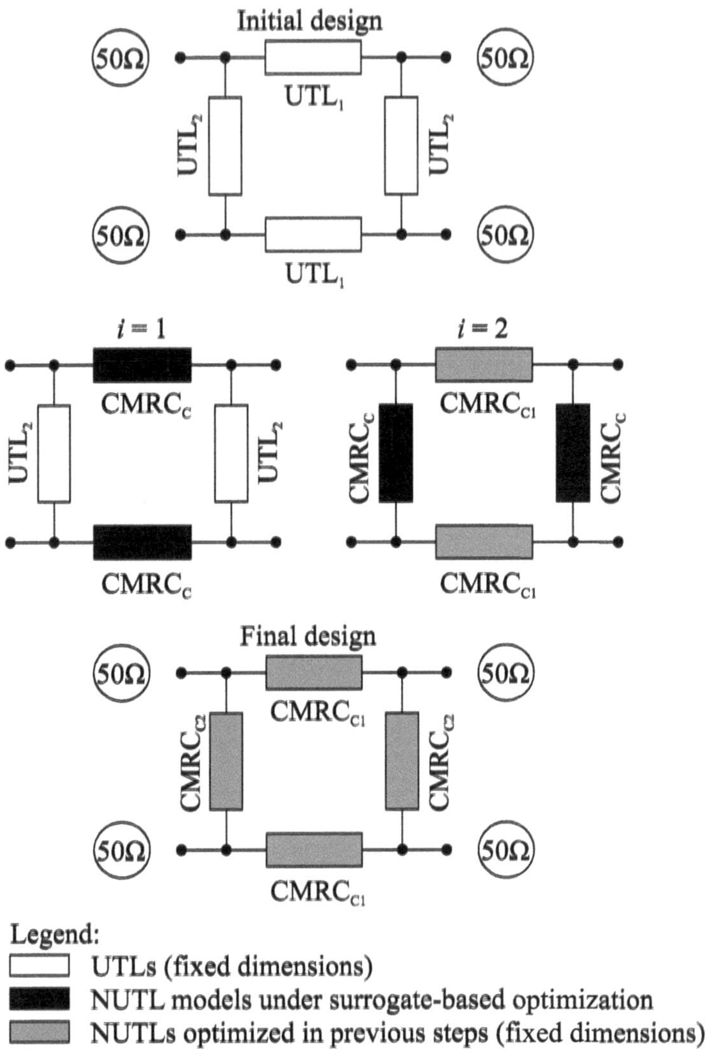

Fig. 9 Schematic description of a BLC under iterative construction [80]

64-point EM model evaluation is approximately 26 s for the atomistic SBO approach and 29 s for the holistic EMBO approach (both methods used i7 2,600 k 8 GB RAM PC), resulting in a total EM design cost of 2.16 min in case of the former approach and 20.3 min in case of the latter one.

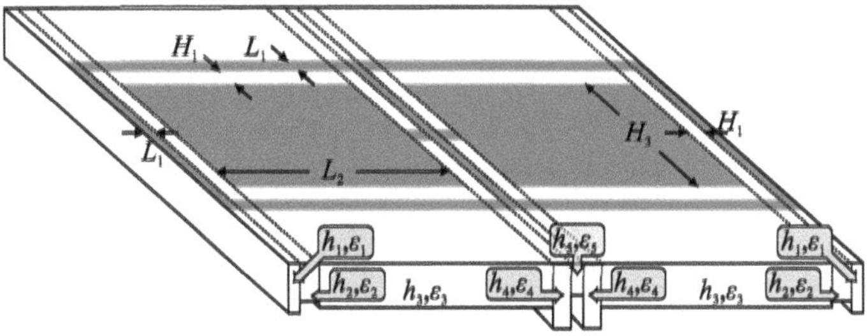

Fig. 10 Surrogate model layout representation of a $CMRC_C$ [80]. Model includes geometrical parameterization as well as preassigned parameter description

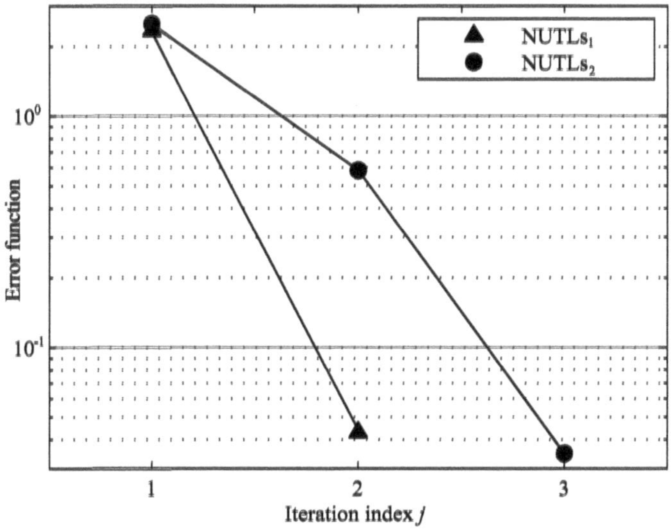

Fig. 11 ISM algorithm convergence plot obtained for the miniaturized BLC design example. Iteration index corresponds to the number of high-fidelity model evaluations [80]

4 Experimental Results

All final design examples discussed in the previous section have been manufactured and measured (see Figs. 12 and 13). One should notice that the abbreviated MT has been fabricated in a back-to-back configuration (see Fig. 12b) for the source and load impedance to be 50 Ω. Theoretical characteristics of the designed circuits, obtained by means of EM simulations, have been included for comparison purposes. It can be observed that the measured MT performance presents a reflection

Table 3 Final designable geometric parameters

Final design dimensions		CMRC$_C$	
		NUTLs$_1$	NUTLs$_2$
x^* [mm]	L_1	0.15	0.15
	H_1	0.15	0.15
	L_2	2.55	3.05
	H_2	1.8	0.65

Table 4 Miniaturized BLC: total design cost

Total design cost	Optimization method			
	Holistic		Atomistic	
	Direct	SBO	Direct	SBO
Convergence	Yes	No	Yes	Yes
Total number of high-fidelity model evaluations	42	7	15	5

coefficient $|S_{11}| \leq -15$ dB in the specified frequency band. Furthermore, measured characteristics of the fabricated MT demonstrate a 8 % bandwidth enlargement in comparison to the theoretical performance. Insertion loss $|S_{21}|$ is smaller than 0.7 dB in 0.85–1.9 GHz frequency range and smaller than 1 dB in 1.9–2.65 GHz band. Lossless conductor used during EM simulations as well as the fabrication tolerance is accounted for differences in frequency characteristics between simulated and measured MT responses. The comparison between theory and experiment also reveals that ΔS_{21} is ranged between 0.3 and 0.6 dB in the predefined frequency range. In case of the miniaturized BLC, an agreement between theoretical and experimental characteristics has been found (see Fig. 13). It is important to emphasize that the 36 % rate of miniaturization has been achieved without major degradation in the performance of the circuit.

5 Conclusion

A computationally efficient design approach to compact microwave circuits with complex topologies has been presented and experimentally validated. The generality of the method discussed makes it suitable for a wide class of N-port unconventional microwave circuits conceived by a sequential alternation of a conventional design solution. The introduction of the atomistic optimization design approach as a vital alternative to a holistic EM optimization design methodology has been found useful in application to computationally demanding microwave circuits with unconventional topologies. Moreover, a combination of a sequential atomistic optimization

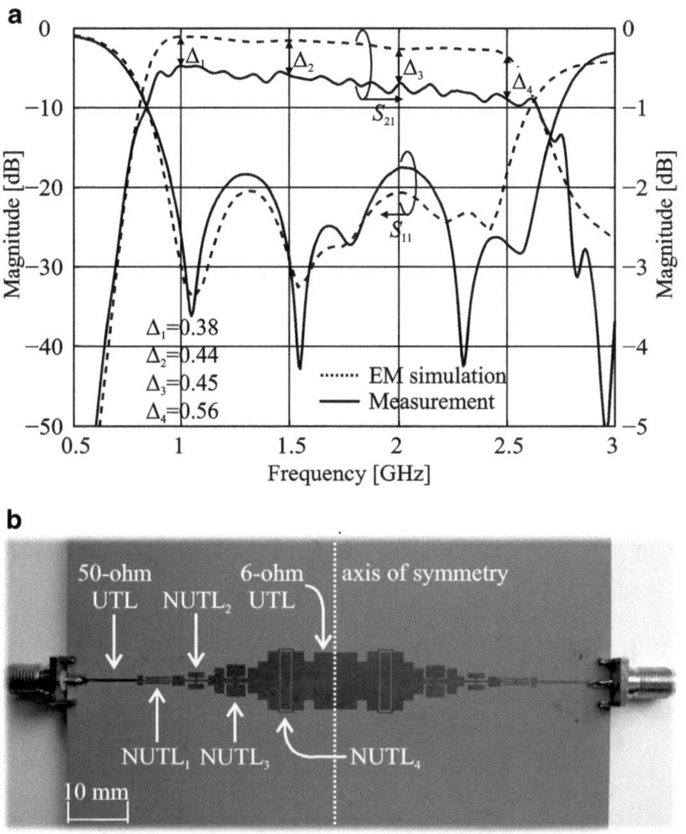

Fig. 12 (**a**) Measured transmission characteristics of the abbreviated MT in comparison to its theoretical (simulated) performance; (**b**) a photograph of the abbreviated MT in a back-to-back configuration [80]

approach and the ISM technique has resulted in a method that outclasses other competitive design methodologies mentioned in this work. The robustness and computational efficiency of the method elaborated in application to circuits with miniaturized footprints has been obtained at the price of finding an approximation of the global optimum, rather than the global optimum itself. The computational gain from the application of the atomistic SBO concept promoted in this work is much more impressive in case of the circuit described by more design variables. The number of design variables, for which the utilization of this method is cost-efficient still remains an open issue.

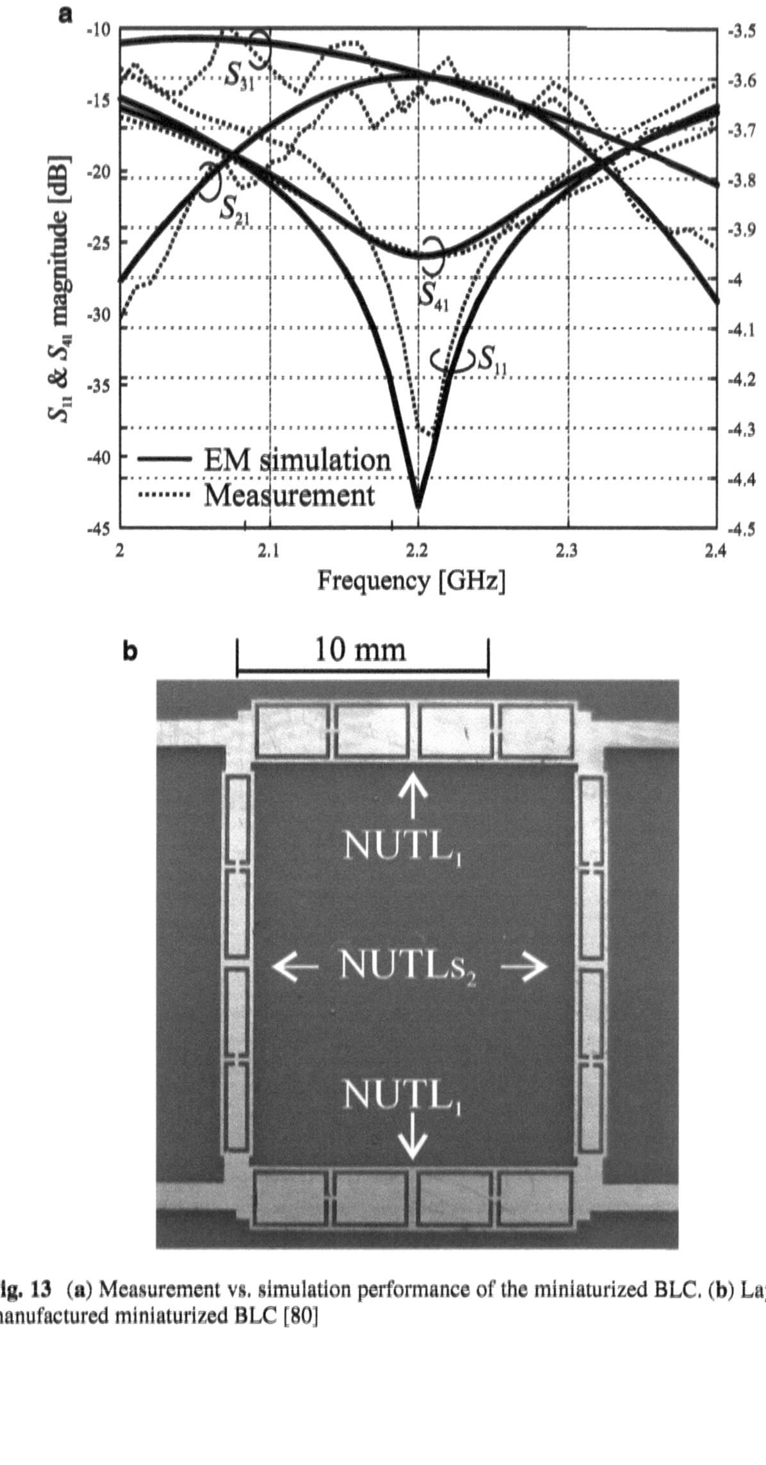

Fig. 13 (**a**) Measurement vs. simulation performance of the miniaturized BLC. (**b**) Layout of the manufactured miniaturized BLC [80]

References

1. Gilmore, R., Besser, L.: Practical RF Circuit Design for Modern Wireless Systems. Artech House, Norwood (2003)
2. Ahn, H.-R.: In: Chang, K. (ed.) Asymmetric Passive Components in Microwave Integrated Circuits. Series: Wiley Series in Microwave and Optical Engineering, pp. 56–151. Wiley, New Jersey (2006)
3. Xu, H.-X., Wang, G.-M., Lu, K.: Microstrip rat-race couplers. IEEE Microw. Magazine **12**, 117–129 (2011)
4. Ahn, H.-R., Bumman, K.: Toward integrated circuit size reduction. IEEE Microw. Magazine **9**, 65–75 (2008)
5. Zatloukal, P., Heddle, R.M., Dabrowski, C.J.: Wireless mobile phone including a headset. US Patent no. 7373182 (2008)
6. Morimoto, H.: Personal digital assistant, wireless communication system and method of link establishment. US Patent no. 2004/0203372 (2004)
7. Ahn, S.-H.: Electronic smart meter enabling demand response and method for demand response. US Patent no. 8234017 (2012)
8. Harris, L.C., Smith, R.L.: Interferometric switched beam radar apparatus and method. US Patent no. 7755533 (2010)
9. Lewis, A.D., Mousseau, G.P., Gilhuly, B.J., Patterson, I.M., Banh, V.T., Rogobete, A., Burns, A.G., Lazaridis, M.: Wireless router system and method. US Patent no. 7529230 (2009)
10. Judd, M.D., Lovinggood, B.W., Tennant, D.T., Maca, G.A., Kuiper, W.P., Alford, J.L., Thomas, M.D., Veihl, J.C.: Repeaters for wireless communication systems. US Patent no. 8630581 (2014)
11. Pozar, D.M.: Microwave Engineering, 2nd edn, pp. 351–498. Wiley, New York (1998)
12. Opozda, S., Kurgan, P., Kitlinski, M.: A compact seven-section rat-race hybrid coupler incorporating PBG cells. Microw. Opt. Technol. Lett. **51**, 2910–2913 (2009)
13. Di Paolo, F.: Networks and Devices Using Planar Transmission Lines. CRC Press, Boca Raton (2000)
14. Staras, S., Nartavicius, R., Skudutis, J., Urbanavicius, V., Daskevicius, V.: Wide-Band Slow-Wave Systems. Taylor & Francis Group, Boca Raton (2012)
15. Hirota, T., Minakawa, A., Muraguchi, M.: Reduced-size branch-line and rat-race hybrids for uniplanar MMICs. IEEE Trans. Microw. Theory Tech. **38**, 270–275 (1990)
16. Gillick, M., Robertson, I.D., Joshi, J.S.: Coplanar waveguide two-stage balanced MMIC amplifier using impedance-transforming lumped-distributed branchline couplers. IEEE Proc. Microw. Antennas Propag. **141**, 241–245 (1994)
17. Chiang, Y.-C., Chen, C.-Y.: Design of a wide-band lumped-element 3-dB quadrature coupler. IEEE Trans. Microw. Theory Tech. **49**, 476–479 (2001)
18. Hong, J.-S., Lancaster, M.J.: Microstrip Filters for RF/Microwave Applications. Wiley, New York (2001)
19. Li, Y., Zhang, Z., Li, Z., Zheng, J.: High-permittivity substrate multiresonant antenna metallic cover of laptop computer. IEEE Antennas Wirel. Propag. Lett. **10**, 1092–1095 (2011)
20. Chen, Y.-C., Hsu, C.-H.: Inverted-E shaped monopole on high-permittivity substrate for application in industrial, scientific, medical, high-performance radio local area network, unlicensed national information infrastructure, and worldwide interoperability for microwave access. IET Microw. Antennas Propag. **8**, 272–277 (2014)
21. Awai, I., Kubo, H., Iribe, T., Wakamiya, D., Sanada, A.: An artificial dielectric material of huge permittivity with novel anisotropy and its application to a microwave BPF. In: IEEE MTT-S Int. Microw. Symp. Dig., pp. 1085–1088 (2003)
22. Wu, H.-S., Tzuang, C.-K.C.: Artificially integrated synthetic rectangular waveguide. IEEE Trans. Microw. Theory Tech. **53**, 2872–2881 (2005)
23. Elek, F., George, E.: On the slow wave behavior of the shielded mushroom structure. In: IEEE MTT-S Int. Microw. Symp. Dig., pp. 1333–1336 (2008)

24. Seki, S., Hasegawa, H.: Cross-tie slow-wave coplanar waveguide on semi-insulating GaAs substrate. Electron. Lett. **17**, 940–941 (1981)
25. Xue, Q., Shum, K.M., Chan, C.H.: Novel 1-D microstrip PBG cells. IEEE Microw. Guid. Wave Lett. **10**, 403–405 (2000)
26. Shum, K.M., Xue, Q., Chan, C.H.: A novel microstrip ring hybrid incorporating a PBG cell. IEEE Microw. Wirel. Compon. Lett. **11**, 258–260 (2001)
27. Sun, K.-O., Ho, S.-J., Yen, C.-C., Weide, D.: A compact branch-line coupler using discontinuous microstrip lines. IEEE Microw. Wirel. Compon. Lett. **8**, 519–520 (2005)
28. Eccleston, K.W., Ong, S.H.M.: Compact planar microstripline branch-line and rat-race couplers. IEEE Trans. Microw. Theory Tech. **51**, 2119–2125 (2003)
29. Gandini, E., Ettorre, M., Sauleau, R., Grbic, A.: A lumped-element unit cell for beam-forming networks and its application to a miniaturized Butler matrix. IEEE Trans. Microw. Theory Tech. **61**, 1477–1487 (2013)
30. Mongia, R., Bahl, I., Bhartia, P.: RF and Microwave Coupler-Line Circuits. Artech House, Boston (1999)
31. Hou, J.-A., Wang, Y.-H.: Design of compact 90° and 180° couplers with harmonic suppression using lumped-element bandstop resonators. IEEE Trans. Microw. Theory Tech. **58**, 2932–2939 (2010)
32. Brillouin, L.: Wave Propagation in Periodic Structures. McGraw-Hill Book Company, Inc., New York (1946)
33. Caloz, C., Itoh, T.: Electromagnetic Metamaterials: Transmission Line Theory and Microwave Applications: The Engineering Approach. Wiley, Hoboken (2006)
34. Yang, F.-R., Ma, K.-P., Qian, Y., Itoh, T.: A uniplanar compact photonic-bandgap (UC-PBG) structure and its applications for microwave circuits. IEEE Trans. Microw. Theory Tech. **47**, 1509–1614 (1999)
35. Kurgan, P., Kitlinski, M.: Novel doubly perforated broadband microstrip branch-line coupler. Microw. Opt. Technol. Lett. **51**, 2149–2152 (2009)
36. Zhou, C., Yang, H.Y.D.: Design considerations of miniaturized least dispersive periodic slow-wave structures. IEEE Trans. Microw. Theory Tech. **56**, 467–474 (2008)
37. Chun, Y.-H., Hong, J.-S.: Compact wide-band branch-line hybrids. IEEE Trans. Microw. Theory Tech. **54**, 704–709 (2006)
38. Wang, J., Wang, B.-Z., Guo, Y.-X., Ong, L.C., Xiao, S.: A compact slow-wave microstrip branch-line coupler with high performance. IEEE Microw. Wirel. Compon. Lett. **17**, 501–503 (2007)
39. Kurgan, P., Kitlinski, M.: Doubly miniaturized rat-race hybrid coupler. Microw. Opt. Technol. Lett. **53**, 1242–1244 (2011)
40. Kurgan, P., Kitlinski, M.: Slow-wave fractal-shaped compact microstrip resonant cell. Microw. Opt. Technol. Lett. **52**, 2613–2615 (2010)
41. Kurgan, P., Bekasiewicz, A.: A robust design of a numerically demanding compact rat-race coupler. Microw. Opt. Technol. Lett. **56**, 1259–1263 (2014)
42. Abdi, H., Williams, L.J.: Principal component analysis. Wiley Interdiscip. Rev. Comput. Stat. **2**, 433–459 (2010)
43. Kurgan, P., Filipcewicz, J., Kitlinski, M.: Design considerations for compact microstrip resonant cells dedicated to efficient branch-line coupler miniaturization. Microw. Opt. Technol. Lett. **54**, 1949–1954 (2012)
44. Bekasiewicz, A., Kurgan, P.: A compact microstrip rat-race coupler constituted by nonuniform transmission lines. Microw. Opt. Technol. Lett. **56**, 970–974 (2014)
45. Radtke, K., Kurgan, P., Bekasiewicz, A., Kitlinski, M.: Zminiaturyzowane, planarne filtry pasmowo-przepustowe o nowej topologii. Wiadomości Elektrotechniczne **11**, 34–36 (2012) (in polish)
46. Koziel, S., Kurgan, P.: Low-cost optimization of compact branch-line couplers and its application to miniaturized Butler matrix design. In: Eur. Microw. Conf. (2014, to appear)
47. Liao, S.-S., Sun, P.-T., Chin, N.-C., Peng, J.-T.: A novel compact-size branch-line coupler. IEEE Microw. Wirel. Compon. Lett. **15**, 588–590 (2005)

48. Liao, S.-S., Peng, J.-T.: Compact planar microstrip branch-line couplers using the quasi-lumped elements approach with nonsymmetrical and symmetrical T-shaped structure. IEEE Trans. Microw. Theory Tech. **54**, 3508–3514 (2006)
49. Tang, C.-W., Chen, M.-G.: Synthesizing microstrip branch-line couplers with predetermined compact size and bandwidth. IEEE Trans. Microw. Theory Tech. **55**, 1926–1934 (2007)
50. Jung, S.-C., Negra, R., Ghannouchi, F.M.: A design methodology for miniaturized 3-dB branch-line hybrid couplers using distributed capacitors printed in the inner area. IEEE Trans. Microw. Theory Tech. **56**, 2950–2953 (2008)
51. Ahn, H.-R.: Modified asymmetric impedance transformers (MCCTs and MCVTs) and their application to impedance-transforming three-port 3-dB power dividers. IEEE Trans. Microw. Theory Tech. **59**, 3312–3321 (2011)
52. Tseng, C.-H., Chang, C.-L.: A rigorous design methodology for compact planar branch-line and rat-race couplers with asymmetrical T-structures. IEEE Trans. Microw. Theory Tech. **59**, 2085–2092 (2012)
53. Kuo, J.-T., Wu, J.-S., Chiou, Y.-C.: Miniaturized rat-race coupler with suppression of spurious passband. IEEE Microw. Wirel. Compon. Lett. **17**, 46–48 (2007)
54. Mondal, P., Chakrabarty, A.: Design of miniaturized branch-line and rat-race hybrid couplers with harmonics suppression. IET Microw. Antennas Propag. **3**, 109–116 (2009)
55. Tseng, C.-H., Chen, H.-J.: Compact rat-race coupler using shunt-stub-based artificial transmission lines. IEEE Microw. Wirel. Compon. Lett. **18**, 734–736 (2008)
56. Wang, C.-W., Ma, T.-G., Yang, C.-F.: A new planar artificial transmission line and its applications to a miniaturized Butler matrix. IEEE Trans. Microw. Theory Tech. **55**, 2792–2801 (2007)
57. Tsai, K.-Y., Yang, H.-S., Chen, J.-H., Chen, Y.-J.: A miniaturized 3 dB branch-line hybrid coupler with harmonics suppression. IEEE Microw. Wirel. Compon. Lett. **21**, 537–539 (2011)
58. Ahn, H.-R., Nam, S.: Compact microstrip 3-dB coupled-line ring and branch-line hybrids with new symmetric equivalent circuits. IEEE Trans. Microw. Theory Tech. **61**, 1067–1078 (2013)
59. Chuang, M.-L.: Miniaturized ring coupler of arbitrary reduced size. IEEE Microw. Wirel. Compon. Lett. **21**, 16–18 (2005)
60. Ahn, H.-R., Kim, B.: Small wideband coupled-line ring hybrids with no restriction on coupling power. IEEE Trans. Microw. Theory Tech. **61**, 1806–1817 (2009)
61. Lee, H.-S., Choi, K., Hwang, H.-Y.: A harmonic and size reduced ring hybrid using coupled line. IEEE Microw. Wirel. Compon. Lett. **17**, 259–261 (2005)
62. Ahn, H.-R., Nam, S.: Wide band microstrip coupled-line ring hybrids for high power-division ratios. IEEE Trans. Microw. Theory Tech. **61**, 1768–1780 (2013)
63. Collin, R.E.: Foundations for Microwave Engineering. Wiley, New York (2001)
64. Queipo, N.V., Haftka, R.T., Shyy, W., Goel, T., Vaidynathan, R., Tucker, P.K.: Surrogate-based analysis and optimization. Prog. Aerosp. Sci. **41**, 1–28 (2005)
65. Yelten, M.B., Zhu, T., Koziel, S., Franzon, P.D., Steer, M.B.: Demystifying surrogate modeling for circuits and systems. IEEE Circuits Syst. Magazine **12**, 45–63 (2012)
66. Koziel, S., Ogurtsov, S.: Simulation-driven design in microwave engineering: methods. In: Koziel, S., Yang, X.S. (eds.) Computational Optimization, Methods and Algorithms. Series: Studies in Computational Intelligence, vol. 356. Springer, Berlin (2011)
67. Koziel, S.: Efficient optimization of microwave circuits using shape-preserving response prediction. In: IEEE MTT-S Int. Microw. Symp. Dig., pp. 1569–1572 (2009)
68. Bandler, J.W., Cheng, Q.S., Dakroury, S.A., Mohamed, A.S., Bakr, M.H., Madsen, K., Sondergaard, J.: Space mapping: the state of the art. IEEE Trans. Microw. Theory Tech. **52**, 337–361 (2004)
69. Liu, X., Wang, G., Liu, J.: A wideband model of on-chip CMOS interconnects using space-mapping technique. Int. J. RF Microw. Comput. Aid. Eng. **21**, 439–445 (2011)
70. Bandler, J.W., Biernacki, R.M., Chen, S.H., Grobelny, P.A., Hemmers, R.H.: Space mapping technique for electromagnetic optimization. IEEE Trans. Microw. Theory Tech. **42**, 2536–2544 (1994)

71. Bakr, M.H., Bandler, J.W., Biernacki, R.M., Chen, S.H., Madsen, K.: A trust region aggressive space mapping algorithm for EM optimization. IEEE Trans. Microw. Theory Tech. **46**, 2412–2425 (1998)
72. Bakr, M.H., Bandler, J.W., Ismail, M.A., Rayas-Sanchez, J.E., Zhang, Q.-J.: Neural space-mapping optimization for EM-based design. IEEE Trans. Microw. Theory Tech. **48**, 2307–2315 (2000)
73. Koziel, S., Bandler, J.W.: A space-mapping approach to microwave device modeling exploiting fuzzy systems. IEEE Trans. Microw. Theory Tech. **55**, 2539–2547 (2007)
74. Koziel, S., Meng, J., Bandler, J.W., Bakr, M.H., Cheng, Q.S.: Accelerated microwave design optimization with tuning space mapping. IEEE Trans. Microw. Theory Tech. **57**, 383–394 (2009)
75. Xu, H.-X., Wang, G.-M., Zhang, C.-X., Yu, Z.-W., Chen, X.: Composite right/left-handed transmission line based on complementary single-split ring resonator pair and compact power dividers application using fractal geometry. IET Microw. Antennas Propag. **6**, 1017–1025 (2012)
76. Smierzchalski, M., Kurgan, P., Kitlinski, M.: Improved selectivity compact brand-stop filter with Gosper fractal-shaped defected ground structures. Microw. Opt. Technol. Lett. **52**, 227–229 (2010)
77. Bekasiewicz, A., Kurgan, P., Kitlinski, M.: New approach to a fast and accurate design of microwave circuits with complex topologies. IET Microw. Antennas Propag. **6**, 1616–1622 (2012)
78. Koziel, S., Echeverría-Ciaurri, C., Leifsson, L.: Surrogate-based methods. In: Koziel, S., Yang, X.-S. (eds.) Computational Optimization, Methods and Algorithms, pp. 33–59. Springer, Berlin (2011)
79. Kurgan, P., Filipcewicz, J., Kitlinski, M.: Development of a compact microstrip resonant cell aimed at efficient microwave component size reduction. IET Microw. Antennas Propag. **6**, 1291–1298 (2012)
80. Koziel, S., Bandler, S.W., Madsen, K.: A space mapping framework for engineering optimization: theory and implementation. IEEE Trans. Microw. Theory Tech. **54**, 3721–3730 (2006)
81. Bandler, J.W., Cheng, Q.S., Nikolova, N.K., Ismail, M.A.: Implicit space mapping optimization exploiting preassigned parameters. IEEE Trans. Microw. Theory Tech. **52**, 378–385 (2004)
82. Koziel, S., Cheng, Q.S., Bandler, J.W.: Space mapping. IEEE Microw. Magazine **9**, 105–122 (2008)
83. Agilent ADS, Version 2011: Agilent Technologies, 395 Page Mill Road, Palo Alto, CA, 94304 (2011)

Knowledge Based Three-Step Modeling Strategy Exploiting Artificial Neural Network

Murat Simsek

Abstract Artificial Neural Network (ANN) is an important technique for modeling and optimization in engineering design. It is very suitable in modeling as it needs only the information based on relationship between the input and the output related to the problem. For further improvement in modeling, a priori knowledge about the problem such as an empirical formula, an equivalent circuit model, and a semi-analytical equation is directly embedded in ANN structure through a knowledge based modeling strategy. Three-step modeling strategy that exploits knowledge based techniques is developed to improve some properties of conventional ANN modeling such as accuracy and data requirement. All these improvements ensure better accuracy with less time consumption compared to conventional ANN modeling. The necessary knowledge in this strategy is generated in the first step through conventional ANN. Then this knowledge is embedded in the new ANN model for the second step. Final model is constructed by incorporating the existing knowledge obtained by the second step. Therefore each model generates better accuracy than previous model. Conventional ANN, prior knowledge input, and prior knowledge input with difference techniques are used to improve accuracy, time consumption, and data requirement of the modeling in three-step modeling strategy. The efficiency of three-step modeling strategy is demonstrated on the nonlinear function modeling and the high dimensional shape reconstruction problem.

Keywords Neural Network Modeling • Knowledge Based Modeling • Three-step modeling strategy • Nonlinear function modeling • Inverse scattering problem

1 Introduction

Artificial Neural Network has been extensively preferred as a modeling technique to obtain surrogate model instead of a fine model which has high computational burden. Surrogate based modeling [6, 17] is required to overcome this computa-

---

M. Simsek (✉)
Faculty of Aeronautics and Astronautics, Istanbul Technical University, Istanbul, Turkey
e-mail: simsekmu@itu.edu.tr

S. Koziel et al. (eds.), *Solving Computationally Expensive Engineering Problems*,
Springer Proceedings in Mathematics & Statistics 97,
DOI 10.1007/978-3-319-08985-0_9

tional burden of the fine model. Surrogate based models can be fundamentally developed in two ways. First way only requires input or output mapping without any change in the computationally cheap coarse model. Space mapping based modeling [1, 5, 10, 12–14] is developed considering this approach. Second way is based on updating the coarse model during modeling process for the coarse model. ANN is very suitable to obtain this kind of coarse model.

ANN provides an efficient strategy to solve modeling and optimization problems which are essential in engineering where only input–output data are available instead of mathematical formulations [2, 4, 7, 18, 19]. ANN modeling is generally used to construct a mapping from the input to the output depending on data obtained from detailed physical/EM simulation models or measurements (fine model) and generate approximate results depending on some tunable parameters such as training set, topological structure, and complexity of the fine model.

Since ANN technique constitutes input–output mapping highly depending on the training set, when the points outside of the training range (extrapolation) are used as inputs for final model after training process, responses of the model are probably unsatisfactory compared to the points inside of the training set (interpolation). ANN and the existing knowledge about the fine model should be combined in the same modeling process in order to reduce complexity of the fine model, while improving extrapolation performance or lowering data requirements for training process.

Knowledge based modeling techniques have been developed to embed existing knowledge in conventional ANN modeling [3, 10, 11, 15, 16, 18]. Knowledge based models utilize less training data compared to the need of conventional ANN. The knowledge provides coarse information for modeling and ANN completes rest of the information using less training data. This modeling approach provides more accuracy and better extrapolation performance than ANN models and offers less computational burden compared to the detailed physical/EM simulation models.

In some cases, modeling involves numerous training data to satisfy specific design purposes such as good accuracy, better extrapolation, and less computational burden. However training process takes long time and modeling accuracy cannot be good enough with respect to design purposes. To get over this problem, Knowledge Based ANN (KBANN) techniques emerged to generate an efficient model.

During KBANN modeling, empirical formulas, equivalent circuit models, and semi-analytical equations are exploited as the existing knowledge (coarse model). This coarse model that is less accurate but fast than the fine model facilitates to reduce the complexity of ANN model. If the coarse model does not exist for the fine model, knowledge based modeling is not suitable under this condition.

Three-step modeling strategy is developed to generate required knowledge without any extra data besides training data. Therefore gradual improvement can be obtained by applying knowledge based techniques during modeling process of three-step strategy. Required knowledge that is obtained by conventional ANN in the first step is exploited as the coarse model for Prior Knowledge Input (PKI) technique [18] in the second step. Last step utilizes knowledge come from the second step in order to satisfy narrow output interval and reduce the complexity of ANN model. Since the last step exploits more accurate coarse model, output correction between fine model and second step responses provides better accuracy than conventional

ANN technique. Considering all steps, Prior Knowledge Input with Difference (PKI-D) technique [8, 10, 11, 16] that is used in the third step completes the gap between the fine model and the coarse model.

Considering general approach and detailed information about each step, three-step strategy will be discussed in Sect. 2. The modeling performance will be presented in Sect. 3 through Branin function modeling in Sect. 3.1 and the shape reconstruction of inverse scattering problem in Sect. 3.2. The less complex and more complex models of Branin function will be discussed in Sect. 3.1. Finally the shape reconstruction problem will be considered with different number of data in Sect. 3.2 to demonstrate the efficiency of three-step strategy with less data.

2 Three-Step Modeling Strategy

Conventional ANN modeling is not convenient when the numerous training data is required to obtain sufficient accuracy. More training data involve more number of iterations to satisfy stopping condition. Three-step modeling strategy totally utilizes the same number of iterations and neurons like conventional ANN. Main contribution of the new strategy is that former model improves latter model via knowledge based modeling techniques. This contribution changes according to knowledge based technique. For example, PKI only uses extra input to reduce ANN complexity, hence more accurate coarse model is not necessary to obtain more accurate result than the coarse model. PKI constitutes general correction instead of detail one. Each step of three-step modeling strategy will be discussed in detail following three subsections.

2.1 Step-1: Generating Knowledge via Conventional ANN (M-1)

In this step, required knowledge is obtained by Multi Layer Perceptron (MLP) after training process. Final model is called $M-1$ after training process is completed. Number of neurons and iterations are reduced to one third of conventional ANNs. This process guarantees same total number of neurons and iterations usage when three-step modeling is completed.

Training process for $M-1$ is given in (1). Error value for i. iteration is given in (2). Final model $M-1$ response is shown in (3). In Fig. 1, training process and final model $M-1$ are shown dotted line and bold box, respectively.

$$w^* = \arg\min_{w} \left\| \cdots e^{(i)^T} \cdots \right\| \tag{1}$$

$$e^{(i)} = f_f\left(x_f^{(i)}\right) - f_{ANN}(x_f^{(i)}) \tag{2}$$

$$Y_{M-1} = f_{ANN}\left(x_f\right) \tag{3}$$

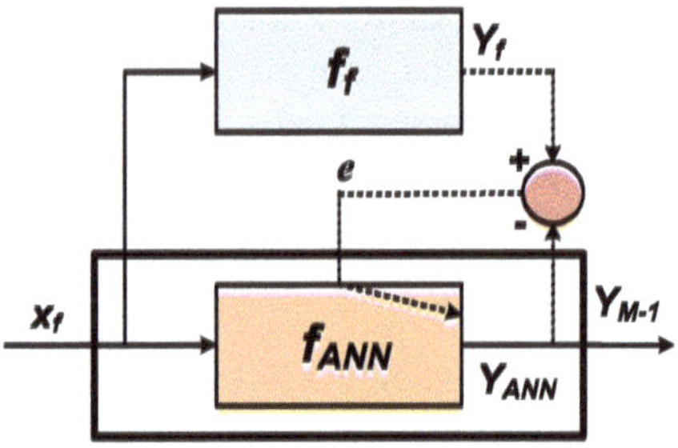

Fig. 1 Block diagram of step-1 via conventional ANN technique for three-step modeling strategy

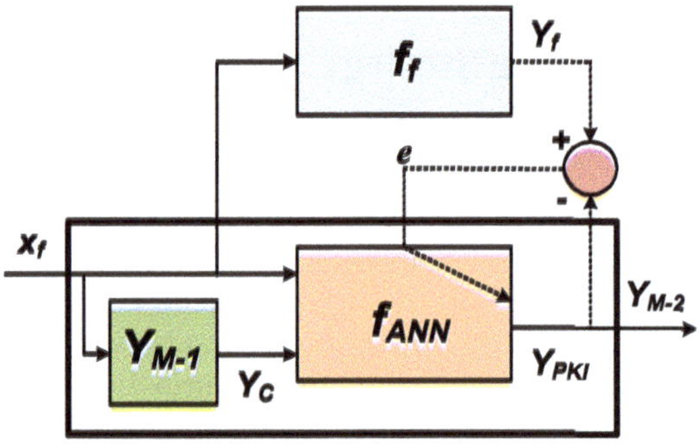

Fig. 2 Block diagram of step-2 via knowledge based PKI technique for three-step modeling strategy

2.2 *Step-2: Using Knowledge to Reduce Complexity via Prior Knowledge Input Model (M-2)*

Knowledge obtained by step-1 is used as an extra input to constitute PKI model. Final model is called $M-2$ after training process is completed. Number of neurons and iterations are same as step-1. Extra input provides extra knowledge to reduce complexity of ANN structure. This model creates better accuracy than $M-1$ via knowledge obtained by $M-1$.

Training process for $M-2$ is given in (4). Error value for i. iteration is given in (5). Final model $M-2$ response is shown in (6). In Fig. 2, training process and final model $M-2$ are shown dotted line and bold box, respectively. This model needs $M-1$ responses to improve $M-2$ responses. $M-1$ response which is shown in (3) is used in (5) and (6)

$$w^* = \arg\min_{w} \left\| \cdots e^{(l)T} \cdots \right\| \tag{4}$$

$$e^{(l)} = f_f\left(x_f^{(l)}\right) - f_{ANN}(x_f^{(l)}, Y_{M-1}^{(l)}) \tag{5}$$

$$Y_{M-2} = Y_{PKI} = f_{ANN}\left(x_f, Y_{M-1}\right) \tag{6}$$

2.3 *Step-3: Learning Difference to Improve Model via Prior Knowledge Input with Difference Model (M-3)*

Knowledge obtained by step-2 is used as an extra input to constitute PKI-D model. In addition to this usage, difference between fine model response and $M-2$ model response is used as a target for ANN structure in $M-3$ model. Final model is called $M-3$ after training process is completed. Number of neurons and iterations are same as step-1 and step-2. Extra input not only provides extra knowledge to reduce complexity of ANN structure but also provides narrow interval via difference usage. This model creates better accuracy than $M-2$ via knowledge obtained by $M-2$.

Training process for $M-3$ is given in (7). Error value for i. iteration is given in (8). Final model $M-3$ response is shown in (9). In Fig. 3, training process and final model $M-3$ are shown dotted line and bold box, respectively. This model needs $M-2$ responses to improve $M-3$ responses. $M-2$ response which is shown in (6) is used in (8) and (9)

$$w^* = \arg\min_{w} \left\| \cdots e^{(i)T} \cdots \right\| \tag{7}$$

$$e^{(i)} = \left(\underbrace{f_f\left(x_f^{(i)}\right) - Y_{M-2}^{(i)}}_{Y_d} \right) - f_{ANN}(x_f^{(i)}, Y_{M-2}^{(i)}) \tag{8}$$

$$Y_{M=3} = Y_{PKI-D} = f_{ANN}\left(x_f, Y_{M-2}\right) + Y_{M-2} \tag{9}$$

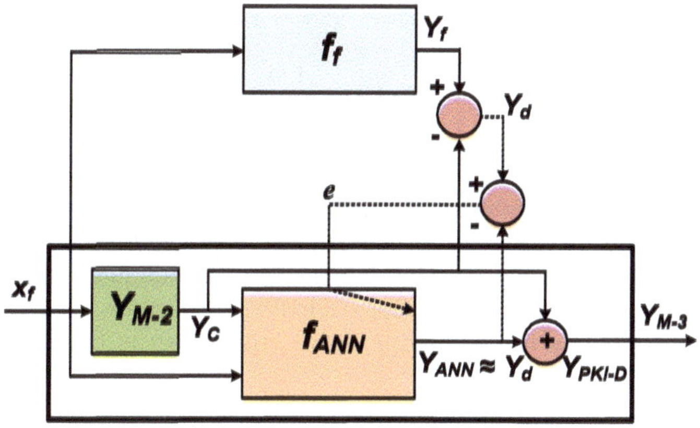

Fig. 3 Block diagram of step-3 via knowledge based PKI-D technique for three-step modeling strategy

3 Examples for Three-Step Modeling

In this section, knowledge based three-step modeling strategy has been mainly discussed by comparing it with conventional ANN technique. Conventional ANN is easy to apply to modeling problems and three-step strategy is developed to improve its performance, thus ANN trained with different number of data has been considered to demonstrate the accuracy and time efficiency of three-step modeling. The Branin function modeling and the shape reconstruction of high dimensional inverse scattering problem are used to show mathematical and engineering projection for this strategy. In addition different number of training data and two complexity levels that produce distinctive effect for ANN modeling are preferred in order to realize better performance criteria for this new strategy.

3.1 Mathematical Modeling Problem: Branin Function

The Branin function is considered to demonstrate the general performance of three-step modeling strategy and conventional ANN technique in modeling. Since Branin function [8, 10] is a well-known benchmark problem for optimization algorithm, it has highly nonlinear behavior and a wide response range.

Branin function modeling involves numerous training data and conventional ANN model is not sufficient for good accuracy, thus it is chosen to compare all results obtained in modeling. Mathematical formulation of Branin function is given in (10).

$$f_f(x_1, x_2) = \left(x_2 - \frac{5{x_1}^2}{4\pi^2} + \frac{5x_1}{\pi} - 6\right)^2 + 10 \cdot \left(1 - \frac{1}{8\pi}\right) \cdot \cos x_1 + 10 \tag{10}$$

According to valid input interval, it is possible to determine the complexity of Branin function. This complexity can be divided into two classes such as response range and nonlinearity. Three-dimensional figures of Branin function is sufficient to determine nonlinearity of the function. Less complex Branin function is depicted in Fig. 4 and more complex version is depicted in Fig. 5. The ratio of maximum output to the minimum output can be used to define criterion about output range of the function.

The maximum and minimum responses of less complex Branin function as depicted in Fig. 4 are 215.6 for $[X_1 = 6.356, X_2 = 15]$ and 54.51 for $[X_1 = 10, X_2 = 10]$, respectively. The maximum and minimum responses of more complex case as depicted in Fig. 5 are 100.6 for $[X_1 = 10, X_2 = 6.356]$

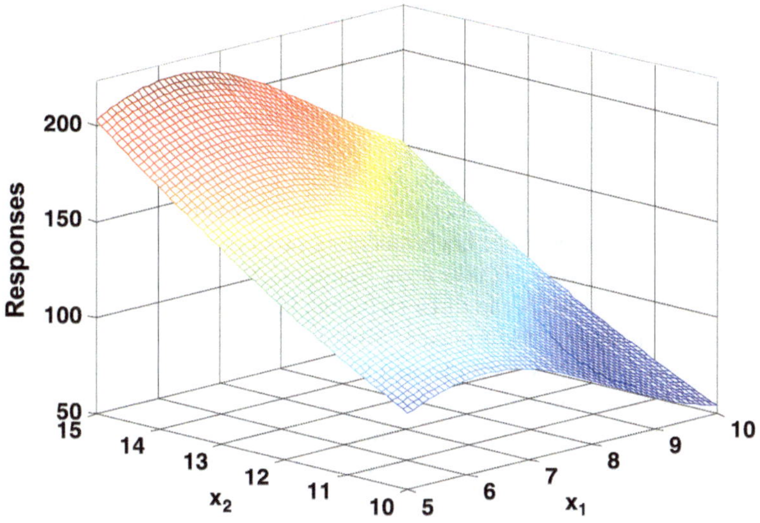

Fig. 4 Less complex fine model for Branin function modeling

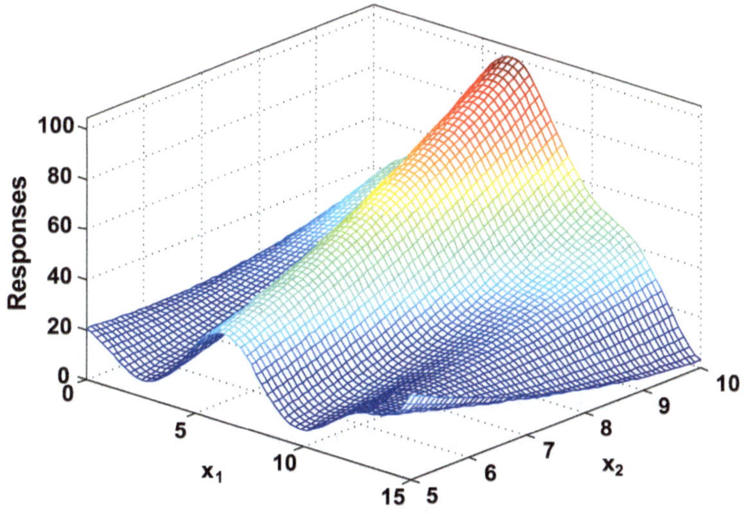

Fig. 5 More complex fine model for Branin function modeling

and 3.094 for $[X_1 = 15, X_2 = 10]$, respectively. $\frac{215.6}{54.51} \Rightarrow 3.96$ for less complex case and $\frac{100.6}{3.094} \Rightarrow 32.51$ for more complex case are obtained using recommended formulations above in order to determine the complexity of output range.

3.1.1 Less Complex Branin Function Modeling

Conventional ANN model is constituted by MLP with two hidden layers for less complex Branin function. Each layer has 30 neurons. Training process takes 300 iterations and utilizes 10,000 data to constitute final model. Each model in three-step modeling strategy is constituted MLP with two hidden layer as well. Each layer has 10 neurons. Therefore total number of neurons are equal for conventional ANN and three-step modeling. Each modeling process in three-step strategy uses 100 iterations and utilizes 10,000 data. Therefore total number of iterations are equal for conventional ANN and three-step modeling. Twenty five test data are used to demonstrate modeling performance. Moreover MATLAB m-file is used to run iteration processes of ANN for all techniques.

Error calculation is an important part of comparison. Equation (11) is used for calculation of normalized test error. While N denotes number of test data, (12) and (13) are used for calculation of mean error and maximum error, respectively.

$$Error = \frac{|X_{original} - X_{method}|}{X_{original}} \tag{11}$$

$$Mean\ Error = \frac{1}{N} \times \sum_{i=1}^{N} \frac{|X_{original,i} - X_{method,i}|}{X_{original,i}} \tag{12}$$

$$Max\ Error = \max_{i} \left\{ \frac{|X_{original,i} - X_{method,i}|}{X_{original,i}} \right\} \tag{13}$$

Conventional ANN model and $M-3$ model are compared in two different figures, namely Figs. 6 and 7. Real response of fine, conventional ANN and $M-3$ models are depicted in Fig. 6. As shown in these figures, gradual improvement of three-step modeling strategy can be recognized in detail from error performance of each steps as depicted in Fig. 6b. Final model $M-3$ of three-step modeling has better accuracy than conventional ANN for the same training data as depicted in Fig. 7. For detailed information, Fig. 7b shows how much improvement can be achieved with three-step strategy with regard to the same training data in conventional ANN modeling. Time consumption analysis is very important to compare conventional ANN and three-step modeling. This analysis also gives the complexity of developing three-step modeling and conventional ANN. All results about less complex Branin function are summarized in Table 1. Three-step model is constructed as a combination of $M-1$, $M-2$, and $M-3$. These results demonstrate that three-step modeling strategy provides efficient modeling performance with regard to accuracy and time consumption for modeling of less complex Branin function.

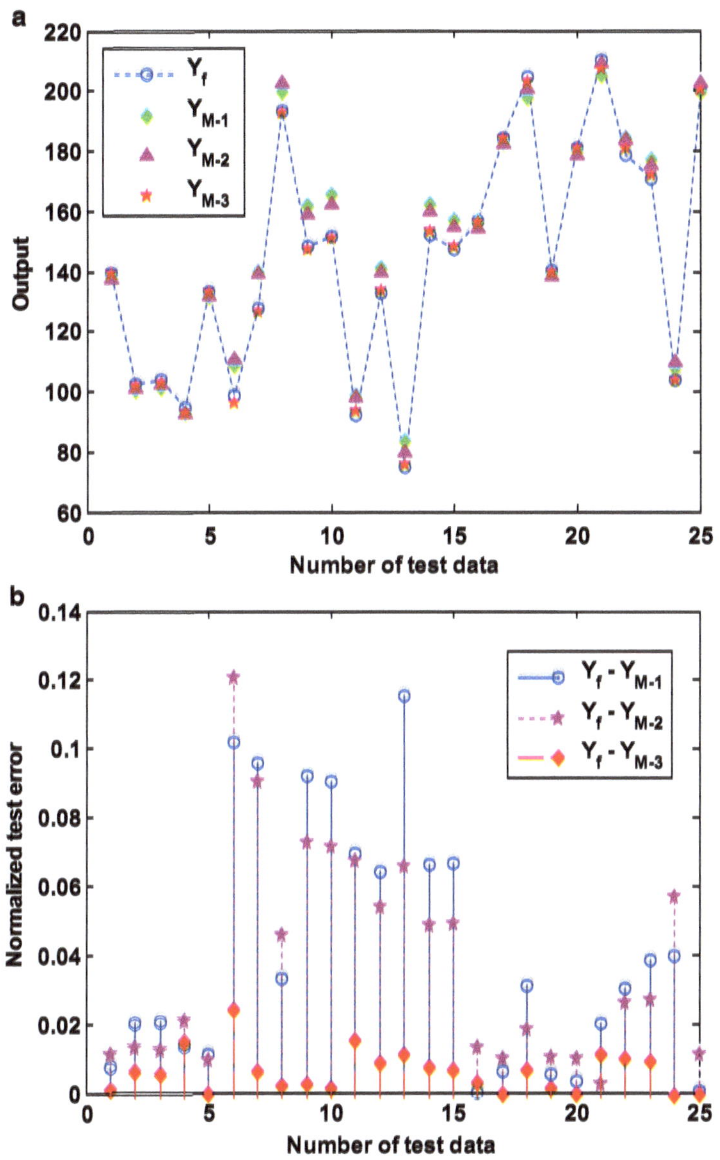

Fig. 6 Modeling results of the Branin function for less complex fine model: (**a**) Normalized test error for M-1 model, M-2 model, and M-3 model (**b**) Output of less complex fine model, M-1 model, M-2 model, and M-3 model for test data

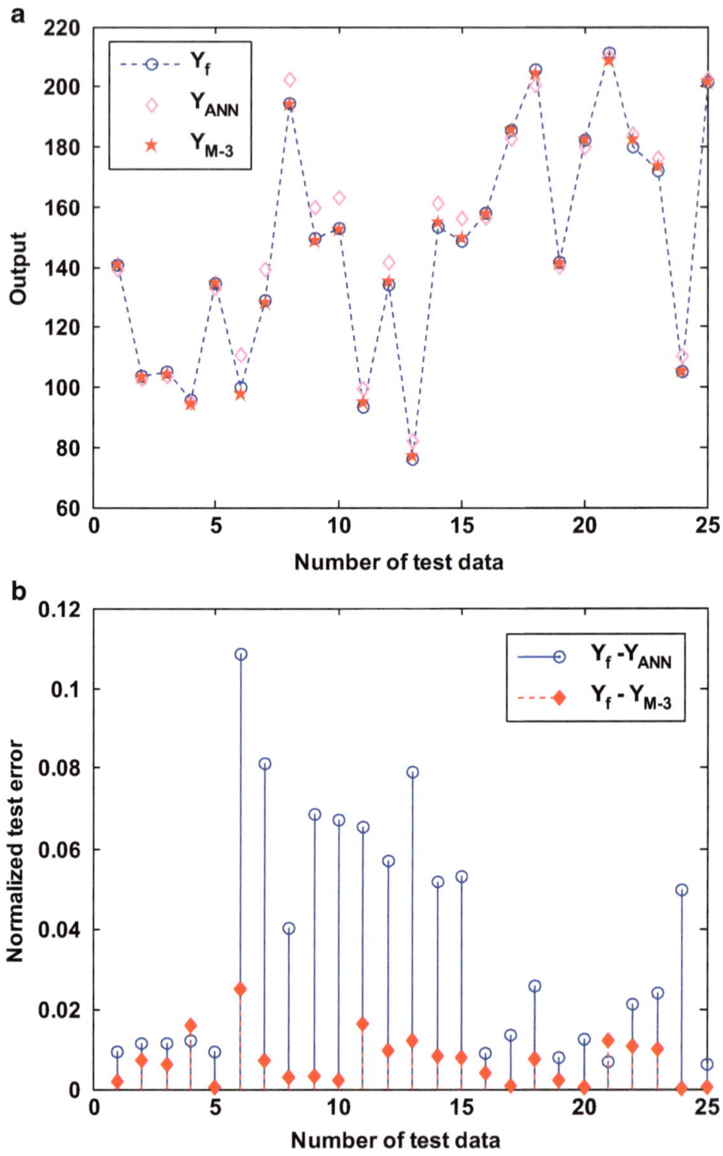

Fig. 7 Modeling results of the Branin function for less complex fine model: (**a**) Normalized test error for ANN model and M-3 model (**b**) Output of fine model, ANN model, and M-3 model for test data

3.1.2 More Complex Branin Function Modeling

Conventional ANN model is constituted by MLP with two hidden layers for more complex Branin function. Each layer has 60 neurons. Training process takes 600

Table 1 Comparing all techniques for modeling of less complex Branin function

Methods	Iteration number	Neuron number	Time consumption	Max error	Mean error
ANN	300	30-30	859.80 (s)	1.086 e-01	3.611 e-02
$M-1$	100	10-10	180.65 (s)	1.166 e-01	4.271 e-02
$M-2$	100	10-10	204.05 (s)	1.219 e-01	3.854 e-02
$M-3$	100	10-10	190.71 (s)	2.501 e-02	7.041 e-03
Three-step	300	30-30	575.41 (s)	2.501 e-02	7.041 e-03

iterations and utilizes 20,000 data to constitute final model. Each model in three-step modeling strategy is constituted MLP with two hidden layer as well. Each layer has 20 neurons. Therefore total number of neurons in model is same for conventional ANN and three-step modeling. Each modeling process in three-step strategy takes 200 iterations and utilizes 20,000 data like conventional ANN. Therefore total number of iterations in model is same for conventional ANN and three-step modeling. Twenty five test data are used to demonstrate modeling performance of two kinds of techniques.

$M-1$, $M-2$, and $M-3$ models are also compared in two different figures. Real response of fine model, $M-1$, $M-2$, and $M-3$ model are depicted in Fig. 8a. Normalized test error for three methods is depicted in Fig. 8b.

Real response of fine, conventional ANN, and $M-3$ models are depicted in Fig. 9a. The difference between three-step model and conventional ANN can be recognized from Fig. 9b. As shown in these figures, three-step modeling generates more accurate results gradually for the same training data and number of iterations.

All results for more complex Branin function modeling are summarized in Table 2. Three-step model consists of $M-1$, $M-2$, and $M-3$. These results demonstrate that three-step modeling strategy provides efficient modeling performance with regard to accuracy and time consumption for modeling of more complex Branin function.

3.1.3 Results for Branin Function Modeling

Considering Branin function modeling, it is necessary to summarize all the results about less complex and more complex fine models. Normalized test errors are depicted in Fig. 10a considering different number of data. This figure indicates that the accuracy of three-step modeling gradually improves starting from model 1 and arrives final value with model 3. In addition three-step strategy provides better accuracy that increases according to number of training data than conventional ANN. Time consumptions for different kinds of fine models are depicted in Fig. 10b.

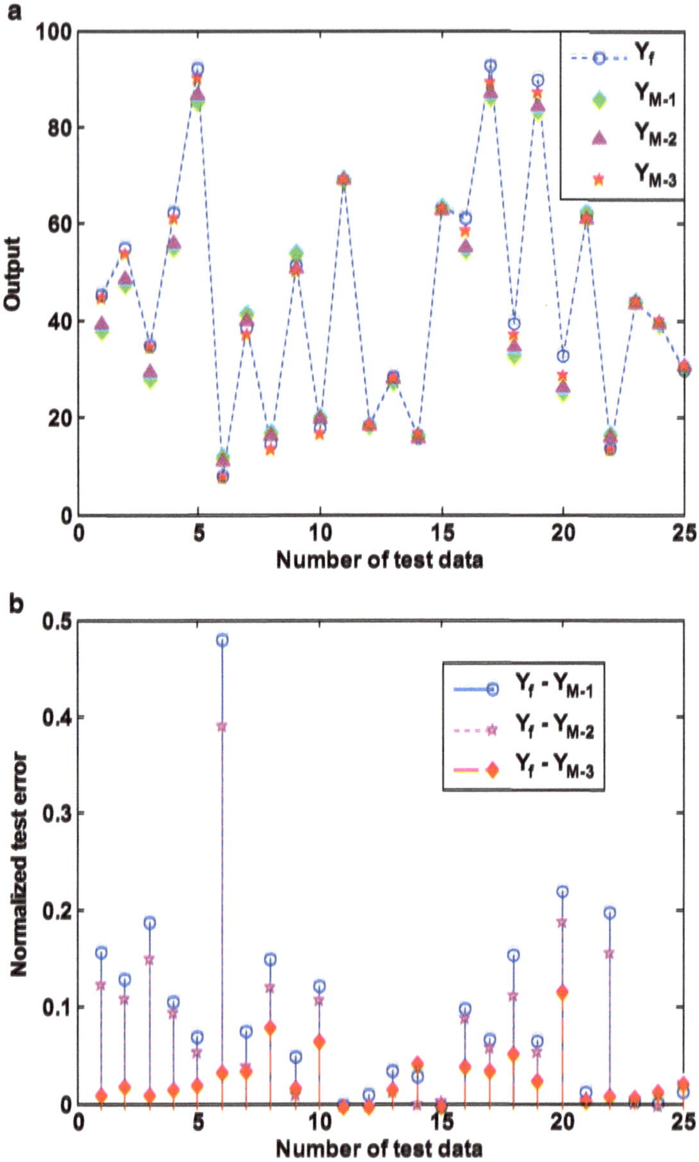

Fig. 8 Modeling results of the Branin function for more complex fine model: (**a**) Normalized test error for M-1 model, M-2 model and M-3 model (**b**) Output of more complex fine model, M-1 model, M-2 model and M-3 model for test data

Time consumption increases according to number of training data, while three-step modeling still preserves better time consumption performance towards to conventional ANN.

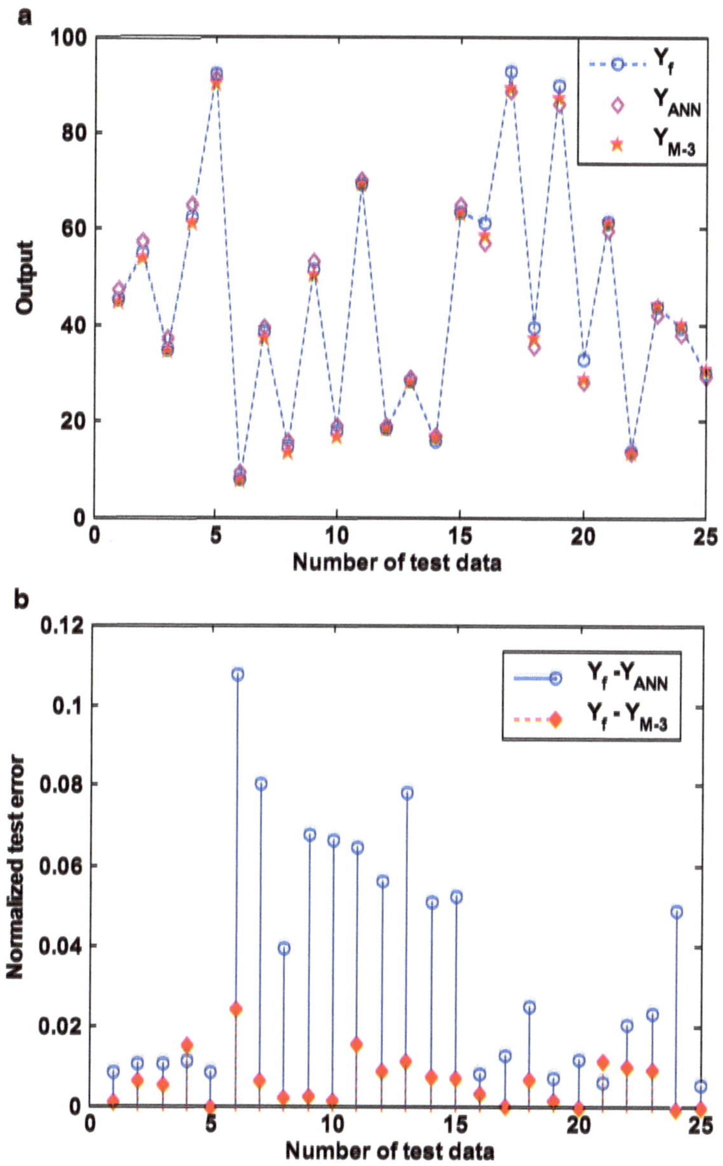

Fig. 9 Modeling results of the Branin function for more complex fine model: (**a**) Normalized test error for ANN model and M-3 model (**b**) Output of fine model, ANN model, and M-3 model for test data

Table 2 Comparing all techniques for more complex Branin function modeling

Methods	Iteration number	Neuron number	Time consumption (s)	Max error	Mean error
ANN	600	60-60	7,216.61	1.698 e-01	5.124 e-02
$M-1$	200	20-20	882.59	4.825 e-01	9.986 e-02
$M-2$	200	20-20	908.51	3.943 e-01	7.837 e-02
$M-3$	200	20-20	899.78	1.197 e-01	2.961 e-02
Three-step	600	60-60	2,690.88	1.197 e-01	2.961 e-02

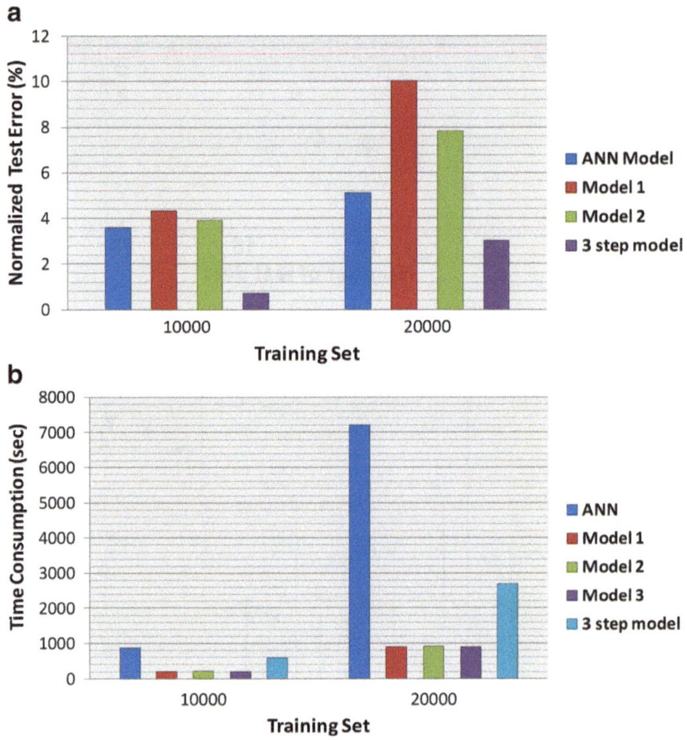

Fig. 10 Comparing results of the Branin function for conventional ANN, M-1, M-2, and three-step models: (**a**) Normalized test error (**b**) Time consumption

3.2 Engineering Modeling Problem: The Shape Reconstruction of Inverse Scattering

We consider the direct scattering problem [14] depicted in Fig. 11. The arbitrary shaped infinitely long impedance cylinder in free space is illuminated by plane wave whose polarization is cylinder axis (z axis). Cylinder contour can be expressed by means of Fourier series as

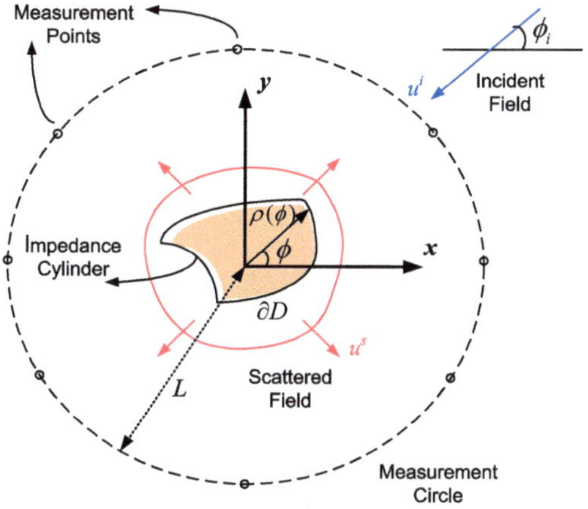

Fig. 11 Scattering problem geometry [14]

$$\rho(\phi) = \sum_{p=-P}^{P} a_p e^{ip\phi}, \qquad a_{-p} = a_p^* \tag{14}$$

where a_p is Fourier coefficients satisfying $a_{-p} = a_p^*$ and obtained as in (15)

$$a_p = \frac{1}{2\pi} \int_0^{2\Pi} \rho(\phi) e^{-ip\phi} d\phi \tag{15}$$

The fine model which is used to calculate measured electric field via direct scattering formulations constitutes a relation between the Fourier coefficients and the measured electric field as depicted as in Fig. 12.

3.2.1 The Shape Reconstruction of High Dimensional Inverse Scattering Problem

The shape reconstruction of the conducting cylinder using electromagnetic field measurements [9, 12, 14] will be used to demonstrate efficiency of three-step strategy. For this application, data efficiency that is provided by three-step strategy will be exhibited using different geometries of conducting cylinder. The conducting cylinder is illuminated by TMz wave with frequency 33 MHz and angle of incidence $\phi_i = 0$. The scattered field data are measured at ten points on measurement circle with radius 100λ as indicated in Fig. 11. The shape of conducting cylinder is represented by one real and four complex Fourier coefficients as indicated in Fig. 12.

Conventional ANN is used not only as one method for solving the problem but also as a way of building the coarse model for three-step modeling strategy. The considered conventional ANN structure is feed-forward MLP with two hidden layers.

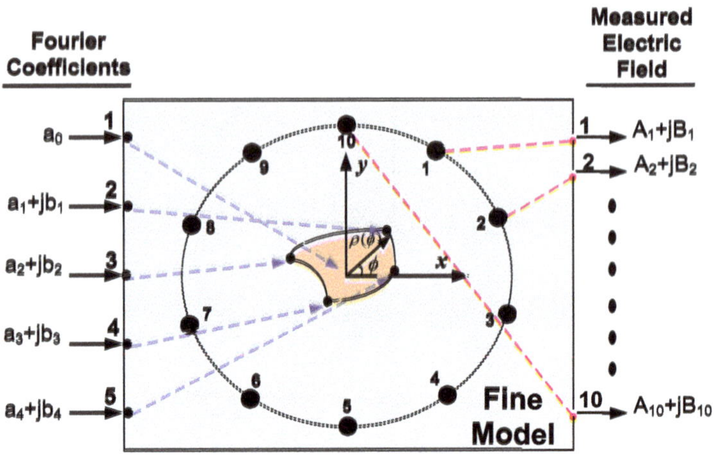

Fig. 12 Fine model for scattering problem [13]

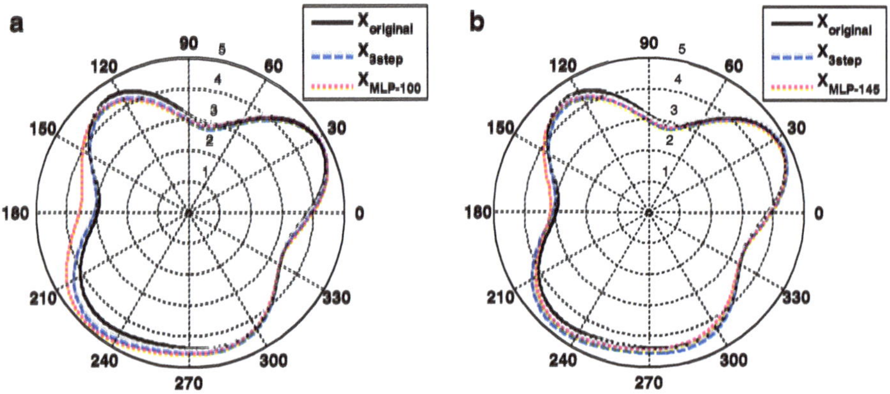

Fig. 13 Shape reconstruction of geometry-1 obtained from original (fine model), three-step trained with 100 data, MLP-100 and MLP-145 models

Error measurement is required to compare all methods. The mean and maximum errors are determined by (12) and (13). The efficiency of three-step strategy is tested via five different geometry. Only two of them can be shown in the figures. In addition the results obtained from three-step strategy for 100 and 200 training data compare to both $MLP - 100$ (ANN trained with 100 data) and $MLP - 200$ (ANN trained with 200 data) in order to demonstrate time consumption performance for same training data. To show data efficiency, same results obtained from three-step strategy compare to both $MLP - 145$ and $MLP - 245$ with respect to the accuracy.

All results of $geometry - 1$ for 100 and 145 training data are depicted in Fig. 13. Normalized test errors are 5 % ($MLP - 100$), 2.4 % ($MLP - 145$) and 2 % (three-step for $MLP - 100$). The similar results for more training data are also

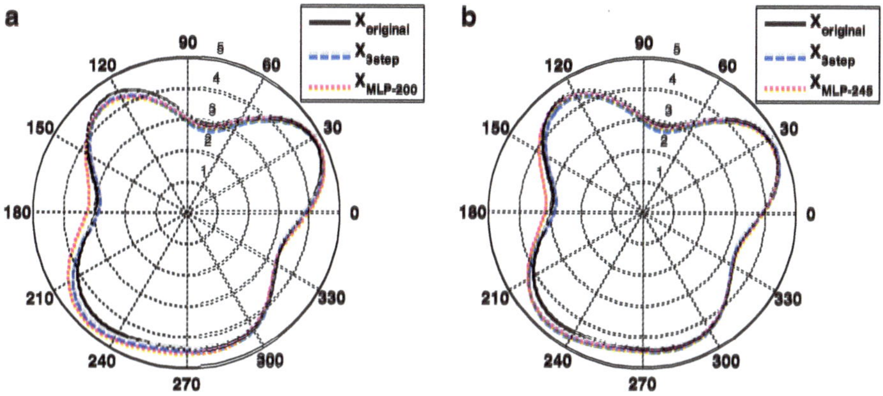

Fig. 14 Shape reconstruction of geometry-1 obtained from original (fine model), three-step trained with 200 data, MLP-200 and MLP-245 models

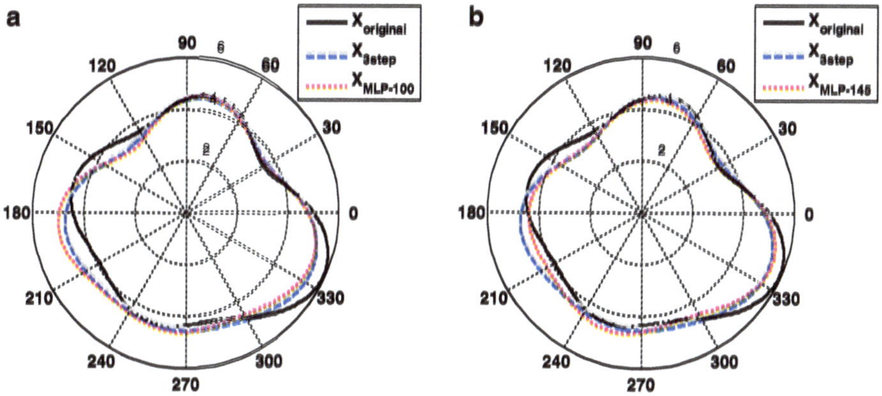

Fig. 15 Shape reconstruction of geometry-2 obtained from original (fine model), three-step trained with 100 data, MLP-100 and MLP-145 models

depicted in Fig. 14. In this case, normalized test errors are 3.2 % ($MLP-200$), 2.3 % ($MLP-245$) and 1.7 % (three-step for $MLP-200$). All results of $geometry-2$ for 100 and 145 training data are depicted in Fig. 15. Normalized test errors are 6 % ($MLP-100$), 4.4 % ($MLP-145$) and 4.9 % (three-step for $MLP-100$). The similar results for more training data are also depicted in Fig. 16. In this case, normalized test errors are 3.7 % ($MLP-200$), 3.1 % ($MLP-245$), and 2.5 % (three-step for $MLP-200$). All these figures show that three-step modeling strategy improves accuracy and time consumption using same training data and also provides more accuracy and similar time consumption using less training data than conventional ANN. Mean errors and maximum errors for five different geometry and other results are summarized for inverse scattering problem in Table 3.

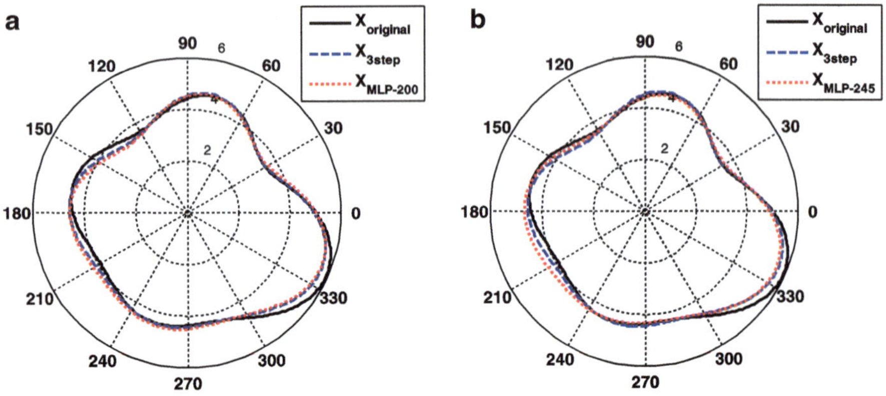

Fig. 16 Shape reconstruction of geometry-2 obtained from original (fine model), three-step trained with 200 data, MLP-200 and MLP-245 models

3.2.2 Results for Shape Reconstruction of Inverse Scattering Problem

Considering shape reconstruction of inverse scattering problem, it is necessary to summarize all the results in terms of accuracy and required number of data. Normalized test errors are depicted in Fig. 17a in terms of different number of data. This figure indicates that the accuracy of three-step modeling gradually improves starting from model 1 and arrives final value with model 3. In addition three-step strategy provides better accuracy that increases according to the number of training data than conventional ANN. Time consumptions for different kinds of fine models are depicted in Fig. 17b. Time consumption increases according to number of training data, while three-step modeling still preserves better time consumption performance towards to conventional ANN. Three-step strategy also provides better accuracy with less number of training data. For example, three-step strategy trained with 100 data that can provide more accurate results than conventional ANN trained with 145 data also has less time consumption than conventional ANN as depicted in Fig. 17.

4 Conclusion

In this work, three-step modeling strategy is considered to improve modeling in terms of accuracy, data use and time consumption and the results are compared with conventional ANN technique. Three-step modeling is constituted gradually using knowledge based techniques. First step is used to create required knowledge for the second step. Second step improves model response of the first step. Final step is performed with prior knowledge input with difference using response of the second step. Main advantage of final step is that coarse model obtained by the second

Table 3 Comparing all techniques for the shape reconstruction of inverse scattering problem

Methods	Iteration number	Neuron number	Time consumption training set + training	Mean of max error	Mean of mean error
$MLP - 100$	300	60-60	$\underbrace{10.031 \times 100 + 56.862}_{1059.962(\text{sec})}$	0.136375 (13.6 %)	0.040893 (4.1 %)
$MLP - 145$	300	60-60	$\underbrace{10.031 \times 145 + 81.682}_{1536.177(\text{sec})}$	0.088649 (8.9 %)	0.031881 (3.2 %)
Three-step for $MLP - 100$	300	60-60	$\underbrace{10.031 \times 100 + 172.364}_{1175.464(\text{sec})}$	0.081963 (8.2 %)	0.030998 (3.1 %)
$MLP - 200$	300	60-60	$\underbrace{10.031 \times 200 + 112.788}_{2118.988(\text{sec})}$	0.086136 (8.6 %)	0.028124 (2.8 %)
$MLP - 245$	300	60-60	$\underbrace{10.031 \times 245 + 172.406}_{2630.001(\text{sec})}$	0.083554 (8.4 %)	0.027061 (2.7 %)
Three-step for $MLP - 200$	300	60-60	$\underbrace{10.031 \times 200 + 356.492}_{2362.692(\text{sec})}$	0.062246 (6.2 %)	0.021382 (2.1 %)

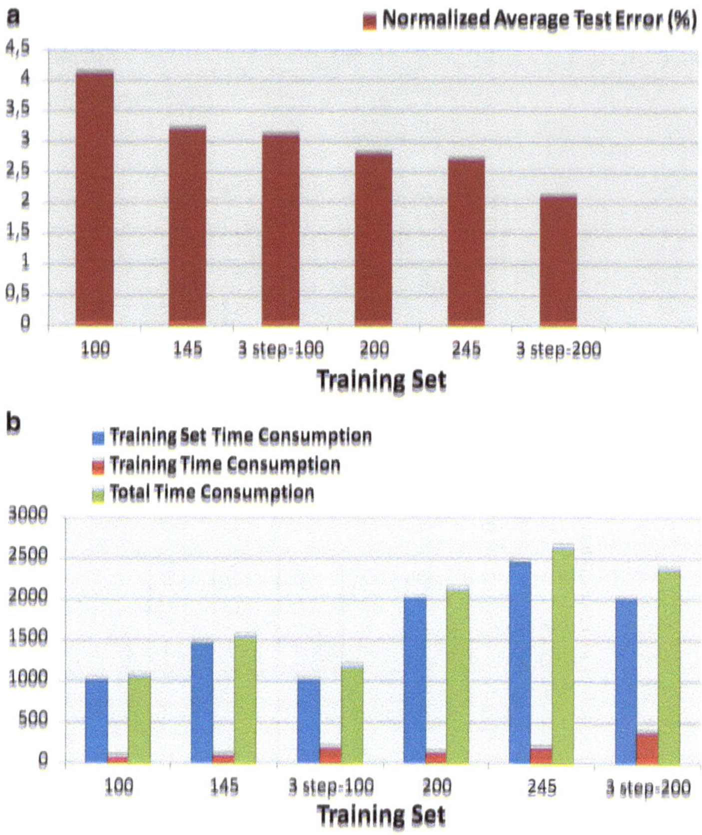

Fig. 17 Comparing results of the shape reconstruction for conventional ANN, M-1, M-2 and three-step models: (**a**) Normalized average test error (**b**) Time consumption

step is used twice in the modeling process. Therefore three-step model accuracy is better than conventional ANNs. Time consumption performance of three-step modeling strategy is another reason to prefer this strategy. To demonstrate efficiency of time consumption, total number of training data, total number of neurons and total number of iterations are fixed for both three-step strategy and conventional ANN during the Branin function modeling. In this case, three-step strategy provides more accuracy with less time consumption than conventional ANN. In order to demonstrate data efficiency, total number of neurons and total number of iterations are fixed for both each part of three-step strategy and conventional ANN during the shape reconstruction of inverse scattering problem. Although time consumption is same for same training data, it is possible to obtain more accuracy for less training data from three-step strategy with respect to conventional ANN.

References

1. Bandler, J.W., Cheng, Q.S., Dakroury, S.A., Mohamed, A.S., Bakr, M.H., Madsen, K., Sondergaard, J.: Space mapping : The state of the artc. IEEE Trans. Microw. Theory Tech. **52**(1), 337–361 (2004)
2. Burrascano, P., Fiori, S., Mongiardo, M.: A review artificial neural networks applications in microwave computer-aided design. Int. J. RF Microw. Comput. Aided Eng. **9**(3), 158–174 (1999)l
3. Devabhaktuni, V.K., Chattaraj, B., Yagoub, M.C.E., Zhang, Q.J.: Advanced microwave modeling framework exploiting automatic model generation, knowledge neural networks, and space mapping. IEEE Trans. Microw. Theory Tech. **51**(7), 1822–1833 (2003)
4. Haykin, S.: Neural network - a comprehensive foundation, 2nd edn. Printice Hall, New Jersey (1999)
5. Koziel, S., Bandler, J.W.: Modeling of microwave devices with space mapping and radial basis functions. Int. J. Numer. Model Electron. Netw. Dev. Fields **21**(1–2), 187–203 (2008)
6. Koziel, S., Leifsson, L. (ed.): Surrogate-Based Modeling and Optimization. Springer, New York (2013)
7. Rayas-Sanchez, J.E.: Em-based optimization of microwave circuits using artificial neural networks: the state-of-the-artc. IEEE Trans. Microw. Theory Tech. **52**(1), 420–435 (2004)
8. Simsek, M.: Developing 3-step modeling strategy exploiting knowledge based techniques. In: The 20th European Conference on Circuit theory and Design, Linkoping, 29–31 August 2011
9. Simsek, M.: The reconstruction of shape with 3-step modeling strategy. In: Scientific Computing in Electrical Engineering, ETH Zurich and ABB Corporate Research, Switzerland, 11–14 September 2012
10. Simsek, M., Sengor, N.S.: A knowledge-based neuromodeling using space mapping technique: compound space mapping-based neuromodeling. Int. J. Numer. Model Electron. Netw. Dev. Fields **21**(1–2), 133–149 (2008)
11. Simsek, M., Sengor,N.S.: An efficient inverse ANN modeling approach using prior knowledge input with difference method. In: The European Conference on Circuit theory and Design, Antalya, 23–27 August 2009
12. Simsek, M., Sengor, N.S.: Solving inverse problems by space mapping with inverse difference method. In: Roos, J., Costa, L.R.J. (eds.) Scientific Computing in Electrical Engineering SCEE 2008, Mathematics in Industry, vol. 14, pp. 453–460. Springer, Berlin (2010)
13. Simsek, M., Sengor, N.S.: The efficiency of difference mapping in space mapping-based optimization. In: Koziel, S., Leifsson, L. (eds.) Surrogate-Based Modeling and Optimization, pp. 99–120. Springer, New York (2013)
14. Simsek, M., Tezel, N.S.: The reconstruction of shape and impedance exploiting space mapping with inverse difference method. IEEE Trans. Antenna Propag. **60**(4), 1868–187 (2012)
15. Simsek, M., Zhang, Q.J., Kabir, H., Cao, Y., Sengor, N.S.: The recent developments in knowledge based neural modeling. Proc. Comput. Sci. **1**(1), 1315–1324 (2010)
16. Simsek, M., Zhang, Q.J., Kabir, H., Cao, Y., Sengor, N.S.: The recent developments in microwave design. Int. J. Math. Model. Numer. Optim. **2**(2), 213–228 (2011)
17. Sondergard, J.: Optimization using surrogate models by the space mapping technique. PhD thesis, Informatics and Mathematical Modelling, Technical University of Denmark (2003)
18. Zhang, Q.J., Gupta, K.C.: Neural Networks for RF and Microwave Design. Artech House, Boston (2000)
19. Zhang, Q.J., Gupta, K.C., Devabhaktuni, V.K.: Artificial neural networks for RF and microwave design—from theory to practice. IEEE Trans. Microw. Theory Tech. **51**(4), 1339–1350 (2003)

Large-Scale Global Optimization via Swarm Intelligence

Shi Cheng, T.O. Ting, and Xin-She Yang

Abstract Large-scale global optimization (LSGO) is a challenging task with many scientific and engineering applications. Complexity, nonlinearity and size of the problems are the key factors that pose significant challenges in solving such problems. Though the main aim of optimization is to obtain the global optimal solutions with the least computational costs, it is impractical in most applications. Thus, a practical approach is to search for suboptimal solutions and good solutions, which may not be easily achievable for large-scale problems. In this chapter, the challenges posed by LSGO are addressed, followed by some potential strategies to overcome these difficulties. We also discuss some challenging topics for further research.

Keywords Large-Scale Global Optimization • Swarm Intelligence Optimization • Population Diversity • Exploration/Exploitation

1 Introduction

Optimization problems can be challenging to solve, especially for highly nonlinear problems. Finding solutions to such problems becomes even more challenging when the problem size becomes large, and in this case, we have to deal with large-scale

S. Cheng (✉)
International Doctoral Innovation Centre, The University of Nottingham, Ningbo, People's Republic of China, & Division of Computer Science, The University of Nottingham, Ningbo, People's Republic of China
e-mail: shi.cheng@nottingham.edu.cn

T.O. Ting
Department of Electrical & Electronic Engineering, Xi'an Jiaotong-Liverpool University, Suzhou, People's Republic of China
e-mail: toting@xjtlu.edu.cn

X.-S. Yang
School of Science and Technology, Middlesex University, The Burroughs, London NW4 4BT, UK
e-mail: x.yang@mdx.ac.uk

S. Koziel et al. (eds.), *Solving Computationally Expensive Engineering Problems*, Springer Proceedings in Mathematics & Statistics 97,
DOI 10.1007/978-3-319-08985-0_10

optimization problems. Even for problems of small and moderate sizes, finding the best feasible solutions is not straightforward. Sometimes, it may be useful to consider the problem of interest in terms of its modality. In the context of modality, optimization problems can be divided into two categories: unimodal problems and multimodal problems. As indicated by its name, a unimodal problem has a single optimum solution whereas a multimodal problem may have more than one global solution, together with potentially many local solutions.

For many large-scale, complex optimization problems, there are no efficient algorithms at all. In many cases, heuristic algorithms such as evolutionary algorithms (EA) are the main alternatives, and they can be very useful. However, they may be inefficient in tackling global optimization problems in the case of multimodal problems, due to the possible occurrence of the premature convergence, and such issue occurs when the solution is trapped in local optima [1–3]. The good news is that for a given type of problem, their efficiency can be increased by tuning their parameter settings, usually determined based on empirical results [4].

In essence, an optimization problem in $\mathbb{R}^n$ is a mapping of $f : \mathbb{R}^n \rightarrow \mathbb{R}^k$, where $\mathbb{R}^n$ is known as a decision space [5], also known as search space [6], and $\mathbb{R}^k$ is the objective space [7]. Further, optimization problems can be sub-divided into two categories according to the value of k. Thus, when $k = 1$ for a given problem, it can be categorized as Single Objective Problem (SOP), and when $k > 1$, this is known as multi-objective optimization (MOO), or multicriteria optimization [8–10]. The evaluation function in optimization, $f(\mathbf{x})$, maps the decision variables to objective vectors. Each solution in decision space is associated with a fitness value with respect to relevant objective space. This situation is illustrated in Fig. 1 for the case of $n = 3$ and $k = 2$.

However, as our emphasis here is on the swarm intelligence in the context of large-scale optimization problems, we will focus on the single objective formulation without the loss of generality. Thus, we have

$$\text{Minimize} \quad f(\mathbf{x})$$

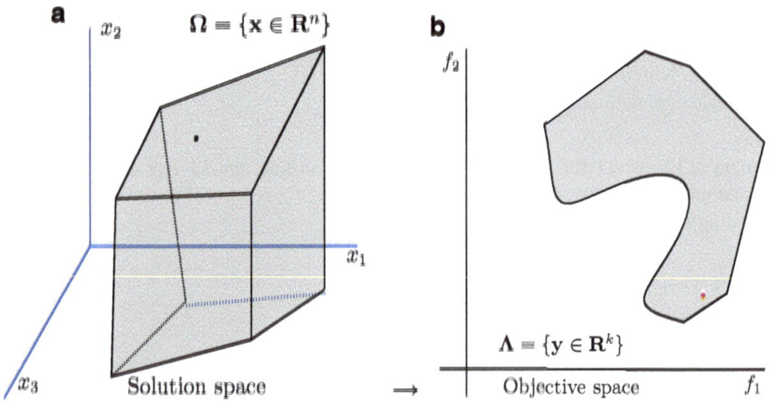

Fig. 1 The mapping from solution space to objective space

where

$$\mathbf{x} = (x_1, x_2, \ldots, x_n) \in X.$$

Here, $X \subset \mathbb{R}^n$ represents the decision space with n dimensions. In addition, $f : X \to \mathbb{R}$ denotes the continuous real-valued objective function mapped from an n-dimensional decision space to one fitness value $F(\mathbf{x})$. Usually, a problem with $n \geq 1,000$ is referred to as a continuous LSGO problem [11]. The single objective formulation of LSGO problem is given here; however, the LSGO problem also can be a multi-objective problem.

For large-scale optimization problems, there are many additional challenges, and this chapter will focus on these challenging issues concerning the solution methods. Therefore, it is organized as follows. Section 2 reviews the basic concepts in swarm intelligence. Section 3 outlines the challenges in the large-scale optimization problems. The techniques concerning swarm intelligence adopted for solving large-scale problems are discussed in Sect. 4. Some applications of swarm intelligence in real-world large-scale problems are briefly reviewed in Sect. 5, followed by the brief conclusions in Sect. 6.

2 Swarm Intelligence

Many real-world applications can be transformed into optimization problems, but this does not mean that it would be easy to solve such problems. In reality, problems can be quite complex, as the design variables can be continuous, discrete, or even mixed, while the functions may not be differentiable. Therefore, to find even a feasible solution may not be easy. Obviously, in special cases of unimodal, continuous problems, traditional methods such as gradient-based methods (e.g., hill-climbing) can be used to find the global solutions. But for multimodal problems, gradient-based methods do not work well, and their results will largely depend on the initial starting points, even for smooth problems. In reality, real-world problems are rarely unimodal or differentiable, and therefore, significant challenges arise.

Among all the methods that are designed to tackle large-scale problems, metaheuristic algorithms are among the most promising types [12]. One class of the metaheuristic algorithms is the so-called Swarm Intelligence (SI). SI-based algorithms typically use a population of multiple agents so as to mimic the successful characteristics of natural systems such as ants, fish, birds, bats, fireflies and others [12–17]. In essence, multiple interacting agents may evolve and thus may lead to self-organized behaviour, the so-called collective intelligence. Contemporary algorithms such as particle swarm optimization, ant colony optimization, bat algorithm, firefly algorithm, cuckoo search, bee algorithms as well as bacterial foraging optimization all have demonstrated some unique characteristics and potential for solving nonlinear optimization problems [12, 18].

The wide use and popularity of SI-based algorithms may be attributed to their simplicity, parellelism and flexibility. Most SI-based algorithms are relatively simple and easy to implement, and the use of multiple agents makes them natural to be parallelized. In addition, such seemingly simple algorithms tend to be very flexible in dealing with a wide range of problems. All these features make swarm intelligence based algorithms quite efficient in many applications [12, 16]. Despite such success, there are still many problems. This chapter will focus on the challenges and potential improvements in solving large-scale problems.

3 Challenges in Large-Scale Global Optimization

The size of a problem is sometimes a deciding factor that affects the solution strategies. However, the combination of size with nonlinearity often causes the major difficulty. For example, if the problems are linear, then linear programming techniques such as the simplex method can often deal with large-scale problems (up to a few million design variables) easily. However, for nonlinear problems, these techniques will not work. Nonlinearity with moderately large-scale sizes can pose substantial challenges.

Though there is no agreed exact definition for large-scale problems, typically nonlinear problems with a large number of variables, e.g., more than thousands variables, can be called LSGO problems [19, 20]. In practice, the performance of many algorithms deteriorates quickly as the dimension of the problem increases. For example, the nearest neighbour approaches are very effective in categorization, but they will be very ineffective for high-dimensional problems. The computational complexity often increases dramatically as the problem size increases. It is well known that the travelling salesman problems (TSP) can have exponential complexity, and thus no efficient algorithms for the TSP class of problems.

3.1 Large-Scale Optimization

Large-scale problems typically have more than 1,000 design variables [21], and they can have many issues, including algorithms, data structures, memory problem and performance issues. One of the main issue is the choice of algorithms, and in many cases, there is no good algorithm at all. Even if there is a good algorithm available, data structures and memory management can also be very important. Even with all these problems sorted, there still exists some performance issue. Whether the computational time is acceptable, and in most cases, the computational costs are too high and thus impractical. In addition, proper performance measures are needed to ensure a fair comparison of different algorithms.

For LSGO, many optimization methods suffer from the "curse of dimensionality" [22–26], which may imply that their performance deteriorates quickly as the dimension of the search space increases [27]. There are several reasons that cause this phenomenon. Firstly, as mentioned above, the complexity of the problem increases exponentially with respect to the number of dimensions. The "empty space phenomenon" is a good example to this scenario [24, 28, 29]. With a number of m possible solutions with n dimensions, the fraction of the feasible search space becomes negligible. In addition, the bias can be accumulated. For example, in particle swarm optimization, the solution update depends on the combination of several vectors, i.e., the current value, the difference between current value and previous best value. In a low-dimensional space, the direction of the vector combination has the high probability to direct towards the global optimum. However, the distance metric for the low dimension space may not be effective in a high-dimensional space. The search direction may be far away from the global optimum due to the bias accumulation.

For traditional methods, Benson et al. compared three major methods: the interior-point method, a trust-region algorithm and the quasi-Newton methods that works for 10,000 dimensions with promising results [21].

From the implementation point of view, to obtain a good approximate solution quickly may be more useful than to find an accurate solutions very slowly. Fortunately, in many SI-based algorithms, good feasible solutions can be found with various improvements and strategies, even for high-dimensional problems [6, 27, 30–36]. However, different degrees of success exist and more extensive research is highly needed.

3.2 *Good Solution or Good Convergence?*

One of the challenges for large-scale optimization is that there is no guarantee for global optimality, except for special cases such as linear problems and convex problems. In general, in order to get a good set of solutions in a practically acceptable timescale, one has to sacrifice the possibility of finding the true global optimality. Therefore, there is some compromise between speed and global optimality. In order to obtain good solutions, a good convergence rate is needed. However, if the convergence rate is high, it can usually lead to premature convergence. In most case, a prematurely converged solution is usually not a good solution because it can be stuck in any point in the search if the convergence speed is too high, because premature convergence is essentially stagnation with almost no diversity in the solution population. Consequently, there is a trade-off between getting a truly optimal solution and good algorithm convergence.

Even so, from a practical point of view, LSGO requires a fast convergence on the large search space, and thus any algorithm that can find the "good enough" solution(s) within a limited time is preferred.

For convergence in the context of computational intelligence, there is no well-accepted definition, and the traditional definition of sequence convergence cannot be used directly. In fact, convergence in probability and convergence in distribution are acceptable. Mathematical speaking, convergence with probability 1 can be defined as follows: If an objective function f is measurable in a feasible solution space Ω in the measurable subset of $\Re^n$ (i.e., $\Omega \subset \Re^n$), an algorithm A produces a search sequence $\{x_k\}_{k=0}^{\infty}$. Then, the convergence with probability 1 is expressed as

$$\lim_{k \to \infty} P(x_k \in R_{opt}) = 1,$$

where R_{opt} is the optimal set of the solutions in the search space [37].

However, even with this definition, the proof of convergence often requires complex mathematical tools such as dynamical systems and Markov chain theory [12, 36]. Unless theory proves otherwise, most convergence analyses have focused on the convergence to the best solution of the population during the iterative search process, there is no guarantee that the best solution found by an algorithm is truly the global optimum for the problem of interest.

Even without theoretical analysis, convergent behaviour and characteristics can be observed in practice when running an algorithm. In fact, premature convergence and stagnation can also be frequently observed in many algorithms such as particle swarm optimization and genetic algorithms.

3.3 *Performance Measures*

For the purpose of comparing different algorithms, the performance measure can be very important. After all, the performance is measured against a criterion. In the literature, the following two main criteria are in use [19]: functional evaluations and accuracy. To compare two algorithms for solving a given problem, a fixed accuracy or tolerance is usually chosen, then the aim is to compare the number of function evaluations. If algorithm A uses fewer evaluations than B, then A tends to be better than B. However, care must be taken when drawing such conclusions, as a single run is normally inconclusive. In practice, multiple independent runs are needed so that meaningful statistics such as the means and standard deviations can be calculated.

On the other hand, another measure is to compare the accuracy of the solution found for a fixed number of functional evaluations. Again the accuracy should be calculated based on multiple independent runs. In addition, a third measure is to compare computational times for other things being fixed. However, extreme care must be taken for this approach as computational time can depend on the details of implementation, computer configurations and data structures even for efficient algorithms. In addition, run times may vary on the same computer when running at different times due to the potential hidden processes in the system, especially on Windows operating systems.

Even if the above measures are properly implemented, multiple runs on a diverse range of problems are required to ensure fair comparison and sensible conclusion. Statistical testing and hypotheses should be carried out to draw meaningful conclusions.

For a more thorough performance measure, performance profiling is a very good measure [21, 38]. The performance profile developed by Dolan and Moré works well to compare m algorithms on the same set of N problems [38]. If $t_{Q,A}$ is the runtime required by algorithm A to solve problem Q, then the performance ratio is defined by

$$\rho_{Q,A} = \frac{t_{Q,A}}{\min\{t_{Q,A} : 1 \le A \le m\}},$$

which means the ratio of the current solver (algorithm) to the best time of all algorithms.

In a special case when an algorithm cannot find a solution, $\rho_{Q,A} = \infty$ [21], The performance profile $P_A \in [0, 1]$ of an algorithm on the whole set of problems can be defined as

$$P_A(\tau) = \frac{1}{N} \text{size}\,\{Q : 1 \le Q \le N, \rho_{Q,A} \le \tau\},$$

which essentially represents the cumulative distribution function of $\rho_{Q,A}$. Here τ is a constant. Statistically speaking, $P(\tau)$ can be considered as efficiency for $\tau = 1$, while $\lim_{\tau \to \infty} P(\tau)$ becomes the probability of success.

4 Techniques and Potentials for Solving Large-Scale Problems

Many effective strategies have been proposed for high-dimensional optimization problems, including problem decomposition and subcomponents cooperation, parameter adaptation and surrogate-based fitness evaluations [6, 21]. In the context of swarm intelligence, there are some strategies that try to improve the performance of swarm intelligence based algorithms for large-scale global optimization.

The literature in this area is expanding, and thus we do not intend to be complete in reviewing the relevant literature. Instead, we will highlight a few useful approaches or strategies for solving large-scale optimization problems in the context of nature-inspired algorithms. These strategies are diversity promotion, exploitation enhancement, decomposition and adaptation. As a matter of fact, there is no clear distinct between these strategies. Most algorithms use one or a combination of several approaches so as to be effective in solving large-scale problems.

4.1 Diversity Promotion

The diversity in the population of any SI-based algorithms can be important for sufficient exploration of the search space. In fact, diversity is almost equivalent to exploration [20,39]. Premature convergence is more likely to happen in a population with low diversity. In many cases, the population variance can be a good measure of solution diversity. If the variance gradually reduces, then the population will converge, but it can also be an indicator of premature convergence. Therefore, maintaining a good diversity in the population can avoid potential premature convergence. However, how to maintain the diversity is still a challenging question, though there are some approaches that can be useful. The aim of the diversity is to ensure the population maintains the capability of jumping out of local optima [40]. Some algorithms use initialization as a part of diversity control.

- **Random partial re-initialization**: As its name indicates, random partial re-initialization means reserving particles by means of a random approach. This strategy is able to achieve a great ability of exploration as a majority of the particles are re-initialized.
- **Elitist partial re-initialization**: This strategy keeps a better half of the population, with the other half being re-initialized. In this case, the algorithm increases the ability of exploration and this is essentially the fitness-proportional randomization technique.

Though the above approaches use the term "re-initialization", they are in fact randomization techniques. Some parts of the population are just generated by randomization, and this does not re-start the search process. However, the term "re-initialization" does have a meaning of indicating reset of some of the solutions so as to keep them afresh.

4.2 Exploitation Enhancement

In almost all algorithms, the convergence is a result of the appropriate use of local information such as gradients. Such use of updated information is referred to exploitation. Thus, it is natural to understand that exploitation enhancement may lead to improved convergence if used wisely. In fact, exploitation enhancement contradicts the diversity promotion. Therefore, some trade-off or balance between diversity/exporation and exploitation. Despite the importance, there is no good strategy in maintaining this balance [16].

Exploitation can be static by using some local information, and it can also be dynamical by using updated information found in the search so far. It seems that dynamical exploitation can be a promising approach and should be investigated further.

4.3 Decomposition

For large-scale problems, a "divide and conquer" approach can be very useful. A good example is the dynamical programming which uses a dynamical re-use of the results by solving a series of subproblems. Another example is sequential quadratic programming [19]. Based on the idea of "Divide and Conquer", a large-scale problem can be decomposed into a series of sub-problems in lower dimensions. For example, the separation of the whole decision variable set into a number of groups. Each group forms a sub-space of solutions, a certain evolutionary algorithm is applied on each sub-space, and after that, useful search information is shared among different groups. The advantage of this approach is that it can reduce the problem size, however, the optimality may be affected because the optimal solutions to each sub-problems do not guarantee the optimality for the whole problem. However, some studies can be useful in producing good solutions [33, 41].

4.4 Adaptation

The "No Free Lunch" (NFL) theorems for optimization, proved by Wolpert and Macready [35, 42], suggest that no algorithm is better than another algorithm in the absolute sense when measured in terms of average performance for all problems. However, NFL theorems are not valid for the cases of co-evolution [43]. This may mean that large-scale problems can be solved by coevolutionary methods or adaptive methods. Potentially, an algorithm that can adapt according to the problem landscape can be advantageous to solve LSGO problems. Several studies have investigated such possibility, including adaptive coevolutionary differential evolution algorithm [44], scalability of generalized adaptive differential evolution [34], and self-adaptive mixed distribution based univariate estimation of distribution algorithm [45].

4.5 Multi-Stage Strategy

Another potentially effective way to deal with large-scale problems is to use multi-stage strategies. A simple and yet very powerful two-stage strategy is called the Eagle Strategy [46], which uses a combination of two different stages in iterative manner so as to reduce the computational efforts and thus increase the efficiency [36]. It can be expected that a good combination of multi-stage strategies with adaptation, while maintaining a good diversity and properly enhancement of exploitation, would be a very effective approach. Certainly, future research should explore this further.

5 Applications

The applications of swarm intelligence based algorithms are very diverse, while their application for large-scale optimization is sparse in contrast. However, the good news is that more applications in this area start to expand.

Among SI-based algorithms, it seems that PSO is relatively thoroughly investigated due to its relative long history, and thus more case studies and variants have emerged [47]. In the context of large-scale numerical optimization, existing studies include particle swarm optimization [47], covariance matrix adaptation evolution strategy [48], self-adaptive mixed distribution based univariate estimation of distribution algorithm [45], dynamic multi-swarm particle swarm optimizer with local search algorithm [49, 50], cooperative co-evolution particle swarm optimization (CCPSO) [51,52] and velocity divergence of CCPSO [53]. In addition, harmony search has also been applied to relatively large-scale problems such as water distribution networks [54].

New algorithms have also demonstrated promising efficiency in dealing with large-scale problems. For example, the travelling salesman problem has been solved by cuckoo search [31], while computationally extensive structural optimization problems have been solved by the firefly algorithm [32].

In the area of telecommunications and wireless communications, more and more applications are becoming increasingly large-scale problems [55,56]. Massive data are being generated from the long-term and/or large-scale applications as driven by information technology and social networks. Consequently, many large-scale optimization now concerns data mining applications and massive data sets.

6 Conclusions

Many real-world applications are large-scale problems and thus pose special challenges to solve. Though swarm intelligence based algorithms can be promising, challenging issues still exist. In this chapter, we have briefly reviewed the main challenges associated with swarm intelligence in the context of large-scale global optimization.

Strategies for improving solution diversity, exploitation enhancement, decomposition, adaptation and multistage strategies are discussed. It can be expected that an effective approach requires a good combination of all these strategies.

Obviously, other issues also exist. Even applications of about 1,000–10,000 dimensions exist (but rare), real-world applications such as business optimization and structural optimization can have millions of design variables, which can be truly large-scale. It is not clear at the moment if the existing algorithms can be scaled up to deal with these truly large-scale problems. In addition, algorithms are just one part of the challenges. As the speed of computers increases, distributed and parallel computing as well as cloud computing facilities can be good tools for solving large-scale global optimization problems.

Acknowledgements This work was carried out at the International Doctoral Innovation Centre (IDIC). The authors acknowledge the financial support from Ningbo Education Bureau, Ningbo Science and Technology Bureau, China's MOST and The University of Nottingham. The work is also partially supported by National Natural Science Foundation of China (NSFC) under grant No.60975080, 61273367; and Ningbo Science & Technology Bureau (Project No.2012B10055).

References

1. De Jong, K.A.: An analysis of the behavior of a class of genetic adaptive systems. PhD thesis, Department of Computer and Communication Sciences, University of Michigan (1975)
2. Mauldin, M.L.: Maintaining diversity in genetic search. In: Proceedings of the National Conference on Artificial Intelligence (AAAI 1984), pp. 247–250 (1984)
3. Goldberg, D.E.: Genetic Algorithms in Search, Optimization and Machine Learning. Addison-Wesley, Boston (1989)
4. Eiben, Á.E., Hinterding, R., Michalewicz, Z.: Parameter control in evolutionary algorithms. IEEE Trans. Evol. Comput. **3**, 124–141 (1999)
5. Adra, S.F., Dodd, T.J., Griffin, I.A., Fleming, P.J.: Convergence acceleration operator for multiobjective optimization. IEEE Trans. Evol. Comput. **12**, 825–847 (2009)
6. Jin, Y., Sendhoff, B.: A systems approach to evolutionary multiobjective structural optimization and beyond. IEEE Comput. Intell. Mag. **4**, 62–76 (2009)
7. Sundaram, R.K.: A First Course in Optimization Theory. Cambridge University Press, Cambridge (1996)
8. Purshouse, R.C., Fleming, P.J.: On the evolutionary optimization of many conflicting objectives. IEEE Trans. Evol. Comput. **11**, 770–784 (2007)
9. Adra, S.F., Fleming, P.J.: Diversity management in evolutionary many-objective optimization. IEEE Trans. Evol. Comput. **15**, 183–195 (2011)
10. Yang, X.S.: Multiobjective firefly algorithm for continuous optimization. Eng. Comput. **29**, 175–184 (2013)
11. Nemhauser, G.L.: The age of optimization: Solving large-scale real-world problems. Oper. Res. **42**, 5–13 (1994)
12. Yang, X.-S., Cui, Z.H., Xiao, R.B., Gandomi, A.H., Karamanoglu, M. (eds.): Swarm Intelligence and Bio-Inspired Computation: Theory and Applications. Elseveir, Waltham (2013)
13. Eberhart, R.C., Shi, Y.: Guest editorial special issue on particle swarm optimization. IEEE Trans. Evol. Comput. **8**, 201–203 (2004)
14. Engelbrecht, A., Li, X., Middendorf, M., Gambardella, L.M.: Editorial special issue: Swarm intelligence. IEEE Trans. Evol. Comput. **13**, 677–680 (2009)
15. Panigrahi, B.K., Shi, Y., Lim, M.-H.: Handbook of Swarm Intelligence: Concepts, Principles and Applications. Adaptation, Learning, and Optimization, vol. 8. Springer, Berlin (2011)
16. Yang, X.S.: Review of meta-heuristics and generalised evolutionary walk algorithm. Int. J. Bioinspired Comput. **3**(2), 77–84 (2011)
17. Yang, X.S.: Efficiency analysis of swarm intelligence and randomization techniques. J. Comput. Theor. Nanosci. **9**(2), 189–198 (2012)
18. Fister, I., Fister Jr., I., Yang, X.S., Brest, J.: A comprehensive reivew of firefly algorithms. Swarm and Evol. Comput. **13**, 34–46 (2013)
19. Yang, X.-S.: Engineering Optimization: An Introduction with Metaheuristic Applications. Wiley, Hoboken (2010)
20. Cheng, S.: Population Diversity in Particle Swarm Optimization: Definition, Observation, Control, and Application. PhD thesis, Department of Electrical Engineering and Electronics, University of Liverpool (2013)

21. Benson, H.Y., Shanno, D.F., Vanderbei, R.J.: A comparative study of large-scale nonlinear optimization algorithms. In: Pillo, G.D., Murli, A. (eds.) High Performance Algorithms and Software for Nonlinear Optimization, pp. 95–127. Kluwer Academic, Dordrecht (2003)
22. Bellman, R.: Adaptive Control Processes: A guided Tour. Princeton University Press, Princeton (1961)
23. Donoho, D.L.: Aide-Memoire. High-Dimensional Data Analysis: The Curses and Blessings of Dimensionality, tech. rep., Stanford University (2000)
24. Lee, J.A., Verleysen, M.: Nonlinear Dimensionality Reduction. Information Science and Statistics. Springer, Berlin (2007)
25. Hastie, T., Tibshirani, R., Friedman, J.: The Elements of Statistical Learning: Data Mining, Inference, and Prediction. Springer Series in Statistics, 2nd edn. Springer, New York (2009)
26. Domingos, P.: A few useful things to know about machine learning. Commun. ACM **55**, 78–87 (2012)
27. Cheng, S., Shi, Y., Qin, Q.: Dynamical exploitation space reduction in particle swarm optimization for solving large scale problems. In: Proceedings of 2012 IEEE Congress on Evolutionary Computation, (CEC 2012), pp. 3030–3037, IEEE, 2012
28. Scott, D.W., Thompson, J.R.: Probability density estimation in higher dimensions. In: Gentle, J.E. (ed.) Computer Science and Statistics: Proceedings of the Fifteenth Symposium on the Interface, pp. 173–179, 1983
29. Verleysen, M.: Learning high-dimensional data. In: Ablameyko, S., Gori, M., Goras, L., Piuri, V. (eds.) Limitations and Future Trends in Neural Computation. NATO Science Series, III: Computer and Systems Sciences, vol. 186, pp. 141–162. IOS Press (2003)
30. Cheng, S., Shi, Y., Qin, Q., Bai, R.: Swarm intelligence in big data analytics. In: Yin, H., Tang, K., Gao, Y., Klawonn, F., Lee, M., Weise, T., Li, B., Yao, X. (eds.) Intelligent Data Engineering and Automated Learning - IDEAL 2013. Lecture Notes in Computer Science, vol. 8206, pp. 417–426. Springer, Berlin (2013)
31. Ouaarab, A., Ahiod, B., Yang, X.-S.: Discrete cuckoo search algorithm for the travelling salesman problem. Neural Comput. Appl. **24**, 1–11 (2013)
32. Gandomi, A.H., Yang, X.S., Alavi, A.H.: Mixed variable structural optimization using firefly algorithm. Comput. Struct. **89**, 2325–2336 (2011)
33. Yang, Z., Tang, K., Yao, X.: Differential evolution for high-dimensional function optimization. In: Proceedings of 2007 IEEE Congress on Evolutionary Computation (CEC 2007), pp. 35231–3530, IEEE, 2007
34. Yang, Z., Tang, K., Yao, X.: Scalability of generalized adaptive differential evolution for large-scale continuous optimization. Soft Comput. **15**, 2141–2155 (2011)
35. Yang, X.S.: Free lunch or no free lunch: That is not just a question? Int. J. Artif. Intell. Tools **21**(03) (2012)
36. Yang, X.-S., Karamanoglu, M., Ting, T., Zhao, Y.-X.: Applications and analysis of bio-inspired eagle strategy for engineering optimization. Neural Comput. Appl. 1–10 (2013)
37. Francisco, S.J., Wets, J.B.R.: Minimization by random search techniques. Math. Oper. Res. **6**, 19–30 (1981)
38. Dolan, E.D., Moré, J.J.: Benchmarking optimization software with performance profiles. Math. Program. **91**, 201–214 (2002)
39. Cheng, S., Shi, Y.: Diversity control in particle swarm optimization. In Proceedings of 2011 IEEE Symposium on Swarm Intelligence (SIS 2011), pp. 110–118, (Paris, France), 2011
40. Cheng, S., Shi, Y., Qin, Q.: Promoting diversity in particle swarm optimization to solve multimodal problems. In: Lu, B.-L., Zhang, L., Kwok, J. (eds.) Neural Information Processing. Lecture Notes in Computer Science, vol. 7063, pp. 228–237. Springer, Berlin (2011)
41. Zhang, K., Li, B.: Cooperative coevolution with global search for large scale global optimization. In: Proceedings of 2012 IEEE Congress on Evolutionary Computation, (CEC 2012), pp. 1–7, IEEE, 2012
42. Wolpert, D.H., Macready, W.G.: No free lunch theorems for optimization. IEEE Trans. Evol. Comput. **1**, 67–82 (1997)

43. Wolpert, D.H., Macready, W.G.: Coevolutionary free lunches. IEEE Trans. Evol. Comput. **9**, 721–735 (2005)
44. Yang, Z., Zhang, J., Tang, K., Yao, X., Sanderson, A.C.: An adaptive coevolutionary differential evolution algorithm for large-scale optimization. In: *Proceedings of 2009 IEEE Congress on Evolutionary Computation, (CEC 2009)*, pp. 102–109, IEEE, 2009
45. Wang, Y., Li, B.: A self-adaptive mixed distribution based uni-variate estimation of distribution algorithm for large scale global optimization. In: Chiong, R. (ed.) Nature-Inspired Algorithms for Optimisation. Studies in Computational Intelligence, 193, pp. 171–198. Springer, Berlin (2009)
46. Yang, X.S., Deb, S.: Two-stage eagle strategy with differential evolution. Int. J. Bioinpsired Comput. **4**(1), 1–5 (2012)
47. Hsieh, S.-T., Sun, T.-Y., Liu, C.-C., Tsai, S.-J.: Solving large scale global optimization using improved particle swarm optimizer. In: Proceedings of 2008 IEEE Congress on Evolutionary Computation, (CEC 2008), pp. 1777–1784, IEEE, 2008
48. Omidvar, M.N., Li, X.: A comparative study of cma-es on large scale global optimisation. In: Li, J. (ed.) AI 2010: Advances in Artificial Intelligence. Lecture Notes in Computer Science, vol. 6464, pp. 303–312. Springer, Berlin (2011)
49. Zhao, S.-Z., Liang, J.J., Suganthan, P.N., Tasgetiren, M.F.: Dynamic multi-swarm particle swarm optimizer with local search for large scale global optimization. In: Proceedings of the 2008 IEEE Congress on Evolutionary Computation, (CEC 2008), pp. 3845–3852, 2008
50. Liang, J.J., Qu, B.Y.: Large-scale portfolio optimization using multiobjective dynamic mutli-swarm particle swarm optimizer. In: Proceedings of the 2013 IEEE Symposium on Swarm Intelligence (SIS 2013), pp. 1–6, 2013
51. Li, X., Yao, X.: Tackling high dimensional nonseparable optimization problems by cooperatively coevolving particle swarms. In: Proceedings of the 2009 IEEE Congress on Evolutionary Computation, (CEC 2009), pp. 1546–1553, 2009
52. Li, X., Yao, X.: Cooperatively coevolving particle swarms for large scale optimization. IEEE Trans. Evol. Comput. **16**, 210–224 (2012)
53. Hu, S., Li, B.: Velocity divergence of ccpso in large scale global optimization. In: Yin, H., Tang, K., Gao, Y., Klawonn, F., Lee, M., Weise, T., Li, B., Yao, X. (eds.) Intelligent Data Engineering and Automated Learning - IDEAL 2013. Lecture Notes in Computer Science, vol. 8206, pp. 545–552. Springer, Berlin Heidelberg (2013)
54. Geem, Z.W.: Optimal cost design of water distribution networks using harmony search. Eng. Optim. **38**, 259–280 (2006)
55. Liu, Y., Zhou, G., Zhao, J., Dai, G., Li, X.-Y., Gu, M., Ma, H., Mo, L., He, Y., Wang, J., Li, M., Liu, K., Dong, W., Xi, W.: Long-term large-scale sensing in the forest: recent advances and future directions of greenorbs. Front. Comput. Sci. China **4**(3), 334–338 (2010)
56. Mc Gibney, A., Klepal, M., Pesch, D.: Agent-based optimization for large scale wlan design. IEEE Trans. Evol. Comput. **15**, 470–486 (2011)

Evolutionary Clustering for Synthetic Aperture Radar Images

Chin Wei Bong and Xin-She Yang

Abstract Image segmentation is a multiobjective optimization problem. The aim of this paper is to propose and apply a small population multiobjective evolutionary clustering method for solving segmentation of SAR (synthetic aperture radar) images. The multiobjective optimization method is based on the scatter search, which can usually avoid using many random components, and this method is based on a small population approach, known as the reference set, whose individuals are combined to construct new solutions which are generated systematically. The reference set is initialized from an initial population composed of diverse solutions, and then updated with the solutions resulting from the local search improvement. The proposed method uses fuzzy clustering method to optimize two fitness functions in terms of the global fuzzy compactness of the clusters and the fuzzy separation. The proposed approach incorporates the concepts of Pareto dominance, external archiving, diversification, and intensification of solutions. Experiments for various objective formulations and solution combination methods are tested for syntactic, COINS and SAR images to show the precision of the algorithm. Furthermore, we also compare our proposed method with other multiobjective evolutionary clustering methods such as multiobjective clustering with automatic k-determination (MOCK) and NSGA-II. The performance of the proposed method is encouraging.

Keywords Evolutionary algorithm • Clustering

C.W. Bong (✉)
Mystech Solution Sdn. Bhd., Kuching, Sarawak, Malaysia

Wawasan Open University, Penang, Malaysia
e-mail: bongwendy@gmail.com; cwbong@wou.edu.my

X.-S. Yang
School of Science and Technology, Middlesex University, London, UK
e-mail: xy227@cam.ac.uk; X.Yang@mdx.ac.uk

S. Koziel et al. (eds.), *Solving Computationally Expensive Engineering Problems*,
Springer Proceedings in Mathematics & Statistics 97,
DOI 10.1007/978-3-319-08985-0_11

1 Introduction

SAR sensors can penetrate clouds and they can work in bad weather conditions and at nighttime when optical sensors are inoperable. Thus, SAR images have been widely used by researchers and industrial sponsors in the past decades. Applications of SAR include climate studies, forest resources identification, marine environments inspection, and so on. An important and yet challenging task in SAR image applications is image segmentation. It is defined as the extraction of the important objects from an input image [1]. It partitions the pixels in the image into homogeneous regions, each of which corresponds to some particular landcover type. The process has received much attention and is also considered one of the most difficult low-level tasks because the process performance needs to be adapted to the changes in image quality, which is affected by variations in environmental conditions, imaging devices, time of day, and other factors [2, 3].

There are many approaches available for SAR image segmentation in the literature, including threshold methods [4, 5], clustering algorithms [6, 7], statistic model-based methods [8–11], and morphologic methods [12, 13]. The clustering algorithms are the most popular and the earliest approaches used. Among the existing clustering algorithms for SAR images, nature-inspired algorithms include particle swarm optimization [14, 15], and artificial immune system [16] are relatively recent methods being used. However, each approach poses its own limitation, and most of these studies have optimized a single objective for specific applications [17]. In reality, image segmentation is a multiobjective optimization problem, which is difficult to solve, and there is a significant gap between the nature of image segmentation problems and real-world solutions [18]. Thus, a multiobjective optimization (MO) approach is an appropriate method to use for a real-world application [19–21]. In addition, this approach seems to be promising with the nature of SAR image segmentation [22–25].

Therefore, this chapter examines the approach of nature-inspired clustering-based SAR image segmentation. A new algorithm based on a small population evolutionary algorithm, namely the scatter search, is considered. We propose and apply the proposed multiobjective scatter search to study SAR image segmentation using hybrid scatter search (HSS). This method incorporates the concepts of Pareto dominance, external archiving, diversification, and intensification of the solutions. The rest of this chapter is organized as follows. Section 2 presents related studies concerning MO problems and scatter search. Section 3 describes our proposed method in detail. Experimental results, comparing HSS with other standard image segmentation methods are presented in Sect. 4. Finally, conclusions are drawn in Sect. 5.

2 Literature Review

The majority of existing clustering algorithms are based on only one internal evaluation function, which is a single-objective function that measures intrinsic properties of a partitioning such as the spatial separation between the clusters or the compactness of the clusters. However, it is sometimes difficult to reflect the quality of partitioning reliably with only one internal function evaluation which may be violated for certain datasets [26]. In this paper, we use a multiobjective optimization approach to overcome the defects of the single-objective clustering algorithms such as Fuzzy C-Mean. Given that the objective functions (no less than two) for clustering are complementary, the simultaneous optimization of several of those objectives may lead to high-quality solutions and improve the robustness towards different data properties. Multiobjective clustering with automatic k-determination (MOCK) proposed by [27] may be the first application of MO clustering in data clustering. Although many methods have been proposed for MO clustering, only a few applications have been reported in image segmentation [28]. Among those applications, NSGA-II [29] was found to be the most frequently used method. It is difficult to apply the current MO clustering technology to image segmentation, owing to an extremely large amount of data need to be handled and thus such handling in an evolutionary algorithm (EA) is tedious and computationally expensive [30].

Nature-inspired techniques have been used with clustering methods, either evolutionary algorithms (EAs) or non-EAs [31]. Although they have been applied in the image segmentation problem, there is still room for more extensive research and the results and performance can be improved further. For the successful operation of an EA with multiobjective clustering (or multiobjective EA clustering) in image segmentation, some of the important issues that should be addressed include the proper size of population, suitable genetic encoding of partitioning, appropriate set of objective functions, and suitable genetic operators (e.g., mutation and crossover). For the methods to work, one of the challenges is to formulate a suitable set of objective functions and the efficient generation of solutions that offer reduced optimization cost. However, the search strategies of the different metaheuristics are highly dependent on the philosophy of these metaheuristic algorithms themselves.

Here, metaheuristics refer to a general algorithmic framework that can be applied to different optimization problems, with relatively few modifications required to adapt them to a specific problem [32, 33]. They are strategies that "guide" the search process. They aim to explore efficiently the search space to find (near-) optimal solutions. Every metaheuristic approach should be designed with the aim of effectively and efficiently exploring a search space. An effective initialization scheme should serve as a good start, without biased towards any unpromising local regions. The search performed by a metaheuristic should be "clever" enough both to explore intensively the areas of the search space with high-quality solutions and to move to unexplored areas of the search space when necessary.

We extend the work of archive-based HSS [34], which follows the scatter search (SS) structure but uses mutation and crossover operators from EA. Scatter search

[35] is an EA in the sense that it incorporates the concept of population. Compared to other EAs, it usually uses fewer random components, and it uses a small population, known as the reference set, whose individuals are combined to construct new solutions which are generated systematically. The reference set is initialized from an initial population, composed of diverse solutions, and they are updated with the solutions resulting from the local search improvement. Scatter search has been found to be successful in a wide variety of optimization problems because it offers reduced optimization cost. Until recently it had been extended to deal with MO problems [34] and also SAR image segmentation problem [36]. According to [26], SS can serve as a powerful local search engine for tasks such as generating missing parts of an approximate Pareto front because of its flexibility and ease of use. In [24], the authors first tested the algorithm in gray-scale images. Now, we provide a more in-depth analysis for the segmentation of SAR image segmentation.

3 Multiobjective Clustering with HSS

The objective function defined in MO clustering can be formulated based on the validity index in clustering algorithms [17]. The optimization of the validity index usually aims for the optimal number of clusters with the optimal clustering output. Although there are many available indices, none of them perform satisfactorily for a wide range of datasets [37]. Therefore, MO clustering methods should be used to optimize the validity indices of two to three clusters to complement their strengths and compensate their weaknesses [38].

The proposed approach consists of three main phases: (1) the features of the image are extracted; (2) HSS optimizes two complementary clustering objective functions using a multiobjective clustering method. It outputs a set of mutually dominant clustering solutions, corresponding to different tradeoffs between the two objectives; (3) all the dominant clustering solutions are combined together to generate the final best solution and to assign each pixel in the sample dataset to one of the clusters. Subsequently, all the remaining pixel data are assigned to one of the clusters, according to the relationship with the assigned sample dataset.

The image segmentation problem can be formulated as clustering the pixels of the images in the intensity space [39, 40]. Here, a fuzzy clustering algorithm produces a $K \times n$ membership matrix $U(X) = [u_{kj}]$, $k = 1, \ldots, K$ and $j = 1, \ldots, n$, where u_{kj} denotes the membership degree of pattern x_j to cluster C_k.

Here, the individuals are made up of real numbers which represent the coordinates of the cluster centers. If individual i encodes the centers of K_i clusters in a p dimensional space, its length l_i will be $p \times K_i$. In the initial population, each string i encodes the centers of K_i of clusters, such that

$$K_i = \left(rand\,(\,)\,\%K^*\right) + 2 \qquad (1)$$

where *rand*( ) is a random integer generator, and K^* is a soft estimate of the upper bound of the number of clusters. Therefore, the number of clusters will vary from 2 to $K^* + 1$. The K_i centers encoded in an individual of the initial population are randomly selected with distinct points from the input dataset.

In computing the first objective functions, the centers $V = \{v_1, v_2, \ldots v_k\}$ encoded in a given individual are extracted. The fuzzy membership values $u_{ik} = 1, 2, \ldots, K$, where $k = 1, 2, \ldots, n$ are computed using

$$u_{ik} = \sum_{j=1}^{K} \left(\frac{D(v_i, x_k)}{D(v_j, x_k)} \right)^{-\frac{2}{m-1}}, \quad 1 \le i \le K;\ 1 \le k \le n \tag{2}$$

where $D(v_i, x_k)$ denotes the distance between ith cluster center and kth data point and $m \in \{1, \infty\}$ is the fuzzy exponent. In this chapter, the Euclidean distance measure is used. If $D(v_j, x_k)$ is equal to zero for some j, the u_{ik} is set to zero for all $i = 1, 2, \ldots, K$, $i \ne j$, while u_{jk} is set equal to 1. Subsequently, the center of each cluster $v_i = 1, 2, \ldots, K$ is updated by using

$$v_i = \frac{\sum_{k=1}^{n} u_{ik}{}^m x_k}{\sum_{k=1}^{n} u_{ik}{}^m}, \quad 1 \le i \le k \tag{3}$$

Next, the membership values are recomputed. The variation σ_i and fuzzy cardinality n_i of the ith cluster, $i = 1, 2, \ldots, K$, are calculated as follows:

$$\sigma_i = \sum_{k=1}^{n} u_{ik}{}^m D(v_i, x_k), \quad 1 \le i \le K, \tag{4}$$

and

$$n_i = \sum_{k=1}^{n} u_{ik}, \quad 1 \le i \le K. \tag{5}$$

Therefore, the global compactness J of the solution represented by the chromosome is computed as

$$J = \sum_{i=1}^{K} \frac{\sigma_i}{n_i} = \sum_{i=1}^{K} \frac{\sum_{k=1}^{n} u_{ik}{}^m D(v_i, x_k)}{\sum_{k=1}^{n} u_{ik}}. \tag{6}$$

The second fitness function, or fuzzy separation S, is computed as follows: the center v_i of the ith cluster is assumed to be the center of a fuzzy set $\{v_j | 1 \le i \le K, j \ne i\}$. Hence the membership degree of each v_j to v_i, $j \ne i$ is computed as

$$u_{ik} = \sum_{j=1}^{K} \left(\frac{D(v_i, x_k)}{D(v_j, x_k)} \right)^{-\frac{2}{m-1}}, \quad 1 \le i \le K;\ 1 \le k \le n. \tag{7}$$

Subsequently, the fuzzy separation can be defined as

$$S = \sum_{i=1}^{K} \sum_{j=1,\ j \neq i}^{K} u_{ij}{}^{m} D\left(v_i, x_j\right). \tag{8}$$

The third objective function is a newly developed point symmetry distance based cluster validity index, *FSym*-index [23]. It is defined as follows:

$$FSym\ (K) = \left(\frac{1}{K} \times \frac{E_1}{E_k} \times D_K\right) \tag{9}$$

where K is the number of cluster,

$$D_K = max_{i,j=1}^{K} d\left(v_i, v_j\right). \tag{10}$$

Note that $d_{ps}(v_j, x_k)$ is the point symmetry based distance between the cluster center v_j and the data point x_k. If the corresponding $d_{ps}(v_j, x_k) = d\,(v_i, x_j)$ is smaller than a pre-specified value, we update the membership u_{ij} using the following criterion: $u_{ij} = 1$; if $i = k$, $u_{ij} = 0$; if $i \neq k$. Otherwise, we update the membership u_{ij} by using the rule which corresponds to the normal fuzzy c-means algorithm. In short, the index *FSym* is a composition of three factors, namely $\frac{1}{K}$, $\frac{E_1}{E_k}$ and D_K. As the MO problem here is formulated as the minimization of all the three objectives, hence the objectives are to minimize J, $\frac{1}{S}$ and $\frac{1}{FSym\ (K)}$.

In the proposed approach, there are basically five parts:

- *Diversification Generation Procedure*: The procedure is the same as that proposed in [35]. The goal is to generate an initial set P of diverse solutions. This is a straightforward method based on dividing the range of each variable into a number of subranges of equal size; so, the value for each decision variable of every solution is generated in two steps. First, a subrange of the variable is randomly chosen. The probability of selecting a subrange is inversely proportional to its frequency count (the number of times the subrange has already been selected). Second, a value is randomly generated within the selected range. This is repeated for all the solution decision variables.
- *Improvement procedure*: This procedure is to use a *simplex* method to improve new solutions obtained from the diversification generation and solution combination methods. The improvement method takes an individual as a parameter, which is repeatedly mutated with the aim of obtaining a better individual. The term "better" is defined here in a similar way to the constrained-dominance approach used in NSGA-II [29].
- *Reference Set Update procedure*: The reference set is a collection of both high-quality and diverse solutions that are used to generate new individuals. The set itself is composed of two subsets, *RefSet1* and *RefSet2*, of size p and q, respectively. The first subset contains the best quality solutions in $\boldsymbol{P}$, while the second subset should be filled with solutions promoting diversity.

- *Subset Generation and Combination procedure*: This procedure generates subsets of individuals, which will be used to create new solutions with a solution combination method. The strategy used considers all pairwise combinations of solutions in the reference set [34]. Furthermore, this method should avoid producing repeated subsets of individuals, i.e., subsets previously generated. In the subset combination, the procedure is to find linear combinations of reference solutions.
- *External Archive*: The aim of this archive is to store a record of the non-dominated individuals found during the search in order to keep those individuals so as to produce a well-distributed Pareto front. The key issue is the archive management that is to decide whether a new solution should be added to the archive or not. This archive is empty at the beginning. It is continuously updated whenever a new individual or solution is not dominated by the *RefSet1*.

Whenever a new solution is created, it becomes a member of the memory. The size of the memory is kept constant and its worst elements are regularly replaced by the better ones. A data structure called quad trees is used to accelerate the process of updating *External Archive*. The quad tree of points where no point dominates any other point is applied and it is a computationally efficient method to determine whether points in the tree dominate a given new point and to retrieve a point in the tree, which it dominates.

Initially, the diversification generation method is invoked to generate the initial solutions, and each one is passed to the improvement method and the result is the initial set ***P***. Then, a fix number of iterations are performed. At each iteration, the reference set is built, the subset generation method is called, and the main loop of the scatter search algorithm is executed until the subset generation method stops producing new subsets of solutions. There is a restart phase, which consists of three steps. First, the individuals *RefSet1* in are inserted into ***P***; second, the best individuals *n* from the external archive, according to the crowding distance, are also moved to ***P***; and, third, the diversification generation and improvement methods are used to produce new solutions for filling up the set ***P***. The idea of moving *n* individuals from the archive to the initial set is to promote the intensification capabilities of the search towards the Pareto front already found. The intensification degree can vary, depending on the number of chosen individuals.

4 Experiments

The proposed method is evaluated using gray-scale images that include three types of images (synthetic image, coins images, and SAR images as shown in Fig. 1). These SAR images are based on two 3-class SAR images which have been obtained from http://www.sandia.gov/radar/imageryku.html. The SAR image 1 (Fig. 1c) is a Ku-band SAR image with 1 m spatial resolution in the area of Rio Grande River near Albuquerque, New Mexico, USA. This image consists of three types of land

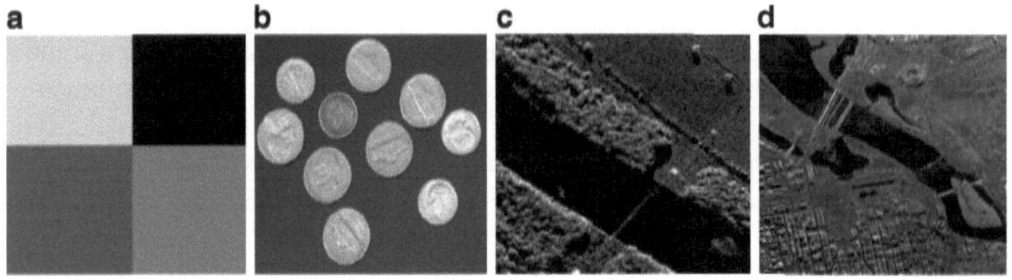

Fig. 1 Images used (**a**) Synthetic image, (**b**) COINS image, (**c**) SAR image 1, (**d**) SAR image 2

Table 1 Two different experiments settings: (1) variation of objective combinations and (2) variation of reference set combination

Setting 1	Variation of objective formulation	
1-A	Minimization of J and $\frac{1}{S}$	
1-B	Minimization of $\frac{1}{S}$ and $\frac{1}{FSym\,(K)}$	
1-C	Minimization of J and $\frac{1}{FSym\,(K)}$	
Setting 2	Reference set variation	Combination method
2-A	RefSet1 ∪ RefSet2	Linear combination
2-B	RefSet1 ∪ RefSet2	SBX combination
2-C	RefSet1, RefSet1	Linear combination
2-D	RefSet1, RefSet1	SBX combination

covers, namely the river, the vegetation, and the crop. Meanwhile, the SAR image 2 (Fig. 1d) is a SAR image in the area of Potomac River in Arlington near Washington. It consists of three types of land covers, namely the buildings, the land, and the water.

Several experiment settings have been conducted to obtain the best combination of the method. Finally, we have evaluated our proposed approach and compared it with two other popular MO methods: NSGA-II [40] and MOCK [27].

4.1 Experiment Settings

To assess the precision of the proposed multiobjective method, we choose to repeat the clustering process that considers all sources of variations in the objective formulations and the reference set with different combination mechanisms. We have conducted experiments based on different settings as shown in Table 1. In Setting 1, we use a different combination of the objective functions with two objectives. Meanwhile in Setting 2, we tested different mechanisms of the subset generation and solution combination.

In the subset generation, we generated all the pairwise combination of individuals belonging to both *RefSet1* and *RefSet2* in Setting 2-A and 2-B (*RefSet1* ∪ *RefSet2*).

In Setting 2-C and 2-D, the subset generation methods produce pairs of individuals belonging only to *RefSet1* or *RefSet2*. The aim is to intensify the search in two directions by reducing the number of combinations with diverse solutions coming from *RefSet2*. Since the lower the number of combinations, the shorter the inner loop of HSS and the higher the number of restarts, promoting feedback of non-dominated solutions from the external archive. The linear combination was used to create new trial solutions in the solution combination mechanism in Setting 2-A and 2-C, while SBX combination was used in Setting 2-B and 2-D.

4.2 Evaluation Method

The performance metrics are to evaluate the closeness to the Pareto front and the diversity in the solutions obtained. First, the Generational Distance (*GD*) was introduced in [30] to measure how far the elements are in the set of non-dominated vectors found from those in the Pareto optimal set and it is defined as

$$GD = \left(\frac{\sqrt{\sum_{l=1}^{n} d_l^2}}{n} \right) \tag{11}$$

where is n the number of vectors in the set of non-dominated solutions found so far and d_l is the Euclidean distance (measured in objective space) between each of these solutions and the nearest member of the Pareto optimal set. All the generated elements are on the Pareto front when $GD = 0$. The non-dominated sets are normalized before this distance measure is calculated to obtain reliable results.

Besides, we use a Spread metric by computing the distance from a given point to its nearest neighbor:

$$\Delta = \frac{\sum_{l=1}^{m} d(e_l, B) + \sum_{X \in B} d(X, B) + \overline{d}}{\sum_{l=1}^{m} d(e_l, B) + B * \overline{d}} \tag{12}$$

where S is a set of solutions, S^* is the set of Pareto optimal solutions, and $e_1 \ldots e_m$ are extreme solutions. In addition, m is the number of objectives,

$$d(X, B) = \min_{X \in B,\, Y \neq X} F\,(X) - F(Y)^2 \tag{13}$$

and

$$\overline{d} = \frac{1}{S^*} \sum_{X \in B} d(X, B)\,. \tag{14}$$

If the solutions are well distributed, including those extreme solutions, $\Delta = 0$. Again, we apply this metric after the normalization of the objective function values.

Table 2 Results of mean values for GD and spread

Mean values for *GD* and spread (subscript)	1-A	1-B	1-C
Synthetic image			
2-A	2.4E−04 $_{2.7E-05}$	**8.7E−05** $_{\mathbf{4.9E-06}}$	4.2E−04 $_{3.2E-04}$
2-B	2.6E−03 $_{8.0E-04}$	1.5E−03 $_{5.1E-04}$	**3.0E−05** $_{\mathbf{1.6E-05}}$
2-C	2.4E−04 $_{8.0E-03}$	8.7E−04 $_{5.1E-04}$	**4.2E−05** $_{\mathbf{1.6E-05}}$
2-D	6.1E−03 $_{7.3E-04}$	1.4E−04 $_{4.1E-04}$	**2.0E−05** $_{\mathbf{3.4E-04}}$
COINS image			
2-A	1.1E−03 $_{1.1E-03}$	8.0E−04 $_{2.1E-04}$	**2.1E−05** $_{\mathbf{3.2E-04}}$
2-B	9.0E−04 $_{7.4E-04}$	4.4E−04 $_{1.1E-03}$	**2.5E−05** $_{\mathbf{3.6E-04}}$
2-C	1.7E−04 $_{1.7E-02}$	1.4E−05 $_{3.4E-04}$	**8.5E−05** $_{\mathbf{2.3E-04}}$
2-D	8.9E−03 $_{3.4E-03}$	3.2E−04 $_{1.1E-03}$	**1.5E−05** $_{\mathbf{3.2E-04}}$

5 Results and Discussions

First, we present the results for segmentation of the syntactic and coins images. The results are considered better when the *GD* and *Spread* values are lower. The comparison of mean values of *GD* and *Spread* values in Table 2 shows that the configuration used in setting 1-B works best with Setting 2-A (highlighted in bold) for syntactic images. Meanwhile, in the setting of 1-C, the setting 2-B, 2-C, and 2-D give the best result (highlighted in bold) for COINS images. This reflects that the minimization of J and $\frac{1}{FSym\,(K)}$ yield better result in general. With the SBX crossover operator, the result for the setting 1-C and 2-D for both images are the best out of all combinations of the bi-objectives. It shows that the diversification/intensification balance is penalized when the subset generation method produces pairs of individuals belonging to *RefSet1* and *RefSet2* in the experiment, thus the method should allow the combinations of individuals belong to the same subset.

Later, the method is compared with two other multiobjective algorithms, namely NSGA-II and MOCK as mentioned earlier. The Pareto fronts obtained with the three algorithms were plotted in Fig. 2 for various settings of bi-objective functions for the COINS image. For this minimization problem, it is demonstrated that the proposed method is able to produce much lower values in all three cases.

Finally, Fig. 3 shows the original SAR images and their segmented results using MOCK, NSGA-II, and the proposed method. In terms of the regional consistency of the water and the buildings region, MOCK and the proposed method are better than NSGA-II. However, all three algorithms cannot achieve the visually correct segmentation in the right top region, i.e. MOCK, and the proposed method misclassifies a big area of the land region as the water. In contrast, the proposed method performs the best in this area where the misclassification is not so serious as MOCK and NSGA-II.

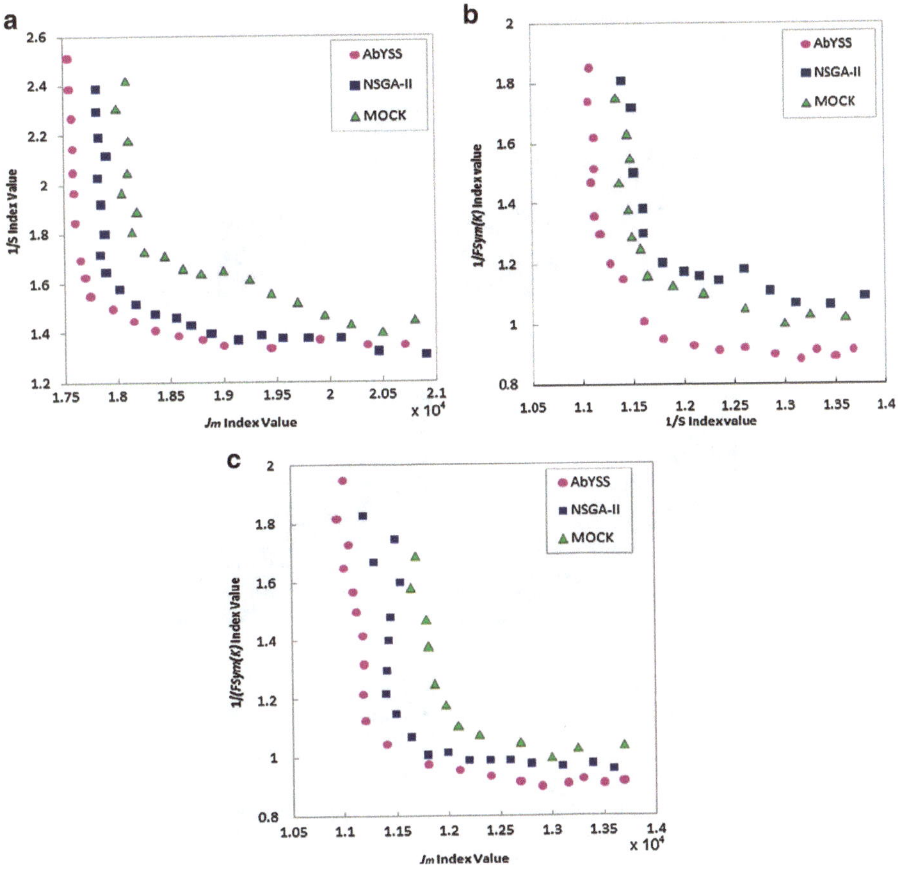

Fig. 2 Pareto front approximation for COINS images for different objective spaces for setting: (**a**) 1-A, (**b**) 1-B, and (**c**) 1-C

6 Conclusion

Most real-world image segmentation problems are challenging, involving simultaneously optimizing multiple criteria with some tradeoffs. Segmentation methods tend to be computationally expensive. Therefore, we have presented a small population multiobjective evolutionary clustering method, based on the scatter search for the SAR image segmentation. This chapter has demonstrated the concepts of Pareto dominance, external archiving, diversification, and intensification of solutions. The performance of the proposed method has been compared with two other methods, and the results of proposed method were encouraging.

Even though the proposed approach achieved the best results, there is still room for improvement. Future work will focus on enhancing the performance speed

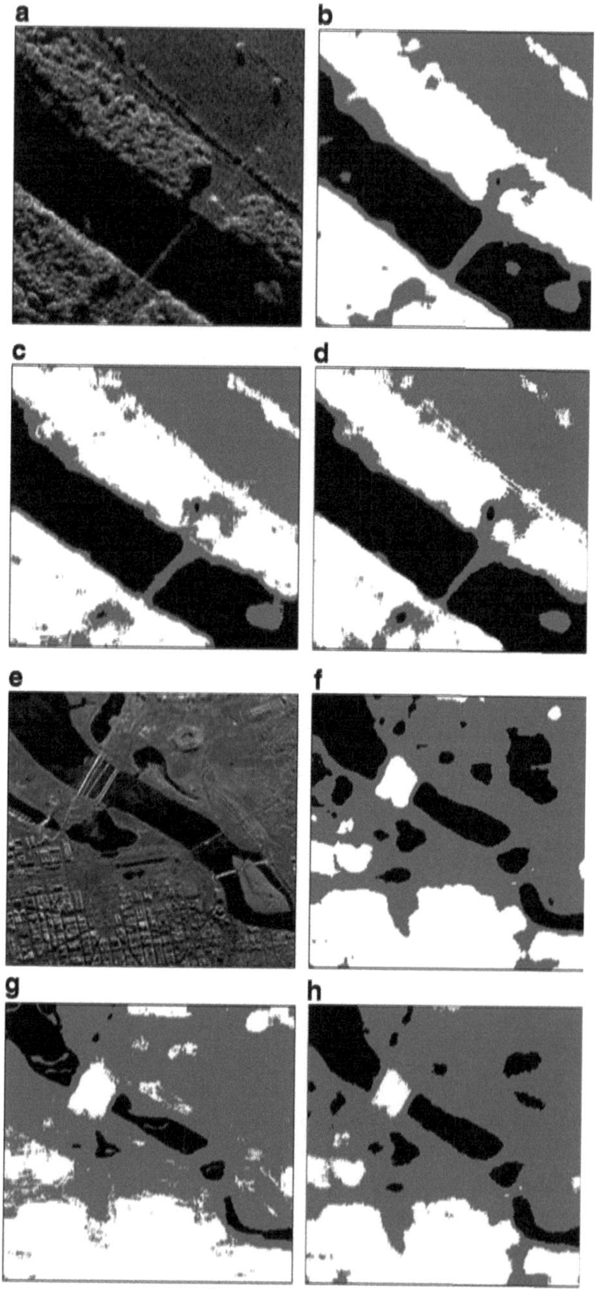

Fig. 3 Two 3-Class SAR images (**a**)(**e**) and their segmentation results for the proposed method (**b**)(**f**), NSGA-II (**c**)(**g**) and MOCK (**d**)(**h**)

of the algorithm. More in-depth research will also be conducted in carrying out the parametric studies in the algorithm. In addition, extension to include more criteria for image segmentation may potentially improve the segmentation quality. Furthermore, there are other promising nature-inspired algorithms such as cuckoo search and firefly algorithm [41], and hybridization with other nature-inspired methods will also be highly useful.

References

1. Gonzalez, R.C., Woods, R.E.: Digital Image Processing. Prentice Hall, Upper Saddle River (2007)
2. Bhanu, B., Lee, S., Das, S.: Adaptive image segmentation using multiobjective evaluation and hybrid search methods. In: Proc. AAAI Fall Symp. on Machine Learning and Computer Vision: What, Why and How? pp. 30–34 (1993)
3. Bhanu, B., Lee, S., Das, S.: Adaptive image segmentation using genetic and hybrid search methods. IEEE Trans. Aerosp. Electron. Syst. **31**(4), 1268–1291 (1995)
4. Zaart, A.E., Ziou, D., Wang, S., Jiang, Q.: Segmentation of SAR images using mixture of gamma distribution. Pattern Recogn. **35**(3), 713–724 (2002)
5. Lee, J.S., Jurkevich, I.: Segmentation of SAR images. IEEE Trans. Geosci. Remote Sens. **27**(6), 674–680 (1989)
6. Kersten, P.R., Lee, J.-S., Ainsworth, T.L.: Unsupervised classification of polarimetric synthetic aperture radar images using fuzzy clustering and EM clustering. IEEE Trans. Geosci. Remote Sens. **43**(3), 519–527 (2005)
7. Chumsamrong, W., Thitimajshima, P., Rangsanseri, Y.: Syntheic aperture radar (SAR) image segmentation using a new modified fuzzy c-means algorithm. In Proc. IEEE Symp. Geosci. Remote Sens., pp. 624–626 (2000)
8. Samadani, R.: A finite mixtures algorithm for finding proportions in SAR images. IEEE Trans. Image Process. **4**(8), 1182–1185 (1995)
9. Deng, H., Clausi, D.A.: Unsupervised segmentation of synthetic aperture radar sea ice imagery using a novel Markov random field model. IEEE Trans. Geosci. Remote Sens. **43**(3), 528–538 (2005)
10. Dong, Y., Forster, B.C., Milne, A.K.: Comparison of radar image segmentation by Gaussian-and Gamma-Markov random field models. Int. J. Remote Sens. **24**(4), 711–722 (2003)
11. Zhao, Q., Li, Y., Liu, Z.: SAR image segmentation using Voronoi tessellation and Bayesian inference applied to dark spot feature extraction. Sensors **13**(11), 14484–14499 (2013)
12. Lemarechal, C., Fjortoft, R., Marthon, P., Cubero-castan, E., Lopes, A.: SAR image segmentation by morphological methods. Proc. SPIE **3497**, 111–121 (1998)
13. Ogor, B., Haese-coat, V., Ronsin, J.: SAR image segmentation by mathematical morphology and texture analysis. In: Proc. IGARSS, pp. 717–719 (1996)
14. Paoli, A., Melgani, F., Pasolli, E.: Clustering of hyperspectral images based on multiobjective particle swarm optimization. IEEE Trans. Geosci. Remote Sens. **47**(12), 4175–4188 (2009)
15. Tian, X., Jiao, L., Zhang, X.: A clustering algorithm with optimized multiscale spatial texture information: application to SAR image segmentation. Int. J. Remote Sens. **34**(4), 1111–1126 (2013)
16. Yang, D., Wang, L., Hei, X., Gong, M.: An efficient automatic SAR image segmentation framework in AIS using kernel clustering index and histogram statistics. Appl. Soft Comput. **16**, 63–79 (2014)
17. Bong, C.W., Rajeswari, M.: Multiobjective clustering with metaheuristic: current trends and methods in image segmentation. IET Image Process. **6**(1), 1–10 (2012)

18. Konak, A., Coit, D.W., Smith, A.E.: Multi-objective optimization using genetic algorithms: a tutorial. Reliab. Eng. Syst. Safety **91**, 992–1007 (2006)
19. Yang, X.S.: Engineering Optimization: An Introduction with Metaheuristic Optimization. Wiley, Hoboken (2010)
20. Jones, D.F., Mirrazavi, S.K., Tamiz, M.: Multi-objective meta-heuristics: an overview of the current state-of the-art. Eur. J. Oper. Res. **137**(1), 1–9 (2002) [1998]
21. Guliashki, V., Toshev, H., Korsemov, C.: Survey of evolutionary algorithms used in multiobjective optimization. Probl. Eng. Cybern. Robot. **60**, 42–54 (2009) [Sofia: Bulgarian Academy of Sciences]
22. Ruochen, L., Wei, Z., Licheng, J., and Fang, L.: A multiobjective immune clustering ensemble technique applied to unsupervised SAR image segmentation. In: Proceedings of the 9th ACM International Conference on Image and Video Retrieval (CIVR 2010), pp. 158–165 (2010)
23. Saha, S., Bandyopadhyay, S.: A new symmetry based multiobjective clustering technique for automatic evolution of clusters. Pattern Recogn. **43**(4), 738–751 (2010)
24. Bong, C.W., Lam, H.Y.: Unsupervised image segmentation with adaptive archive-based multi-objective evolutionary method. In: Kuznetsov, S.O., et al. (eds.) 4th International Conference of Pattern Recognition and Machine Intelligence (PReMI'11), vol. 6744, pp. 92–97. Springer, Heidelberg (2011)
25. Li, Y., Wei, Y., Wang, Y., Jiao, L.: Multi-objective evolutionary for synthetic aperture radar image segmentation with non-local means denoising. Nat. Comput. **13**(1), 39–53 (2013) [2014]
26. Coello, C.A.C.: Evolutionary multi-objective optimization: some current research trends and topics that remain to be explored. Front. Comput. Sci. China **3**(1), 18–30 (2009)
27. Handl, J., Knowles, J.: An evolutionary approach to multiobjective clustering. IEEE Trans. Evol. Comput. **11**(1), 56–76 (2007)
28. Qian, X.: Unsupervised texture image segmentation using multiobjective evolutionary clustering ensemble algorithm. In: IEEE Congress on Evolutionary Computation (CEC '08), pp. 3561–3567. IEEE Press (2008)
29. Deb, K., Pratap, A., Agarwal, S., Meyarivan, T.: A fast and elitist multiobjective genetic algorithm: NSGA-II. IEEE Trans. Evol. Comput. **6**(2), 182–197 (2002)
30. Van Veldhuizen, D.A., Lamont, G.B.: Multiobjective Evolutionary Algorithm Research: A History and Analysis. Air Force Institute, Dayton-Fairborn (1998)
31. Bong, C.W., Rajeswari, M.: Multi-objective nature-inspired clustering and classification techniques for image segmentation. Appl. Soft Comput. **11**(4), 3271–3282 (2011)
32. Blum, C., Roli, A.: Metaheuristics in combinatorial optimization: overview and conceptual comparison. ACM Comput. Surv. **35**(3), 268–308 (2003)
33. Yang, X.S.: Nature-Inspired Metaheuristic Algorithms, 1st edn. Luniver, Bristol (2008)
34. Nebro, A.J., Luna, F., Alba, E., Dorronsoro, B., Durillo, J.J., Beham, A.: AbYSS: adapting scatter search to multiobjective optimization. IEEE Trans. Evol. Comput. **12**(4), 439–457 (2008)
35. Glover, F., Laguna, M., Martí, R.: Scatter search. In: Ghosh, A., Tsutsui, S. (eds.) Advances in Evolutionary Computing: Theory and Applications, pp. 519–537. Springer, Berlin (2003)
36. Bova, N., Ibanez, O., Cordon, O.: Image segmentation using extended topological active nets optimized by scatter search. IEEE Comput. Intell. Magazine **8**(1), 16–32 (2013)
37. Jain, A.K., Murty, M.N., Flynn, P.J.: Data clustering: a review. ACM Comput. Surv. **31**(3), 264–323 (1999)
38. Handl, J., Knowles, J.: Exploiting the trade-off-the benefits of multiple objectives in data clustering. In: Proc. Third Int. Conf. Evolutionary Multi-Criterion Optimization (2005)
39. Mukhopadhyay, A., Maulik, U.: Unsupervised pixel classification in satellite imagery using multiobjective fuzzy clustering combined with SVM classifier. IEEE Trans. Geosci. Remote Sens. **47**(4), 1132–1138 (2009)
40. Mukhopadhyay, A., Maulik, U.: A multiobjective approach to MR brain image segmentation. Appl. Soft Comput. **11**, 872–880 (2011)
41. Yang, X.S.: Cuckoo Search and Firefly Algorithm: Theory and Applications. Springer, Heidelberg (2013)

Automated Classification of Airborne Laser Scanning Point Clouds

Christoph Waldhauser, Ronald Hochreiter, Johannes Otepka, Norbert Pfeifer, Sajid Ghuffar, Karolina Korzeniowska, and Gerald Wagner

Abstract Making sense of the physical world has always been at the core of mapping. Up until recently, this has always dependent on using the human eye. Using airborne lasers, it has become possible to quickly "see" more of the world in many more dimensions. The resulting enormous point clouds serve as data sources for applications far beyond the original mapping purposes ranging from flooding protection and forestry to threat mitigation. In order to process these large quantities of data, novel methods are required. In this contribution, we develop models to automatically classify ground cover and soil types. Using the logic of machine learning, we critically review the advantages of supervised and unsupervised methods. Focusing on decision trees, we improve accuracy by including beam vector components and using a genetic algorithm. We find that our approach delivers consistently high quality classifications, surpassing classical methods.

Keywords Decision Trees • Genetic Algorithm • Full Waveform LIDAR • Floodplain Classification

---

C. Waldhauser (✉) • R. Hochreiter
Institute for Statistics and Mathematics, WU Vienna University of Economics and Business, Welthandelsplatz 1, 1020 Vienna, Austria
e-mail: first.last@wu.ac.at

J. Otepka • N. Pfeifer • S. Ghuffar
Department of Geodesy and Geoinformation, Vienna University of Technology, Karlsplatz 13, 1040 Vienna, Austria
e-mail: first.last@geo.tuwien.ac.at

K. Korzeniowska
Institute of Geography and Spatial Management, Jagiellonian University, ul. Gronostajowa 7, 30387 Kraków, Poland
e-mail: k.korzeniowska@uj.edu.pl

G. Wagner
Vermessung Schmid, Inkustraße 1–7, 3400 Klosterneuburg, Austria
e-mail: wagner@geoserve.co.at

S. Koziel et al. (eds.), *Solving Computationally Expensive Engineering Problems*, Springer Proceedings in Mathematics & Statistics 97,
DOI 10.1007/978-3-319-08985-0_12

1 Introduction

Surveying the very planet we live on has been an ongoing effort since the dawn of mankind. From the early maps of Anatolia to modern geospatial intelligence, the mission of any map was always to make sense of the world around us. Boosting map drawing with the latest advances of machine learning has the potential to largely facilitate the generation of maps and extend their usefullness into application domains beyond path finding.

In this chapter, we present the combined efforts of academia and industry to create a framework for the automated generation of maps. The basis for this project is airborne laser scanning: the systematic recording and digitizing of ground by means of laser emitted from aircraft. The resulting point clouds of the environment are then automatically classified into ground cover types, using supervised learning and evolutionary computation approaches.

This chapter is organized as follows. In the first section we describe the technical background of airborne laser scanning. Section 3 details the work related to develop automated classification models. There we will compare the practical aspects of supervised and unsupervised approaches as well as detail the supervised classification approach we implemented and evolutionary computation extensions to it. That section also features a description of the data set we used to empirically test our approaches. The aspects of implementing our model in industry applications are discussed in the subsequent Sect. 4. We close with some concluding remarks pointing to future research.

2 Airborne Laser Scanning Point Clouds

2.1 *Measurement Principle*

Airborne Laser Scanning (ALS) is a remote sensing method for obtaining geometrical and additional information about objects not in contact with the sensor, i.e. the laser scanner. A laser scanner emits a short pulse of infrared light which travels through the atmosphere and is scattered and partially absorbed by any objects in the instantaneous field of view of the laser beam. If diffuse reflection occurs, which is the standard case for many object surfaces, including, e.g., vegetation, bare ground, and building surfaces, a portion of the incident light is scattered back to the sensor. There, the backscattered signal is detected and recorded. The time lag between emission of the pulse and detection of its echo is the two-way travel time from the sensor to the object. With the known speed of light this time lag is turned into the distance from sensor to object. This is also called laser range finding (LRF). In laser scanning, the beam is scanned across the entire field of view, thus covering a larger extent. Rotating mirrors and comparable devices are used to deflect the laser beam and cover large areas. With the known orientation of the mirror and the known

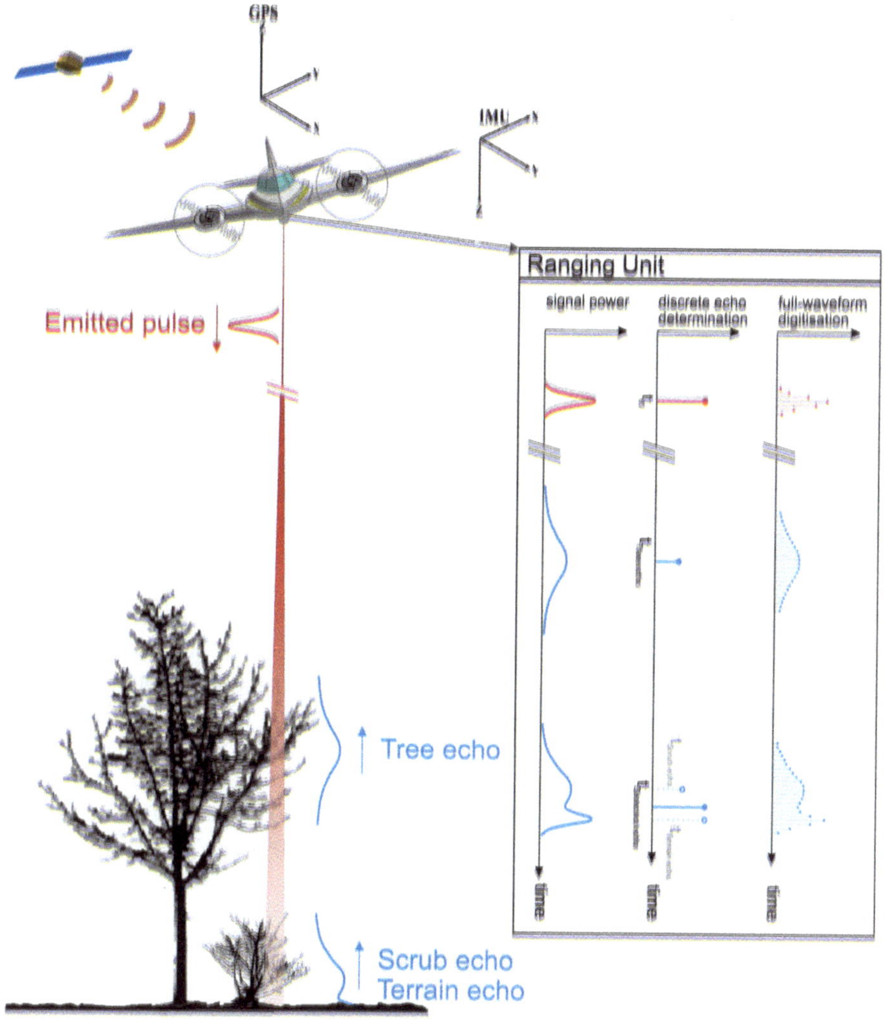

Fig. 1 Diagram explaining the principles of ALS [3]

position of the laser scanner in a global Earth fixed coordinate system (e.g., WGS84, in UTM projection), the location of the objects at which the laser pulse was scattered can be computed. This provides a so-called 3D point cloud: a set of points, each with 3 co-ordinates x, y, z. These points are obtained in the sensor co-ordinate system.

In airborne laser scanning the scanner is mounted on a flying platform (fixed wing or helicopter). Its position is measured with Global Navigation Satellite Systems (GNSS, e.g., GPS). The angular attitude of the sensor platform inside the aircraft is observed with Inertial Measurement Unit (IMU, comprised of accelerometers and gyros). The laser scanner is mounted to look downwards and the beam is scanned at right angles to the flight direction (see Fig. 1).[1] Together with the forward motion of

[1] Full color, high resolution versions of each figure can be found at http://www2.wu.ac.at/alsopt.

the aircraft, larger areas can be scanned. Even larger areas are measured by flying strip-wise above the terrain. This six degree of freedom trajectory defines a moving co-ordinate system for the observation of range and angle from the laser scanner. With an Euclidean transformation the points can be transformed from the sensor co-ordinate system to the global co-ordinate system. Typical results for the accuracy of such points is in the order of 10 cm (single standard deviation in each coordinate).

Besides the observation of distance between the sensor and an object point by the time lag of emission and detection of its echo, also other observations can be retrieved from the received echo. Firstly, it is not always the case that the laser beam hits exactly one object. Due to the diameter of the beam, e.g., 50 cm, multiple objects may be within the beam, but at different heights. Examples include vegetation canopy and ground below. While a part of the signal is reflected at the leaves of the canopy of a tree, other parts of the signal continue traveling downwards until they hit lower vegetation or the ground, from which they are reflected. Thus, each emitted pulse may give rise to several echoes. Other examples, next to vegetation, are power lines and house edges, where a part of the signal is reflected on the roof, while the other part is reflected from the ground.

Furthermore, the backscattered echo can be sampled as a function of time, so-called waveform digitizing. The recorded amplitude depends on the range, on laser scanner device parameters, e.g. the receiving aperture diameter, but also on object properties, i.e., how much of the incident signal is absorbed, scattered diffusely, etc. By means of calibration [26] the parameters of the object like the backscatter cross-section can be determined [20, 27]. The received echo may also be deformed relative to the emitted pulse. An increased echo width [9] is a hint for either hitting a slanted surface or, more often the case, hitting vegetation. Within the footprint a number of leaves may be found which have similar by not identical height. Thus the echoes from all the leaves overlap and form a single widened echo.

2.2 Additional Point Descriptors

A point cloud **P** is a collection of points $p_i = (x, y, z) \in \mathbf{P}$ in a three-dimensional space. The laser scanning point cloud can be analyzed locally to enhance the description of each point further. For instance, given a point density of 4 points/m^2, a local surface model [13] can be computed using, e.g., the ten nearest points. This model may be an inclined plane, with its normal vector being an additional description for the point. The equation of a plane through the point (x_0, y_0, z_0) having a normal vector $n = (a, b, c)$ is given as

$$a\,(x - x_0) + b\,(y - y_0) + c\,(z - z_0) = 0 \tag{1}$$

To compute the three components (a, b, c) of the normal vector, three equations, i.e. three (non-collinear) points are required. To add robustness, generally more than three points are used. A subset of points in the neighborhood are typically selected based on k-nearest neighbors or points within the sphere of a predefined radius. If k nearest neighbors are selected, then there are $k+1$ points and subsequently $k+1$ equations (1). A least squares solution of this overdetermined system of equations estimates an optimal plane by minimizing squared sum of distances between the points and the estimated plane. In the matrix form this equation system is written as

$$A \cdot \beta = 0, \tag{2}$$

where each row of matrix A contains the coordinates of a point relative to the center $[x_n - x_0, y_n - y_0, z_n - z_0]$, here $n = 1..k + 1$ and $\beta = [a, b, c]^T$ is the unknown normal vector. The least squares solution for a system of equations of this (2) form is equivalent to solving the eigenvalue problem of the matrix $A^T A$. The unknown normal vector β of the estimated plane is the eigenvector corresponding to the smallest eigenvalue of $A^T A$. The matrix $A^T A$ is often called structure tensor [7]. The mathematical form of the structure tensor is:

$$A^T A = T = \frac{1}{k} \sum_{i=1}^{k+1} (p_i - \bar{p})^T (p_i - \bar{p}) \tag{3}$$

here $\bar{p} = (x_0, y_0, z_0)$ is the center of the points in the neighborhood.

For house roofs or street surfaces the normal vectors have been shown to reach an accuracy of a few degrees. The normal vector further allows to convert the backscatter cross section into the so-called diffuse reflectance. This value assumes a certain (Lambertian) scattering behavior of the object. This scattering mechanism is described by the reflectance (a unit-less value) and the normal vector of the surface. A surface reflecting all incoming light perfectly diffuse has a reflectance of 1.

The quality of the plane fitting, e.g., the root mean square distances between the optimal plane and the given points indicates the roughness of the surface [11]. The smallest eigenvalue of T gives the variance of the distances between the points and the estimated plane.

The structure tensor T holds plenty more useful information about the distribution of points in the neighborhood. The geometric information encoded in T is essential in the characterization and classification of natural and artificial objects. Three widely used features derived from T are linearity, planarity, and omnivariance. The linearity feature reflects how well the distribution of points can be modeled by a $3D$ line. Points over power lines exhibit such a characteristic, therefore, the linearity feature is essential in classifying power lines and similar structures. The planarity describes the smoothness of the surface which is directly related to the roughness measure and the quality of plane fitting for normal vector estimation. In contrast to power lines and smooth surfaces, laser echoes from trees often spread inhomogeneously across a larger $3D$ volume. This volumetric point distribution is described by the concept of omnivariance. These features are computed using the three eigen values $\lambda_1 \geq \lambda_2 \geq \lambda_3 \geq 0$ of the matrix T:

$$L_T = \frac{\lambda_1 - \lambda_2}{\lambda_1} \tag{4}$$

$$P_T = \frac{\lambda_2 - \lambda_3}{\lambda_1} \tag{5}$$

$$O_T = \sqrt[3]{\lambda_1 \lambda_2 \lambda_3} \tag{6}$$

In addition to L_T, P_T and O_T, features like anisotropy, eigenentropy, and curvature are also derived using the eigenvalues of the structure tensor T [8, 14, 21, 28].

More information about the characteristics of the surfaces can be derived using features like *echo ratio*, *ZRange*, *ZRank*, *NormalizedZ*, and *PointDistance*. *Echo ratio* represents the vertical penetration of the surface [10]. *ZRange* represents the maximum height difference between the points in the neighborhood, while *ZRank* is the rank of the point corresponding to its height in the neighborhood. *NormalizedZ* is the rank of the point (between 0 and 1) multiplied by the height range in the neighborhood. *PointDistance* is the average of all shortest distances between the points in the neighborhood. A more detailed description of these features can be found in [16, 19].

Thus, the point cloud can be augmented by additional parameters besides the coordinates x, y, z: the echo ID (first, second, ... last echo of a sequence of echoes) and overall length of the echo sequence, echo amplitude, echo width, backscatter cross section, diffuse reflectance, roughness (*NormalSigma*), normal vector (*NormalX, NormalY, NormalZ*) echo ratio (*ER*), *ZRange, ZRank, NormalizedZ, PointDensity, PointDistance*, linearity, planarity, and omnivariance.

3 Classification

A major application for the automated processing of point clouds is the classification of points. In this application, every point is assigned a class due to its inherent laser return characteristics and its derived features. If successful, any such endeavor promises massive savings in terms of human resources and time, and thus ultimately in cost.

In the past, the remote sensing community focused on classifying data obtained from satellite measurements [15]. They report that results in general have been only somewhat satisfactory with large portions being continuously misclassified. In contrast we work with air- and not satellite-borne data. This allows for a much higher resolution and considerable less atmospheric interference when measuring. Further, the used full waveform data contains much more information than traditional approaches using laser solely for range measurements. Finally, the method of actively illuminating the ground with a laser beam is superior to passively recording reflections of sunlight.

Classification tasks can be grouped into human and machine based classifications. Machine based classification itself can be split into knowledge- and learning based systems. The former is today's industry standard in ALS point cloud processing, the latter the eventual developmental goal. The main disadvantage of knowledge-based systems over machine learning classifiers is their requirement of explicit definitions of ontologies and classification rules. Machine learning classifiers, on the other hand, base their classifications on rules automatically deduced from the available data with minimal (or no) human intervention. A machine learning classifier with human intervention uses initial human input to deduce automatically classification rules from it, that then can be used to autonomously classify points of previously unseen point clouds.

When charging humans with point classification, a number of factors come into play. Foremost, there is the need for additional data. Usually, this data is provided by means of orthophotos that are (ideally) taken in parallel to the laser scanning. Secondly, the qualification, endurance, and accuracy of the employed human has to be taken into consideration as well. That person needs to be an expert user of geographical information systems and trained to recognize the subtleties of orthophotos.

This confluence of laser scanning data, external data via orthophotos and human experience allows for rather precise classifications of points. So far, human performance has not been surpassed by machines in terms of accuracy. Naturally, human classification is a very time-consuming process. And equally naturally, machines outperform humans in the time domain by many orders of magnitude. Therefore, investigating algorithms for automated point cloud classification is an active area of research.

When turning to learning based classification, two approaches following the classical machine-learning dichotomy of supervised vs. unsupervised learning come to mind. The former requires initial human classifier input to derive a classification of unseen points, while the latter does not. The advantage of supervised classification is that the resulting classes correspond with target classes provided through human input. Since unsupervised classification lacks any human interaction, the classes found may or may not be interpretable or relatable to classes that humans would come up with. In the remainder of this section, we will focus on supervised classification based on initial human interaction and the difficulties that arise from it.

As detailed above and elsewhere [16], point characteristics can be grouped according to the way they were obtained: by direct measurement, by calibrated or spatial improved measurements, by deriving them computationally, by linking with meta data. For the former three groups, problems can arise. Directly measured point features are subject to the specifics of the laser scanner used. The predominant method employed for airborne laser scanning enterprises is a laser that is being deflected off the vertical by a rotating prism or a swinging mirror. This allows to scan a range perpendicular to the flight path and is essential for obtaining complete laser scans. However, this method changes the characteristics of the laser return signal, as the angle of the return signal depends on not only the characteristics of the surface but also the angle of the inbound signal. For instance, when scanning directly below the aircraft only little occlusion will occur, while at extreme scan angles, the laser beam will be obscured by any objects between the aircraft and the ground. This distortion needs to be taken into consideration when working with point cloud data. Section 3.4 below discusses the detection of and compensation for these effects in greater detail.

A further question that needs addressing is rooted in the way derived attributes are being computed. Many such attributes are computed taking in to account a neighborhood of points. Here, neighborhood size becomes a defining factor. Choosing an appropriate neighborhood size is far from trivial. However, neighborhood size theoretically affects the classification quality that can be derived. Further complications arise from different neighborhood sizes that can be chosen for each attribute. In Sect. 3.5 we present a genetic algorithm for finding optimal neighborhood sizes for all neighborhood dependent features involved in the classification.

Before turning to the problems described above, we will briefly introduce the data set we worked with and describe how supervised classification works from human and machine perspectives, respectively.

3.1 Data Set and Example

From the industrial side of view, the motivation for this project was to find a new, fast, and reliable algorithm for the classification of point clouds, which can minimize the manual checking and correction, because every manual manipulation is a very time-consuming task. The scenario described in Sect. 4.4 below was the basis for the development of the models used to automatically classify point clouds.

The data set used was taken from the project *DGM-W Niederrhein* with kind permission of the *Bundesanstalt für Gewässerkunde, Germany*. Four predefined areas have been selected, each not bigger than 60 hectares, with different content like bridge, power lines, houses, coniferous and deciduous trees, concrete, gravel, bare earth, groynes, and water.

The flight was done by airplane with the use of a Riegl LMS-Q560 200 KHz Laser Scanner. Flight speed was 100 knots at an altitude above ground of 600 m. The distance between the flight lines was 300 m. The effective scanning rate was set to 150 KHz with 80 lines per second. The resulting mean point density was about 6 points/m^2 over the whole area (except water areas). A radiometric calibration was computed using asphalt streets in each flight session as calibration reference. The calibration parameters were then applied to compute the reflectance, a normalized intensity value. Roughness shapes (derived from digital orthophotos), which define different ground classes of all areas were known and used as support.

For the classification a list of classes was discussed and defined. First a high-order list with standard classes (level 1), which are most common in the majority of laser scanning projects was generated; then each standard class was refined

Table 1 Defined classes in two levels of granularity

First level		Second level	
Class	Code	Class	Code
Unclassified	0	Unclassified	0
Undefined	1	Undefined	1
		Ground	2
		Sand	18
		Gravel	3
Ground	2	Stone, rock	4
		Asphalt	22
		Cement	21
		River dam, groyne	28
		Deciduous forest	5
Vegetation	5	Coniferous forest	6
		Mixed forest	7
Building	8	Building roof	8
		Wall, building wall	24
water	9	Water	9
		Car, other moving object	10
		Temporary object (under construction)	11
		Bridge	12
Artificial objects	10	Power line	13
		Tower, power pole	14
		Bridge cable	15
		Road protection fence	16
		Bridge construction	17
Technical	23	Technical, e.g. concrete part of a bridge	23
Ground, vegetation	20	Ground, vegetation	20
Error	99	Error	99

into subclasses (level 2) to better represent the different kinds of environment. The classes used for this project can be seen in Table 1.

Following the refined class list and taking the roughness shapes into consideration, a three-dimensional classification was done manually using the TerraScan software package. This was done to provide reference data to generate training and testing data sets for the supervised classification method. Later the manual classification was also used as gold standard for assessing the results of the classification done with both the supervised and unsupervised methods.

3.2 *Reference Data Generation Through Manual Classification*

The term reference data refers to data that is manually classified by humans using external data sources like orthophotos. It serves two purposes: providing training data for supervised classification and representing a gold standard that can be used to test the automatic classification's accuracy.

One method to generate reference data is a manual classification of the data set [12, 24]. This process requires a thorough visual analysis of the data and a labeling of each point. As this process is time consuming, typically only small parts of the data set are manually labeled. Thus, this part should represent the diversity of the terrain surface (flat, hilly, etc.) as well as a large amount of the different target classes with a variety of geometric appearance and distribution of other measured or derived objects. These target classes often comprise natural objects (bare-earth, water, vegetation, etc.) and man-made objects (buildings, roads, bridges, ramps, power and other transmission lines, fences, cars and other moving objects, etc.). Vegetation, as one targeted class for example, can be tall and low and have different density. The variety of vegetation has to be included in the manually labeled part for both, accuracy assessment and machine learning. The diversity of classes depends on the purpose of the classification.

Reference data generation can be performed in a number of different ways. Firstly, an automatic classification based on a selection from available algorithms can be performed, followed by a manual improvement of the results. Secondly, only manual classification of an unclassified data set can be performed. In the case of a large number of classes the second way is recommended. A third option in the generation of reference data by using existing data sets. As those data sets were often acquired at a different time, with a different measurement technology, and often with other applications in mind, the transferability of such a classification is limited. Therefore, the next paragraphs will concentrate on the methods for manual classification.

The most common methods for visualization and reference data generation are described below. The basic and most common method uses a 2D profile (Fig. 2). Profiles are sets of points cut out from the entire point cloud with a vertical rectangular prism, not bound in height. The width of the prism is typically small, e.g. 2 m, whereas its length is larger, e.g. 50 m. These values are sensible when working with point densities ranging from 1 to 20 points/m^2. Profiles allow the user to see a part of the terrain from a side view which enables her to distinguish the points within different classes but also to identify the border between different objects, e.g. building and ground, or vegetation and ground. These borders are harder to identify in a top view. In order to classify larger areas in an organized manner, transects are used. This means that a set of parallel profiles is generated which cover a rectangular area. Advancing in the manual classification from one profile to the next accelerates the entire process. The second method uses a shaded relief map (hillshade) of the surface generated by the points of one class. A hillshade

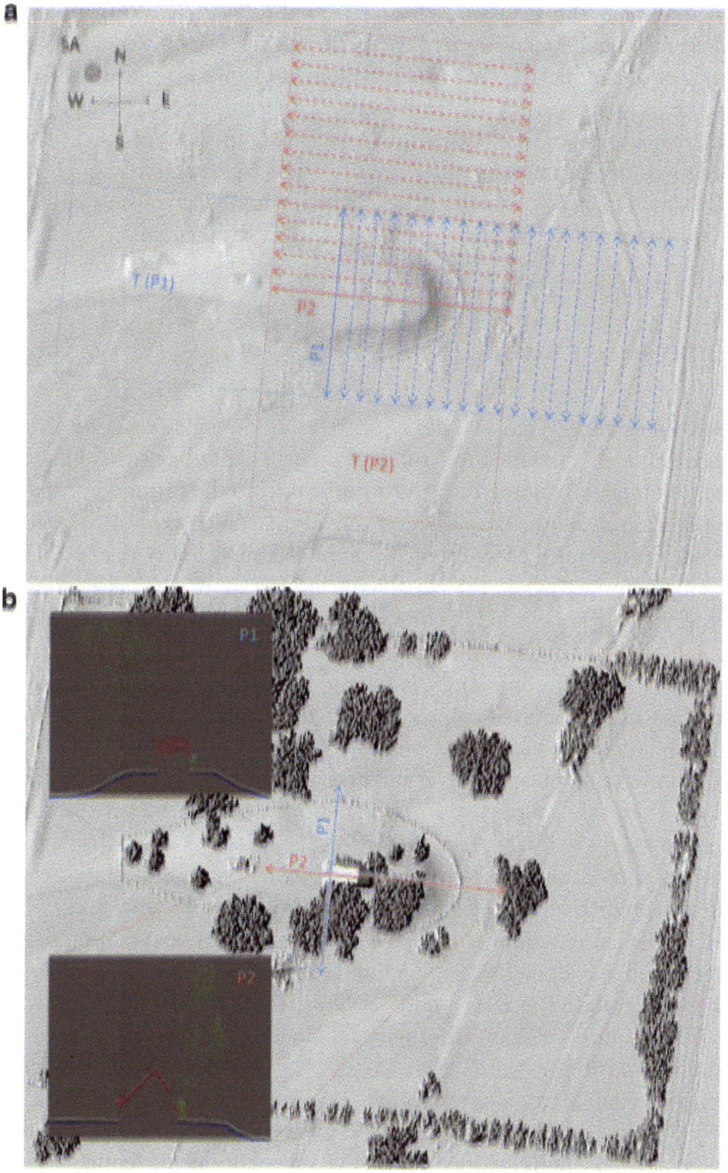

Fig. 2 Methods for visualizing the data during manual classification; (**a**)—shaded relief for DTM; (**b**)—shaded relief for DTM, buildings and vegetation; P1, P2—2D profiles; T(P1), T(P2)—transects for 2D profiles; SA—sun azimuth for shaded relief

requires an artificial illumination source, which is set in a standard manner to an azimuth of 315 degrees, lighting the area from the northwest. Lighting from different directions can substantially help to notice the terrain slope as well as objects located on the ground, especially in the case of mountainous regions. Hillshades can be generated for the bare-earth class, in which the surface represents the digital terrain model (DTM). An example for transects and hillshades can be found in the top panel of Fig. 2. That figure's bottom panel exhibits a DTM. Also combinations of classes, e.g. bare-earth and buildings, can be used. This method can be applied for refining a manual classification, i.e. reclassifying points. This is especially suited to remove small artifacts which occur when close spatial proximity between two classes led to a misclassification in an earlier step.

3.3 Supervised Classification

The idea behind supervised classification is to automatically derive from a small training set enough classification rules, so that a larger, unseen data set can be classified automatically using the model derived from the former. For that purpose, the training data needs to be classified already. Usually, this initial classification is achieved by manually classifying the points. This training data is then used to build a model or equally train the classifying algorithm. In supervised classification, the interpretation of the model comes second, therefore more complex models are favored over simplistic ones that would ease human interpretation; in fact, the boundary to model complexity is dictated only by overfitting avoidance. This model is then used to classify unseen data. To evaluate model performance, true classification information for the unseen data is required as well. However, in production environments, model evaluation for the entire data set is usually not performed. Therefore, supervised classification promises to save a considerable amount of costs.

The method of choice for supervised classification here is classification trees. The tree is a predictive model that links up point features with that point's class. Structurally, the tree consists of leaves and branches. The leaves represent the final class labels and the branches the conjunctions of features that lead up to these class labels. Literature suggests a number of different algorithms for growing a tree [15, 23]. For the purpose of classifying point clouds, we have found Breiman et al.'s Classification and Regression Trees (CART) [1] to strike a good balance between computational complexity and reliability. The implementation we used was that of rpart [25]. In terms of Friedl et al. [5] these trees are univariate classification trees.

Conceptually, a classification tree seeks to partition the entire feature space of a data set, one variable at a time. It does that by selecting a variable and an appropriate splitting value that will contribute maximally to node purity. Node purity is computed using the Gini impurity coefficient:

$$I_G(f) = 1 - \sum_{i=1}^{m} f_i^2 \tag{7}$$

with f_i being the fraction of items labeled to be of class i for a set of m class labels.

This splitting and branch growing continues, until no variable can be found that further increases node purity. The resulting trees can become quite large which hinders interpretation (not a problem for point cloud classification) and are prone to overfitting. This latter limitation can become troublesome when trying to classify point clouds, as the learned model does not generalize well anymore for unseen data. However, using cross-validation and pruning off branches that are not occurring in a significant number of replications proves to be an effective tool against overfitting.

As stated above, the performance of a classification tree can be gauged if not only training but also test data contain true class labels. A measurement statistic of classification performance is the misclassification rate. Let M be a cross-classification matrix between true and predicted class labels and its elements being the counts of the predicted elements and J the number of all points in the point cloud, then

$$MCR = 1 - tr(M)/J \tag{8}$$

is the misclassification rate.

When selecting training data, two factors need consideration: the randomness of the selection process and its stratification. The former factor becomes important once large sets of random numbers need to be created. While computers can always only generate pseudo random numbers, most of them are sufficiently strong for point cloud processing.[2] However, strong random number generation with guaranteed randomness does not suffice to select a suitable training data set, if the classes are not evenly distributed. In that common case, single classes—say temporary construction structures—have only very few points associated with them. When choosing points at random, it is extremely unlikely that many of the rare class points will end up in the training data set. And if a class does not show up in the training data set, the supervised classification algorithm cannot learn the rules required to classify it. Therefore simple random sampling schemes do not work in the presence of rare classes.

To enable the supervised classification of rare classes, stratified sampling needs to be applied. In its simplest form, stratified sampling guarantees that numerous points from each class are selected for the training data set. This, at the expense of having the entire training data set being representative for the point cloud it has been sampled from. The heuristic used for our stratified sampling approach sets the size of the sample for stratum c (s_c) to be either half of the points of that class (S_c) or the overall sample size (k) divided by the number of classes in the point cloud ($|A|$):

$$s_c = \min(\frac{S_c}{2}, \frac{k}{|A|}) \tag{9}$$

[2]We used the R [18] implementation of the Mersenne twister, which has a period of $2^{19937} - 1$.

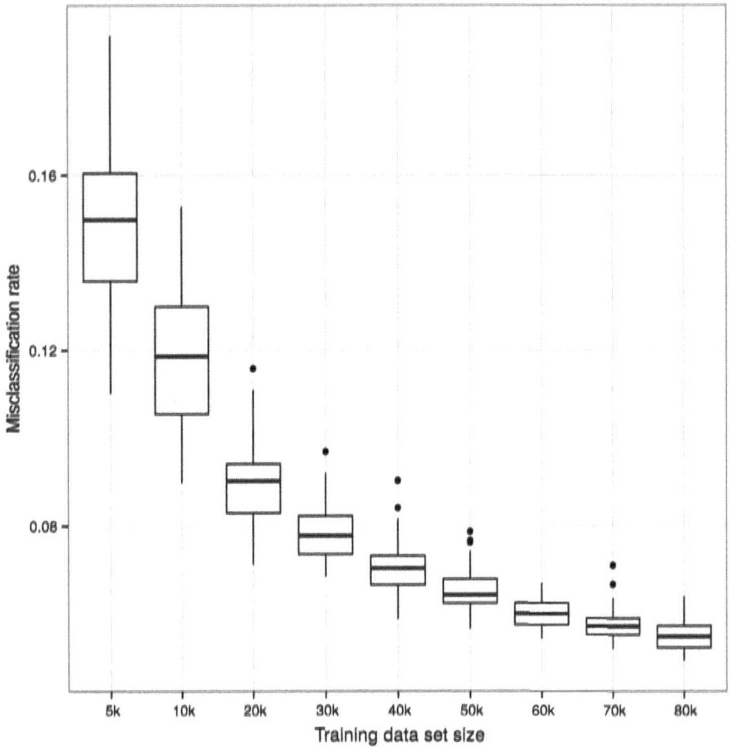

Fig. 3 Misclassification rate as a function of training data size; classification of a 3 million strong point cloud, results bootstrapped with 50 replications

As noted above, the resulting stratified sample is not representative for the entire point cloud anymore: rare classes occur much more often in the training data set than they do in the point cloud. It is therefore necessary to inform the supervised classification algorithm of that misrepresentation.

Perhaps obviously, the performance of a tree depends on the number of data points it is allowed to learn from: the larger the training data set, the better (usually) the classification of test data will be. However, manually classifying points is expensive. Therefore, it is crucial to find a training data set size that is just large enough to produce reliable predictions. Figure 3 depicts this relationship. As can be seen, there is a sharp drop between 10,000 and 20,000 points as training data set size with respect to mean misclassification rate and its dispersion. After about 50,000 points, the improvement gained by adding additional points subsides. We therefore settled for 50,000 points as training data set size. The resulting mean misclassification rate of 0.065 is a usable starting point. In the following, we will discuss aspects of improving this achievement even further.

When classifying point cloud data into predetermined classes, not all classes that appear to be epistemologically justified to humans can be sufficiently identified using laser return signals. For the problem at hand, the points were to be partitioned

Table 2 Classes that were hard to predict. Percentage of points that ended up in that class. Remainder to 100 % is scattered in all classes

True class	Predicted classes
Building, wall	Deciduous forest (67 %)
	Building roof (17 %)
	Building, wall (17 %)
Temporary object	Temporary object (78 %)
	Road protection fence (14 %)
Power pole	Power pole (75 %)
	Road protection fence (16 %)
Error class points	Scattered in all classes

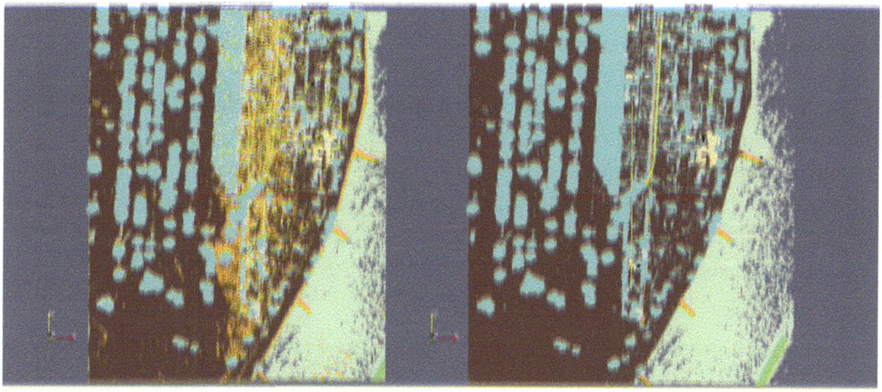

Fig. 4 Predicted (*left*) and true (*right*) classification of a sample area

into 26 classes. Logically, these classes could be broken down into coarsely and finely grained classes. While the coarse classes were successfully classified (MCR: 0.02, $\sigma = 0.002$), the finer classification exhibited the 6.5 % MCR as described above. Table 2 lists the finely grained classes that were notoriously troublesome. Figure 4 shows the differences between automatic and true human classification results.

When casting a more detailed look at these misclassifications, it becomes evident that many of them are conceivably caused by imprecise classifications of humans in the first place. Consider, for example, a road in winter: the asphalt tarmac is at places covered with grit sand to prevent the icing of the road. Grit and asphalt tarmac will differ in texture and material. Therefore, the laser return signal for patches of road that contain more grit sand than others will exhibit different characteristics. In manual classification based on aerial photography, these patches of grit sand are unlikely to be identified and marked as such by the human classifier. To a certain extent, the misclassification rate achieved by supervised classification of finely grained classes can be explained by the algorithm outperforming human classification. This is obviously very dependent on the quality of human classification.

A similar argument holds for the error class. Here, points were classified as errors if some of their measurements exceeded a valid measurement range. The algorithm was informed about these missing values. On the other hand, classification trees are able to cope with missing information by substituting it with the second best split. Therefore, points that a human would not classify because it contained obviously faulty measurements were classified by the algorithm.

Another problem that is rooted in the difficulty of epistemological concepts is the misclassification of many temporary object points as road protection fences. It is difficult for any automated classifier to learn the concept of an object being temporary in nature. While the algorithm successfully classifies almost all temporary objects as some kind of artificial objects, it cannot differentiate between these objects being permanent or temporary (road protection fences).

However, the largest problem in misclassification cannot possibly be rooted in epistemological complexities: Buildings and walls are being classified predominantly as trees. From a geometrical point of view, trees and buildings do indeed share some properties related to their height and volume. On the other hand, distinctive characteristics like texture and material should have been picked up by the algorithm. This type of mistake is also represented in Fig. 5. To some extent, the misclassifications can be explained by snow or leaf covered roofs on top of buildings. Still, this unsatisfactory performance can most likely only be overcome by implementing geometrical shape detection in a post-processing step. This is the focus of ongoing research.

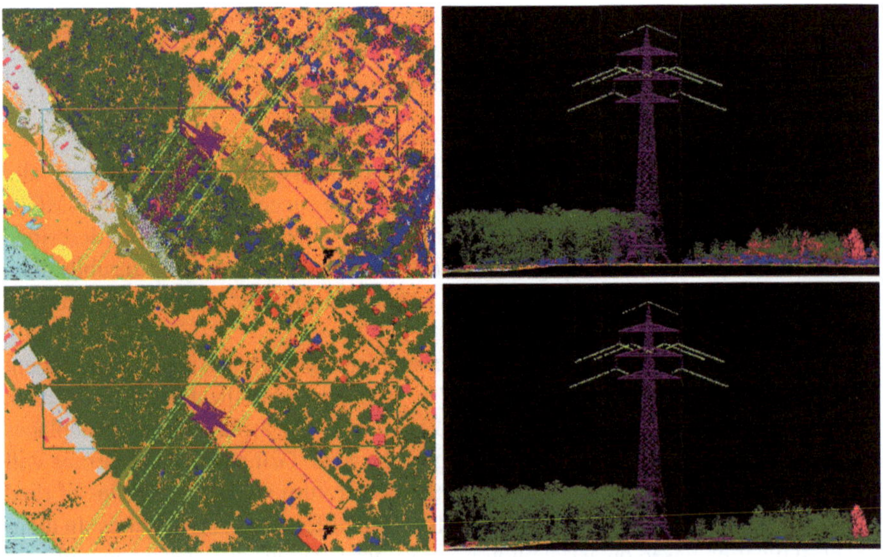

Fig. 5 Misclassification at a power line where pole and vegetation cannot be separated reliably; automatically classified point cloud on *top*, *bottom panel* shows the manually classified one

Table 3 Model quality in mean MCR for models with different kinds of border effect components. Results bootstrapped with 50 replications.

Model type	μ_{MCR}	σ_{MCR}
No border effects	0.081	0.004
Beam vector components	0.065	0.005
Scan angle	0.074	0.005
Beam vector components and scan angle	0.063	0.004

3.4 *Border Effects*

Airborne laser scanning is limited by the principles of optics: dependent on the incident angle, the characteristics of a laser return signal varies. For example, hitting vegetation from the side will produce many more laser echoes than hitting it straight from above. Also, the shape of the beam's cross section depends on that angle. Additional distortion in the characteristics of points may arise from the method of aerial laser scanning. Due to the limited field of view of airborne laser scanners wider areas are scanned by multiple overlapping strips. Typically, these strips overlap to achieve full coverage even in case of wind sheer or minor navigation errors. In these overlapping areas, the properties of the measurement process change (as there are multiple overpasses); a change that needs to be accounted for.

One method to compensate for the different return signal quality/properties is to take the deflection of the laser into account. There are two approaches available. One uses the raw beam vector components (v_x, v_y, v_z) that indicate the deflection of the laser beam for a given point. The other method combines these components to derive the scan angle ϕ:

$$\phi = \arctan(\frac{\sqrt{v_x^2 + v_y^2}}{|v_z|})$$

The following Table 3 shows the effect beam vector components and scan angle have on the misclassification rate. Starting with the simplest model without any compensation for border effects, the mean classification rate lies at 8.1%. Adding the scan angle to the model improves its quality by one, beam vector components by 2 % points. Adding both compensation terms to the model barely improves classification quality with respect to a pure beam vector components model.

3.5 *Scale Space Selection*

A number of point cloud features are not directly measured but computed with respect to any points immediate neighborhood. In general, the local neighborhood of a point can be defined in 2D or 3D. Furthermore, a certain number of closest neighbors, a fixed distance or a combination of both can be used as neighborhood definition. For the following analysis a cylinder (i.e., 2D fixed distance neighborhood) for each point is formed. Obviously, larger radii lead to a stronger averaging effect while smaller ones are prone to overfitting. It, therefore, is important to find the optimal radius for each feature in order to minimize misclassification rate.

To discover the optimal radii for neighborhood-dependent features, a genetic search algorithm [6] was used. In the following we will describe the genetic algorithm used for this optimization and its parameters. We then turn our attention towards evaluating the algorithm's performance in terms of convergence and solution stability. The former examines the relation of improvement achieved due to and time spent on optimization. The latter analyzes the stability of recommended radii across a number of optimizations.

The 13 neighborhood-dependent features were computed each with radii ranging from 1 to 6 m in 0.5 m increments resulting in 11 versions of each feature. The algorithm's genomes were then modeled to be integer vectors of length 13 with each gene being an integer from 1 to 11, encoding the chosen neighborhood size for each feature. The algorithm was initialized with 100 random genomes as starting solutions. The standard genetic operators of single-point cross-over breeding and mutation were employed for evolutionary optimization. Further, pairing genomes for mating was done using tournament selection and a proportion of the top performing solutions was cloned directly into each new generation. To ensure that the gene pool remained fresh and to safeguard against local optima traps, some random genomes were introduced with each generation. Table 4 gives the parameters of the genetic algorithm, which were established by experiment.

The fitness function to be optimized was the misclassification rate as described above. In order to ensure comparability, MCR was computed using the same training–test data split each time. The initial split was generated using a stratified sampling scheme and included 5,137 points in the training data set. Using a random sample of 100,000 points, the algorithm was allowed 500 generations to find the optimum combination of radii for the 13 neighborhood-dependent features. In order to ensure computability within reasonable time, not the entire point cloud could be processed. Therefore, a very large simple random sample of 100,000 points was drawn from the point cloud, and all operations were performed on that sample.

As genetic optimization is essentially stochastic in nature, the optimization was repeated 34 times. Of these 34 replications, 30 reached the same optimum while four stayed behind (by a very small margin). Almost all replications had converged

Table 4 Parameters of the genetic algorithm

Parameter	Value
Population size	100
Tournament size	5
Mutation probability	0.05
Elite proportion	0.1
Reseed proportion	0.1

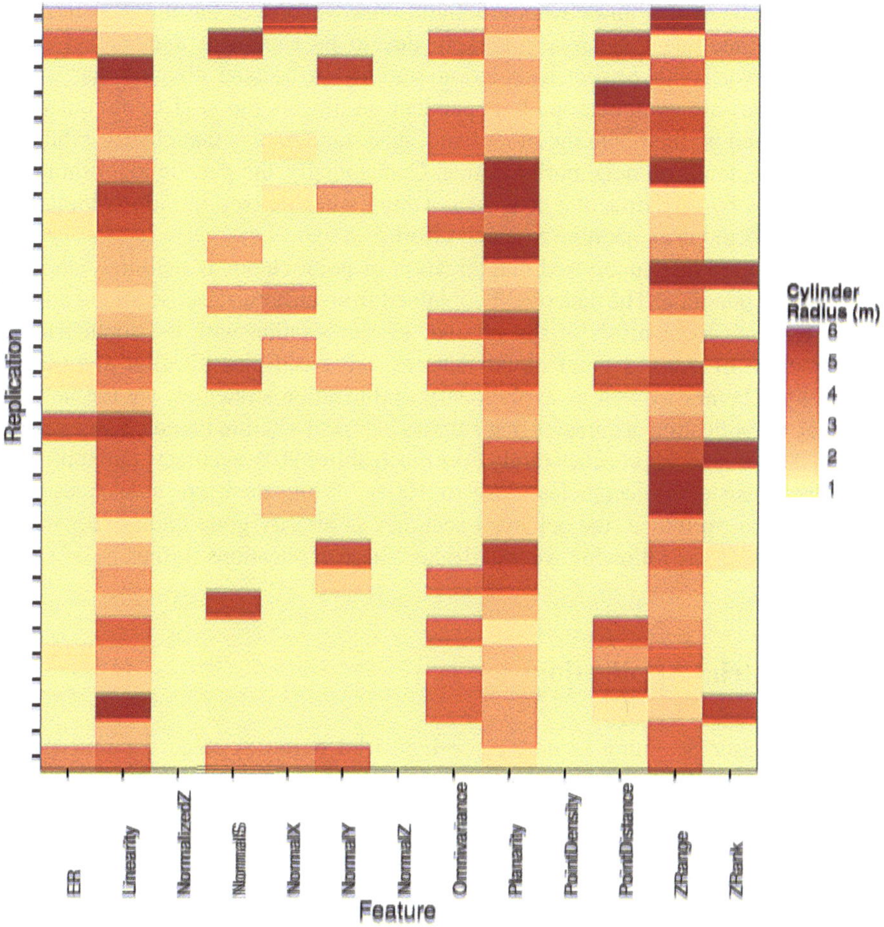

Fig. 6 Solution stability of 30 genetic optimization replications

to the optimum after 50 generations. By generation 75 all 30 successful replications had converged. The optimum discovered implied a misclassification rate of 0.022. When compared to the best misclassification achieved using a constant radius of 6 m (0.065) this is a notable improvement by more than 60 %.

Turning to solution stability, it is of interest whether each replication's terminal solution leads to the same combination of radii or not. Figure 6 displays a heat map of cylinder radii per feature chosen in each (optimal) replication. Features that exhibit the same color shades for the entire column can be considered stable. These are the variables NormalizedZ, NormalZ, and PointDensity. For each of these features, the optimal cylinder radius is at 1 meter. At the other end of the spectrum, very colorful columns, Linearity, Planarity, and Z-Range, are indicative of features whose neighborhood size has no impact on misclassification rate.

The genetic algorithm delivers a definite improvement of the misclassification rate. The remaining 2 % are most likely due to measurement and human classification error. With respect to solution stability, it became obvious that while some features are computationally dependent on neighborhood size, the outcome is not affected by them. On the other hand, there are features that clearly exhibit a strong dependence on neighborhood size. Conceptually, the genetic algorithm can be improved by implementing consensus voting when delivering radii recommendations. This too is an ongoing research effort.

We conclude that supervised classification of point clouds is definitely an idea worthwhile pursuing. The data quality obtained from airborne laser scanning allows for a very precise analysis of the ground. In combination with the sophisticated computation of derived point cloud features, advanced classification algorithms sampling schemes as well as evolutionary optimization strategies, we are able to produce classification accuracies that surpass classical satellite based classification. While the classical approaches rarely ever reach above 90 % accuracy, our approach delivers consistently accuracies close to 100 %. While there are challenges that remain to be overcome, the achieved accuracy is already good enough for many applications. In the following we will discuss these applications further.

4 Industrial Applications

Airborne Laser Scanning is in use for industrial purposes since the mid-1990s and has dramatically improved since then. For example: in the beginning there have been laser scanners with a fixed array of fiber optical conductors, which brought a good point density in the direction of flight, but very poor density in the transverse direction. So a detection of embankments along the flight direction was very hard. Technological advances like the steadily increased measurement rates, improved apertures and new detection algorithms prepared the way for a wide field of applications.

There are different technologies at work in today's laser scanners: they provide sampling rates of up to 600,000 laser pulses per second. Also modern apertures are able to detect more than just one single return per pulse and provide reflectance, echo ID and echo width for each return; some can even penetrate water surfaces and give information on submarine ground and submerged objects.

Higher point densities result in better environment depicting. With today's high point densities, embankments can be well detected by extracting *breaklines* within the point clouds. Normally 4 points/m^2 will be ordered, but customers more often want 8 or more points/m^2. This gives the opportunity to model the ground more precisely. But customers are not only interested in the presence of ground, but they also want to know what kind of ground they are looking at.

Classification is mostly a semiautomatic process, consisting of an automatic step and a manual checking and correction step. One of the aims is to minimize the need of manual correction, due to its cost. Another aim is to improve the automatic detection of more than a standard set of classes to cater to future customer's requirements.

In the following we will present some examples of airborne laser scanning applications.

4.1 Digital Terrain Model

Often a plain model of the ground is needed for planning or research purposes. These models are of great importance, e.g. for road- or railway planning offices, in order to know how much material has to be removed or added for street or railway planning. Therefore the point cloud has to be classified with special emphasis on detecting erroneous echoes. The DTM classes mostly consist of ground, water and unclassified points, which have no influence on the model.

4.2 Digital Surface Model

The DSM features ground, vegetation, buildings, bridges, and sometimes power lines and describes the earth's surface including natural and artificial objects. By subtracting the DTM from the DSM the result will be a normaliuzed DSM. This can then be used, e.g., for easy measurement of building or vegetation heights.

4.3 Avalanche Prediction

In mountainous areas avalanches (snow or boulders) are a common threat, so prediction and subsequently protection is an important task. For aviation purposes it is also necessary to know the position of power lines or cable-cars. Therefore each point needs to be classified along the lines of ground, various vegetation, water, building, power lines, ...

To compute the pathways and probabilities of avalanches in certain areas, one needs to know not only point classes but also inclination, roughness (in this case roughness refers to a parameter, which will tell how fluids will be slowed on a surface), azimuth, ...

All these features can be derived out of the point cloud by classifying using the above algorithm.

4.4 Flooding Prediction

To protect people and environment in areas that are in danger of flooding around rivers, it is vital to know how water is flowing over different types of ground. Therefore ground has to be classified in different roughness classes, that have known properties for flowing or seeping. The classification of roughness areas is normally done by digitizing digital orthophotos [4, 17]. In respect to the classification methods described in Sect. 3, roughness can be set in direct relation with different ground classes. Taking into account the derived DTM together with the digitized breaklines [2], a triangulated surface can be computed.

By combining the DTM surface with information of the different point classes from ground detection, there can be defined areas with varying roughness. This classification is normally done by using digital orthophotos as reference. By classifying the roughness purely from the data contained within a laser point cloud, the high cost of extra orthophotos can be skipped.

4.5 Forestry and Vegetation

The detection of forested areas is an important part of environmental applications. Especially time series analyses, e.g. to estimate deforestation, were often carried out using analog or digital orthophotos so far. However, Airborne Laser Scanning gets more popular for such applications, because it is not restricted to the canopy. The laser beam can often penetrate the vegetation returning multiple echoes. This provides information about the vertical structure of the forest including good knowledge of the ground, which is needed to compute high quality DTMs, tree heights, stem volumes, etc. In urban areas the knowledge of classified vegetation is used in applications for 3D visualizations, urban planning, noise emission charts, etc. [22].

5 Conclusion

In this chapter we presented an overview of advances in processing and automatically classifying point clouds from airborne laser scanning. Particularly, the accuracy of the classification of point clouds can be improved greatly using machine learning based methods like decision trees. There, manually classified training data—a small subset of the entire point cloud—is used to build a classification model. This then in turn can be used to classify the remainder of the point cloud or a fresh one.

These advances in classification accuracy are chiefly due to our making use of the entire full wave form of the laser echoes. Using advanced radiometric and computational methods, for every echo additional properties or features are computed from that echo's wave form, external data and the echo's immediate neighborhood. Using an evolutionary algorithm we were able to identify features where the size of that neighborhood influenced classification accuracy and establish optimal neighborhood size values for these features.

The model presented in this chapter has applications ranging from forestry to avalanche and flooding protection. A more immediate application is the automatic generation of maps. However, this is but the beginning of our journey. We already pointed to the inclusion of shape detection for improving classification accuracy and consensus voting the genetic algorithm to optimize neighborhood size recommendations as current research goals. Further extensions focus on better understanding how the scan angle affects echo properties when analyzing the flights strip-wise. A major issue is the possibility to learn from multiple but possibly unreliably sources. Often, orthophotos related to a point cloud are out-of-date or older maps are used to provide external reference data. Ideally, if we were able to use these data sources to speed up training data and model generation, the entire remote sensing work flow could be revolutionized.

References

1. Breiman, L., Friedman, J., Olshen, R., Stone, C.: Classification and Regression Trees. Wadsworth and Brooks, Pacific Grove (1984)
2. Briese, C., Mandlburger, G., Ressl, C., Brockman, H.: Automatic break line determination for the generation of a DTM along the river Main. In: ISPRS Workshop Laserscanning 2009, pp. 236–241 (2009)
3. Doneus, M., Briese, C., Fera, M., Janner, M.: Archaeological prospection of forested areas using full-waveform airborne laser scanning. J. Archaeol. Sci. **35**(4), 882–893 (2008)
4. Dorninger, P.: Eine praktikable und genaue Methode zur Bestimmung von Wasser-Land-Grenzen aus Laser-Scanner-Daten. BfG-Veranstaltungen **3**, 93–100 (2011)
5. Friedl, M., Brodley, C.: Decision tree classification of land cover from remotely sensed data. Remote Sens. Environ. **61**(3), 399–409 (1997)
6. Goldberg, D., Holland, J.: Genetic algorithms and machine learning. Mach. Learning **3**(2), 95–99 (1988)
7. Gressin, A., Mallet, C., David, N.: Improving 3D LiDAR point cloud registration using optimal neighborhood knowledge. ISPRS Annals of Photogrammetry, Remote Sensing and Spatial Information Sciences I-3, pp. 111–116 (2012)
8. Gross, H., Thoennessen, U.: Extraction of lines from laser point clouds. In: Symposium of ISPRS Commission III: Photogrammetric Computer Vision PCV06. International Archives of Photogrammetry, Remote Sensing and Spatial Information Sciences, vol. 36, pp. 86–91 (2006)
9. Höfle, B., Hollaus, M., Lehner, H., Pfeifer, N., Wagner, W.: Area-based parameterization of forest structure using full-waveform airborne laser scanning data. In: Silvilaser 2008, p. 9, Edinburgh, 2008
10. Höfle, B., Hollaus, M., Hagenauer, J.: Urban vegetation detection using radiometrically calibrated small-footprint full-waveform airborne LiDAR data. ISPRS J. Photogramm. Remote Sens. **67**(1), 134–147 (2012)
11. Hollaus, M., Aubrecht, C., Höfle, B., Steinnocher, K., Wagner, W.: Roughness mapping on various vertical scales based on full-waveform airborne laser scanning data. Remote Sens. **3**(3), 503–523 (2011)

12. Kobler, A., Pfeifer, N., Orginc, P., Todorovski, L., Oštir, K., Džeroski, S.: Repetitive interpolation: A robust algorithm for DTM generation from aerial laser scanner data in forested terrain. Remote Sens. Environ. **108**(1), 9–23 (2007)
13. Kraus, K., Pfeifer, N.: Advanced DTM generation from LiDAR data. In: International Archives of the Photogrammetry, Remote Sensing and Spatial Information Sciences, pp. 23–30, Annapolis, 2001
14. Mallet, C., Bretar, F., Soergel, U.: Analysis of full-waveform LiDAR data for classification of urban areas. Photogramm. Fernerkundung Geoinf. **5**, 337–349 (2008)
15. Mather, P., Tso, B.: Classification Methods for Remotely Sensed Data. CRC press, Boca Raton (2010)
16. Otepka, J., Ghuffar, S., Waldhauser, C., Hochreiter, R., Pfeifer, N.: Georeferenced point clouds: Data model, features and management. ISPRS Int. J. Geoinf. **2**(4), 1038–1065 (2013)
17. Prinz, R.: DGM-W-Modellierung unter Einbeziehung erfasster Buhnen und Buhnenfelder. BfG-Veranstaltungen **3**, 48–57 (2011)
18. R Core Team: R: A Language and Environment for Statistical Computing. R Foundation for Statistical Computing, Vienna, 2013
19. Research Groups Photogrammetry and Remote Sensing: OPALS: Orientation and Processing of Airborne Laser Scanning data. Department of Geodesy and Geoinformation, TU Vienna, Vienna (2013)
20. Roncat, A., Bergauer, G., Pfeifer, N.: B-spline deconvolution for differential target cross-section determination in full-waveform laser scanning data. ISPRS J. Photogramm. Remote Sens. **66**(4), 418–428 (2011)
21. Rusu, R.: Semantic 3D object maps for everyday manipulation in human living environments. Ph.D. thesis, München, Techn. Univ., Diss. (2009)
22. Rutzinger, M., Höfle, B., Pfeifer, N.: Detection of high urban vegetation with airborne laser scanning data. In: Proceedings of ForestSat (2007)
23. Safavian, S., Landgrebe, D.: A survey of decision tree classifier methodology. IEEE Trans. Syst. Man Cybern. **21**(3), 660–674 (1991)
24. Sithole, G., Vosselman, G.: Experimental comparison of filter algorithms for bare-earth extraction from airborne laser scanning point clouds. ISPRS J. Photogramm. Remote Sens. **59**(1–2), 85–101 (2004)
25. Therneau, T., Atkinson, B., Ripley, B.: rpart: Recursive Partitioning (2013). R package version 4.1-3
26. Wagner, W.: Radiometric calibration of small-footprint full-waveform airborne laser scanner measurements: Basic physical concepts. ISPRS J. Photogramm. Remote Sens. **65**(6), 505–513 (2010). ISPRS Centenary Celebration Issue
27. Wagner, W., Ullrich, A., Ducic, V., Melzer, T., Studnicka, N.: Gaussian decomposition and calibration of a novel small-footprint full-waveform digitising airborne laser scanner. ISPRS J. Photogramm. Remote Sens. **60**(2), 100–112 (2006)
28. Weinmann, M., Jutzi, B., Mallet, C.: Feature relevance assessment for the semantic interpretation of 3D point cloud data. In: ISPRS Annals of the Photogrammetry, Remote Sensing and Spatial Information Sciences II-5/W2, pp. 313–318 (2013)

A Novel Approach to the Common Due-Date Problem on Single and Parallel Machines

Abhishek Awasthi, Jörg Lässig, and Oliver Kramer

Abstract This chapter presents a novel idea for the general case of the Common Due-Date (CDD) scheduling problem. The problem is about scheduling a certain number of jobs on a single or parallel machines where all the jobs possess different processing times but a common due-date. The objective of the problem is to minimize the total penalty incurred due to earliness or tardiness of the job completions. This work presents exact polynomial algorithms for optimizing a given job sequence for single and identical parallel machines with the run-time complexities of $O(n \log n)$ for both cases, where n is the number of jobs. Besides, we show that our approach for the parallel machine case is also suitable for non-identical parallel machines. We prove the optimality for the single machine case and the run-time complexities of both. Henceforth, we extend our approach to one particular dynamic case of the CDD and conclude the chapter with our results for the benchmark instances provided in the OR library.

Keywords Scheduling • Common Due Date • Algorithms • Combinatorial optimization • Simulated annealing

1 Introduction

The Common Due-Date scheduling problem involves sequencing and scheduling of jobs over machine(s) against a common due-date. Each job possesses a processing time and different penalties per unit time in case the job is completed before or later than the due-date. The objective of the problem is to schedule the jobs so as to

A. Awasthi (✉) • J. Lässig
Department of Computer Science, University of Applied Sciences Zittau/Görlitz, Görlitz, Germany
e-mail: abhishek.awasthi@hszg.de; joerg.laessig@hszg.de

O. Kramer
Department of Computing Science, Carl von Ossietzky University of Oldenburg, Oldenburg, Germany
e-mail: oliver.kramer@uni-oldenburg.de

S. Koziel et al. (eds.), *Solving Computationally Expensive Engineering Problems*, Springer Proceedings in Mathematics & Statistics 97,
DOI 10.1007/978-3-319-08985-0_13

minimize the total penalty due to earliness or tardiness of all the jobs. In practice, a common due date problem occurs in almost any manufacturing industry. Earliness of the produced goods is not desired because it requires the maintenance of some stocks leading to some expenses to the industry for storage cost, tied-up capital with no cash flow, etc. On the other hand, a tardy job leads to customer dissatisfaction.

When scheduling on a single machine against a common due date, one job at most can be completed exactly at the due date. Hence, some of the jobs will complete earlier than the common due-date, while other jobs will finish later. Generally speaking, there are two classes of the common due-date problem which have proven to be NP-hard, namely:

- Restrictive CDD problem
- Non-restrictive CDD problem.

A CDD problem is said to be *restrictive* when the optimal value of the objective function depends on the due-date of the problem instance. In other words, changing the due date of the problem changes the optimal solution as well. However, in the *non-restrictive* case a change in the value of the due-date for the problem instance does not affect the solution value. It can be easily proved that in the restrictive case, the sum of the processing times of all the jobs is strictly greater than the due date and in the non-restrictive case the sum of the processing times is less than or equal to the common due-date.

In this chapter, we study the restrictive case of the problem. However, our approach can be applied to the non-restrictive case on the same lines. We consider the scenario where all the jobs are processed on one or more machines without pre-emption and each job possesses different earliness/tardiness penalties. We also discuss a particular dynamic case of the CDD on a single machine and prove that our approach is optimal with respect to the solution value.

2 Related Work

The Common due-date problem has been studied extensively during the last 30 years with several variants and special cases [13, 21]. In 1981, Kanet presented an $O(n \log n)$ algorithm for minimizing the total absolute deviation of the completion of jobs from the due date for the single machine, n being the number of jobs [13]. Panwalkar et al. considered the problem of common due-date assignment to minimize the total penalty for one machine [17]. The objective of the problem was to determine the optimum value for the due-date and the optimal job sequence to minimize the penalty function, where the penalty function also depends on the due-date along with earliness and tardiness. An algorithm of $O(n \log n)$ complexity was presented but the special problem considered by them consisted of symmetric costs for all the jobs [17, 21].

Cheng again considered the same problem with slight variations and presented a linear programming formulation [5]. In 1991 Cheng and Kahlbacher and Hall et al. studied the CDD problem extensively, presenting some useful properties for the general case [6, 10]. A pseudo polynomial algorithm of $O(n^2 d)$ (where d is the common due-date) complexity was presented by Hoogeveen and Van de Velde for the restrictive case with one machine when the earliness and tardiness penalty weights are symmetric for all the jobs [11]. In 1991 Hall et al. studied the unweighted earliness and tardiness problem and presented a dynamic programming algorithm [10]. Besides these earlier works, there has been some research on heuristic algorithms for the general common due date problem with asymmetric penalty costs. James presented a tabu search algorithm for the general case of the problem in 1997 [12].

More recently in 2003, Feldmann and Biskup approached the problem using metaheuristic algorithms, namely simulated annealing (SA) and threshold accepting, and presented the results for benchmark instances up to 1,000 jobs on a single machine [4, 7]. Another variant of the problem was studied by Toksari and Güner in 2009, where they considered the common due date problem on parallel machines under the effects of time dependence and deterioration [22]. Ronconi and Kawamura proposed a branch and bound algorithm in 2010 for the general case of the CDD and gave optimal results for small benchmark instances [19]. In 2012, Rebai et al. proposed metaheuristic and exact approaches for the common due date problem to schedule preventive maintenance tasks [18].

In 2013, Banisadr et al. studied the single-machine scheduling problem for the case that each job is considered to have linear earliness and quadratic tardiness penalties with no machine idle time. They proposed a hybrid approach for the problem based upon evolutionary algorithm concepts [2]. Yang et al. investigated the single-machine multiple common due date assignment and scheduling problems in which the processing time of any job depends on its position in a job sequence and its resource allocation. They proposed a polynomial algorithm to minimize the total penalty function containing earliness, tardiness, due date, and resource consumption costs [23].

This chapter is an extension of a research paper presented by the same authors in [1]. We extend our approach for a dynamic case of the problem and for non-identical parallel machines. Useful examples for both the single and parallel machine cases are presented.

3 Problem Formulation

In this section we give the mathematical notation of the common due date problem based on [4]. We also define some new parameters which are necessary for our considerations later on.

Let

n = total number of jobs
m = total number of machines
n_j = number of jobs processed by machine j ($j = 1, 2, \ldots, m$)
M_j = time at which machine j finished its latest job

$W_j^k = k^{th}$ job processed by machine j
P_i = processing time of job i $(i = 1, 2, \ldots, n)$
C_i = completion time of job i $(i = 1, 2, \ldots, n)$
D = the common due date
α_i = the penalty cost per unit time for job i for being early
β_i = the penalty cost per unit time for job i for being tardy
E_i = earliness of job i, $E_i = \max\{0, D - C_i\}$ $(i = 1, 2, \ldots, n)$
T_i = tardiness of job i, $T_i = \max\{0, C_i - D\}$ $(i = 1, 2, \ldots, n)$.

The cost corresponding to job i is then expressed as $\alpha_i \cdot E_i + \beta_i \cdot T_i$. If job i is completed at the due date, then both E_i and T_i are equal to zero and the cost assigned to it is zero. When job i does not complete at the due date, either E_i or T_i is nonzero and there is a strictly positive cost incurred. The objective function of the problem can now be defined as

$$\min \sum_{i=1}^{n} (\alpha_i \cdot E_i + \beta_i \cdot T_i) \ . \tag{1}$$

According to the three-field problem classification introduced by Graham et al. [9], the common due-date scheduling problem on a single machine can be expressed as $1|P_i|\sum_{i=1}^{n}(\alpha_i E_i + \beta_i T_i)$. This three-field notation implies that the jobs with different processing times are scheduled on a single machine to minimize the total earliness and tardiness penalty.

4 The Exact Algorithm for a Single Machine

We now present the ideas and the algorithm for solving the single machine case for a given job sequence. From here onwards we assume that there are n jobs to be processed by a machine and all the parameters stated at the beginning of Sect. 3 represent the same meaning. The intuition for our approach comes from a property presented and proved by Cheng and Kahlbacher for the CDD problem [6]. They proved that the optimal solution for a problem instance with general penalties has no idle time between any two consecutive jobs or in other words, when the schedule is compact. This property implies that at no point of time the machine processing the jobs is left idle till the processing of all the jobs is completed. In our approach we first initialize the completion times of all the jobs without any idle times and then shift all the jobs with the same amount of time.

Let J be the input job sequence where J_i is the ith job in the sequence J. Note that without loss of any generality we can assume $J_i = i$, since we can rank the jobs for any sequence as per their order of processing. The algorithm takes the job sequence J as the input and returns the optimal value for Eq. (1). There are three requirements for the optimal solution: allotment of jobs to specific machines, the order of processing of jobs in every machine, and the completion times for all the jobs.

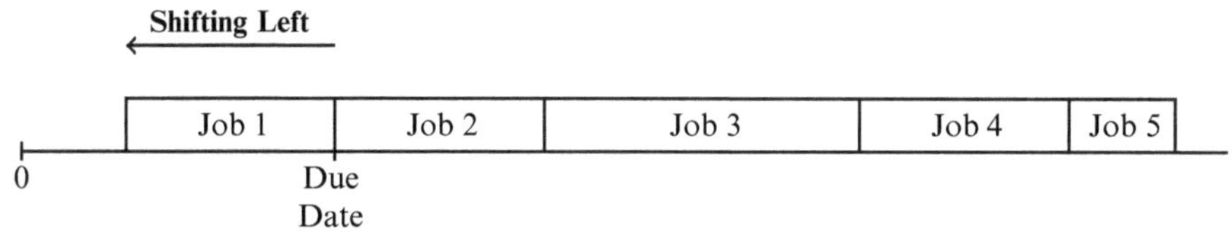

Fig. 1 Left shift (decrease in completion times) of all the jobs towards decreasing total tardiness for a sequence with five jobs. Each reduction is done by the minimum of the processing time of the job which is starting at the due date and the maximum possible left shift for the first job

Using the property of compactness proved by Cheng and Kahlbacher [6], our algorithm assigns the completion times to all the jobs such that the first job is finished at $\max\{P_1, D\}$ and the rest of the jobs follow without any idle time in order to obtain an initial solution which is then improved incrementally. It is quite apparent that a better solution for this sequence can be found only by reducing the completion times of all the jobs, i.e. shifting all the jobs towards decreasing total tardiness penalty as shown in Fig. 1 with five jobs. Shifting all the jobs to the right will only increase the total tardiness.

Hence, we first assign the jobs in J to the machine such that none of the jobs are early and there is no idle time between the processing of any two consecutive jobs, as stated in Eq. (2).

$$C_i = \begin{cases} \max\{P_1, D\} & \text{if} \quad i = 1 \\ C_{i-1} + P_i & \text{if} \quad 2 \leq i \leq n \,. \end{cases} \tag{2}$$

Before stating the exact algorithm for a given sequence for the single machine case, we first introduce some new parameters, definitions, and theorems which are useful for the description of the algorithm. We first define $DT_i = C_i - D, i = 1, 2, \ldots, n$, and $ES = C_1 - P_1$. It is clear that DT_i is the algebraic deviation of the completion time of job i from the due date and ES is the maximum possible shift (reduction of completion time) for the first job.

Definition 1. PL is a vector of length n and any element of PL (PL_i) is the penalty possessed by job i. We define PL, as

$$PL_i = \begin{cases} -\alpha_i, & \text{if} \quad DT_i \leq 0 \\ \beta_i, & \text{if} \quad DT_i > 0 \,. \end{cases} \tag{3}$$

With the above definition we can express the objective function stated by Eq. (1) as $\min(Sol)$, where

$$Sol = \sum_{i=1}^{n} (DT_i \cdot PL_i) \,. \tag{4}$$

The Algorithm 1 mentioned below returns the optimal solution value for any job sequence for the CDD problem on a single machine.

Algorithm 1: Exact Algorithm for Single Machine

1 **Initialize** $C_i \, \forall i$ (Equation (2))
2 **Compute** PL, DT, ES
3 $Sol \leftarrow \sum_{i=1}^{n} (DT_i \cdot PL_i)$
4 $j \leftarrow 2$
5 **while** $(j < n + 1)$ **do**
6 $\quad C_i \leftarrow C_i - \min\{ES, DT_j\}, \forall i$
7 $\quad$ **Update** PL, DT, ES
8 $\quad V_j \leftarrow \sum_{i=1}^{n} (DT_i \cdot PL_i)$
9 $\quad$ **if** $(V_j < Sol)$ **then** $Sol \leftarrow V_j$ **else go to** 11
10 $\quad j \leftarrow j + 1$
11 **return** Sol

5 Parallel Machine Case

For the parallel machine case we first need to assign the jobs to each machine to get the number of jobs and their sequence in each machine. In addition to the parameters explained in Sect. 3, we define a new parameter λ, which is the machine assigned to each job.

Definition 2. We define λ as the machine which has the earliest scheduled completion time of the last job on that machine. Using the notation mentioned in Sect. 3, λ can be mathematically expressed as

$$\lambda = \operatorname*{argmin}_{j=1,2,\ldots,m} M_j \, .$$

Algorithm 2 assigns the first m jobs to each machine, respectively, such that they all finish processing at the due date or after their processing time, whichever is higher. For the remaining jobs, we assign a machine λ to job i since it offers the least possible tardiness. Likewise each job is assigned at a specific machine such that the tardiness for all the jobs is the least for the given job sequence. The job sequence is maintained in the sense that for any two jobs i and j such that job j follows i; the Algorithm 2 will either maintain this sequence or assign the same starting times at different machines to both the jobs. Finally, Algorithm 2 will give us the number of jobs (n_j) to be processed by any machine j and the sequence of jobs in each machine, W_j^k. This is the best assignment of jobs at machines for the given sequence. Note that the sequence of jobs is still maintained here, since Algorithm 2 ensures that any job i is not processed after a job $i + 1$. Once we have the jobs assigned to each machine, the problem then converts to m single machine problems, since all the machines are independent.

Algorithm 2: Exact Algorithm: Parallel Machine

$M_j \leftarrow 0 \; \forall j = 1, 2, \ldots, m$
$n_j \leftarrow 1 \; \forall j = 1, 2, \ldots, m$
$i \leftarrow 0$
for $j \leftarrow 1$ **to** m **do**
  $i \leftarrow i + 1$
  $W_j^1 \leftarrow i$
  $M_j \leftarrow \max\{P_i, D\}$
for $i \leftarrow m + 1$ **to** n **do**
  Compute λ
  $n_\lambda \leftarrow n_\lambda + 1$
  $W_\lambda^{n_\lambda} \leftarrow i$
  $M_\lambda \leftarrow M_\lambda + P_i$
for *each machine* **do**
  Algorithm 1

For the non-identical parallel machine case we need a slight change in the definition of λ in Definition 2. Recall that M_j is the time at which machine j finished its latest scheduled job and λ is the machine which has the least completion time of jobs, among all the machines. In the non-identical machine case we need to make sure that the assigned machine not only has the least completion time but it is also feasible for the particular job(s). Hence, for the non-identical machines case, the definition of λ in Algorithm 2 will change to λ_i where

$$\lambda_i = \underset{j=1,2,\ldots,m}{\text{argmin}} \; M_j \text{ , such that machine } j \text{ is feasible for job } i \text{ .}$$

For the remaining part, the Algorithm 2 works in the same manner as for the identical parallel machines. Algorithm 2 can then be applied to the non-identical independent parallel machine case for the initial allocation of jobs to machines.

6 Illustration of the Algorithms

In this section we explain Algorithms 1 and 2 with the help of illustrative examples consisting of $n = 5$ jobs for both, single and parallel machine cases. We optimize the given sequence of jobs J where $J_i = i$, $i = 1, 2, \ldots, 5$. The data for this example is given in Table 1. There are five jobs to be processed against a common due-date (D) of 16. The objective is to minimize Eq. (4).

Table 1 The data for the exemplary case. The parameters possess the same meaning as explained in Sect. 3

i	P_i	α_i	β_i
1	6	7	9
2	5	9	5
3	2	6	4
4	4	9	3
5	4	3	2

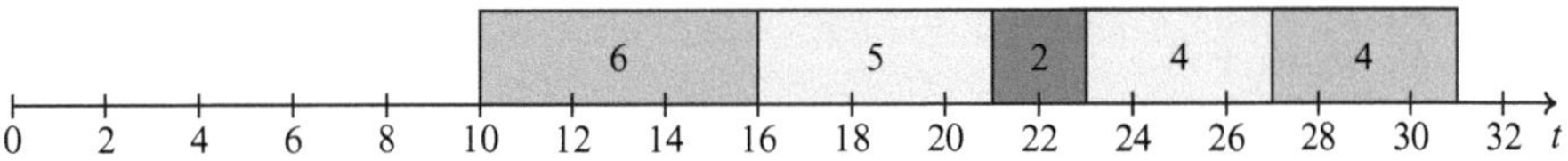

Fig. 2 Initialization of the completion times of all the jobs. The first job completes processing at the due date and the remaining jobs follow without any idle time

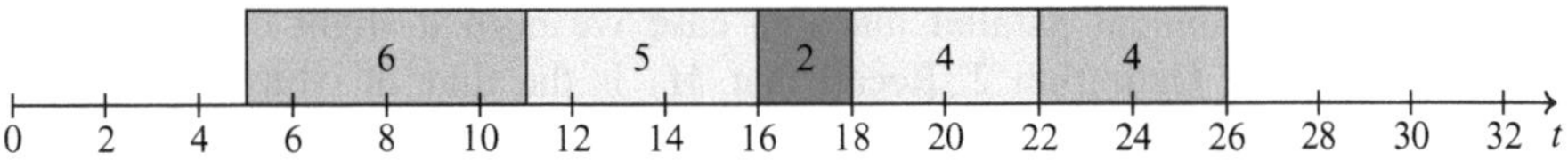

Fig. 3 All the jobs are shifted left by $\min\{ES, DT_j\} = 2$ units processing time

6.1 Single Machine Case

We first initialize the completion times of all the jobs according to Eq. (2) as shown in Fig. 2. The first job is completed at the due-date and possesses no penalty. However, all the remaining jobs from $J_i, i = 2, 3, 4, 5$ are tardy. After the initialization, the total penalty of this schedule is $Sol = \sum_{i=1}^{n}(\alpha_i \cdot E_i + \beta_i \cdot T_i) = (0 \cdot 7 + 0 \cdot 9) + (0 \cdot 9 + 5 \cdot 5) + (0 \cdot 6 + 7 \cdot 4) + (0 \cdot 9 + 11 \cdot 3) + (0 \cdot 3 + 15 \cdot 2)$. Hence, the objective value $Sol = 116$.

After the first left shift of 5 time units, the total penalty of this schedule is $Sol = \sum_{i=1}^{n}(\alpha_i \cdot E_i + \beta_i \cdot T_i) = (5 \cdot 7 + 0 \cdot 9) + (0 \cdot 9 + 0 \cdot 5) + (0 \cdot 6 + 2 \cdot 4) + (0 \cdot 9 + 6 \cdot 3) + (0 \cdot 3 + 10 \cdot 2)$. Hence the objective value $Sol = 81$.

After the third left shift of 2 time units (Fig. 3), the total penalty of this schedule is $Sol = \sum_{i=1}^{n}(\alpha_i \cdot E_i + \beta_i \cdot T_i) = (7 \cdot 7 + 0 \cdot 9) + (2 \cdot 9 + 0 \cdot 5) + (0 \cdot 6 + 0 \cdot 4) + (0 \cdot 9 + 4 \cdot 3) + (0 \cdot 3 + 4 \cdot 2)$. Hence the objective value $Sol = 95$.

Since the new value of the objective function is higher than in the previous step, we have the optimal value and schedule for this problem as shown in Fig. 4 with a total penalty of 81.

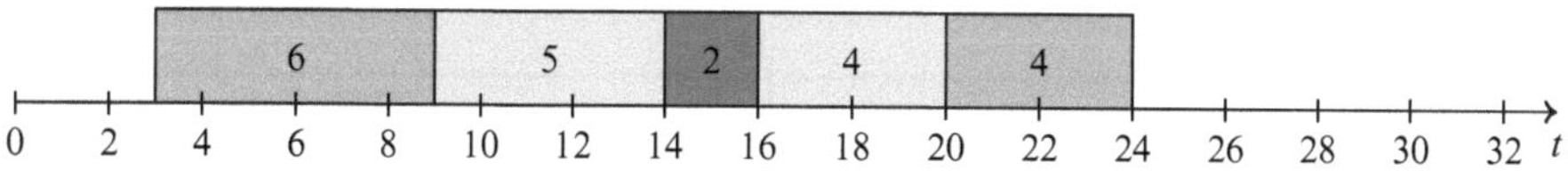

Fig. 4 All the jobs are shifted left by $\min\{ES, DT_j\} = 5$ units processing time

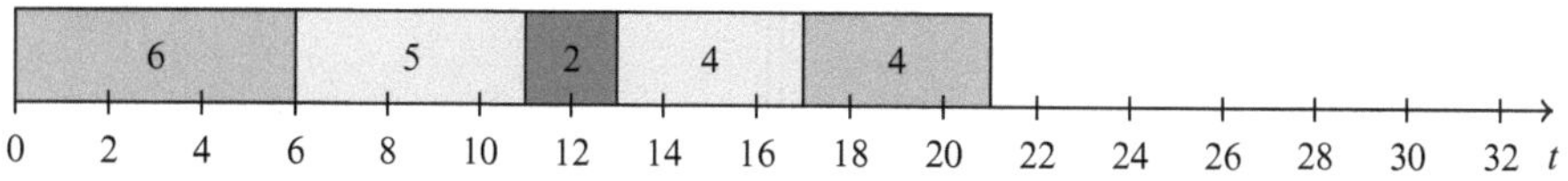

Fig. 5 Final left shift by $ES = 3$ units

For the sake of completeness, Fig. 5 shows the next step if we continue reducing the completion times using the same criterion as before. After the last possible left shift of 3 time units, the total penalty of this schedule is $Sol = \sum_{i=1}^{n}(\alpha_i \cdot E_i + \beta_i \cdot T_i) = (10\cdot 7 + 0\cdot 9) + (5\cdot 9 + 0\cdot 5) + (3\cdot 6 + 0\cdot 4) + (0\cdot 9 + 1\cdot 3) + (0\cdot 3 + 5\cdot 2)$. Hence the objective value $Sol = 146$. The total penalty increases further to a value of 146. Hence, the optimal value for this sequence is 81.

6.2 *Parallel Machine Case*

In the parallel machine case we consider two parallel machines and illustrate how we first assign the jobs in the same job sequence J to the machines and optimize them independently. The data used in this example is the same as in Table 1. The common due-date for the instance is also the same as earlier, $D = 16$.

As shown in Fig. 6a, there are five jobs to be processed on two independent identical parallel machines, against a due-date of 16. Hence, we first assign the jobs to a machine. We start with the first two jobs in the sequence J and assign them to the machines separately at $\max\{P_i, D\}$, Fig. 6b. For the remaining jobs, we subsequently choose a machine which offers least tardiness for each job. The third job in the sequence is assigned to the first machine and the fourth job goes to the second machine on the same lines, as depicted in Fig. 6c. Finally, we have all the jobs assigned to a machine (Fig. 6d) and each machine has a certain number of jobs to process in a given sequence. In this example, the first machine processes three jobs with the processing times of 6, 2, and 4, while the second machine processes two jobs with processing times of 5 and 4, in that order. Once we have this assignment of jobs to machines, we can apply our single machine algorithm to both of them independently to optimize the overall earliness and tardiness penalty. Figure 7 shows the best schedule for both the machines with an overall penalty of 32.

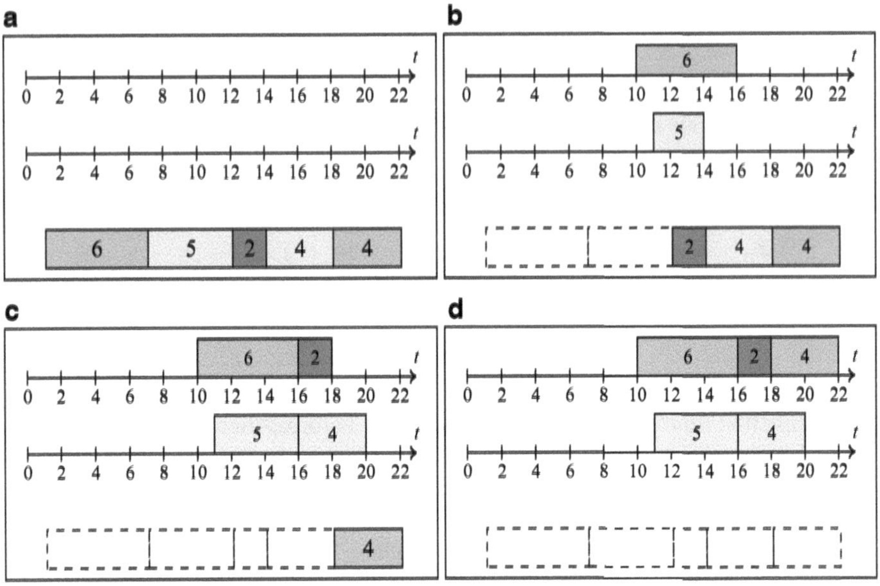

Fig. 6 Illustration of the assignment of jobs to machines. After the assignment, each machine has a certain number of jobs in the given sequence

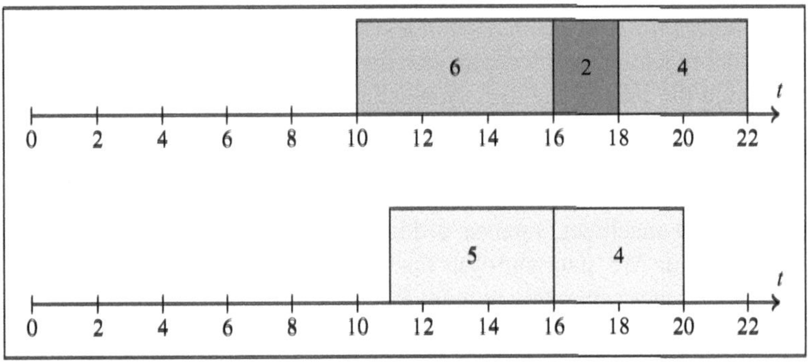

Fig. 7 Final optimal schedule for both the machines for the given sequence of jobs. The overall penalty of 32 is reached, which is the best solution value as per Algorithms 1 and 2.

7 Proof of Optimality

We now prove the optimality of Algorithm 1 with respect to the solution value for the single machine case.

Lemma 1. *If the initial assignment of the completion times of the jobs* (C_i)*, for a given sequence* J *is done according to Eq.* (2)*, then the optimal solution for this sequence can be obtained only by reducing the completion times of all the jobs or leaving them unchanged.*

Proof. We prove the above lemma by considering the two cases of Eq. (2).

Case 1: $D > P_1$
In this case Eq. (2) will ensure that the first job is completed at the due-date and the following jobs are processed consecutively without any idle time. Moreover, with this assignment all the jobs will be tardy except for the first job which will be completed at the due date. The total penalty (say, PN) will be $\sum_{i=1}^{n}(\beta_i \cdot T_i)$, where $T_i = C_i - D, i = 1, 2, \ldots, n$. Now if we increase the completion time of the first job by x units, then the new completion times C_i' for the jobs will be $C_i + x \forall i, (i = 1, 2, \ldots, n)$ and the new total penalty PN' will be $\sum_{i=1}^{n}(\beta_i \cdot T_i')$, where $T_i' = T_i + x$ $(i = 1, 2, \ldots, n)$. Clearly, we have $PN' > PN$ which proves that an increase in the completion times cannot fetch optimality which in turn proves that optimality can be achieved only by reducing the completion times or leaving them unchanged from Eq. (2).

Case 2: $D \leq P_1$
If the processing time of the first job in any given sequence is more than the due-date, then all the jobs will be tardy including the first job as $P_1 > D$. Since all the jobs are already tardy, a right shift (i.e., increasing the completion times) of the jobs will only increase the total penalty and hence worsening the solution. Moreover, a left shift (i.e., reducing the completion times) of the jobs is not possible either, because $C_1 = P_1$, which means that the first job will start at time 0. Hence, in such a case Eq. (2) is the optimal solution. In the rest of the paper we avoid this simple case and assume that for any given sequence the processing time of the first job is less than the due-date. ∎

Theorem 1. *Algorithm 1 finds the optimal solution for a single machine common due date problem, for a given job sequence.*

Proof. The initialization of the completion times for a sequence P is done according to Lemma 1. It is evident from Eq. (2) that the deviation from the due date (DT_i) is zero for the first job and greater than zero for all the following jobs. Besides, $DT_i < DT_{i+1}$ for $i = 1, 2, 3, \ldots, n-1$, since $C_i < C_{i+1}$ from Eq. (2) and DT_i is defined as $DT_i = C_i - D$. By Lemma 1 the optimal solution for this sequence can be achieved only by reducing the completion times of all the jobs simultaneously or leaving the completion times unchanged. Besides, a reduction of the completion times is possible only if $ES > 0$ since there is no idle time between any jobs.

The total penalty after the initialization is $PN = \sum_{i=1}^{n}(\beta_i \cdot T_i)$ since none of the jobs are completed before the due date. According to Algorithm 1 the completion times of all the jobs is reduced by $\min\{ES, DT_j\}$ at any iteration. Since $DT_1 = 0$, there will be no loss or gain for $j = 1$. After any iteration of the *while* loop in line 5, we decrease the total weighted tardiness but gain some weighted earliness penalty for some jobs. A reduction of the completion times by $\min\{ES, DT_j\}$ is the best non-greedy reduction. Let $\min\{ES, DT_j\} > 0$ and t be a number between 0 and $\min\{ES, DT_j\}$. Then reducing the completion times by t will increase the number of early jobs by one and reduce the number of tardy jobs by one. With this operation, if there is an improvement to the overall solution, then a reduction by $\min\{ES, DT_j\}$ will fetch a much better solution (V_j) because reducing the completion times by t will lead to a situation where none of the jobs either start at time 0 (because $ES > 0$) nor any of the jobs finish at the due date since the jobs $1, 2, 3, \ldots, j-1$ are early, jobs $j, j+1, \ldots, n$ are tardy and the new completion time of job j is $C'_j = C_j - t$.

Since after this reduction $DT_j > 0$ and $DT_j < DT_{j+1}$ for $j = 1, 2, 3, \ldots, n-1$, none of the jobs will finish at the due date after a reduction by t units. Moreover, it was proved by Cheng et al. [6] that in an optimal schedule for the restrictive common due date, either one of the jobs should start at time 0 or one of the jobs should end at the due date. This case can occur only if we reduce the completion times by $\min\{ES, DT_j\}$. If $ES < DT_j$, the first job will start at time 0 and if $DT_j < ES$ then one of the jobs will end at the due date. In the next iterations we continue the reductions as long as we get an improvement in the solution and once the new solution is not better than the previous best, we do not need to check any further and we have our optimal solution. This can be proved by considering the values of the objective function at the indices of two iterations; j and $j+1$. Let V_j and V_{j+1} be the value of the objective function at these two indices, then the solution cannot be improved any further if $V_{j+1} > V_j$ by Lemma 2. ■

Lemma 2. *Once the value of the solution at any iteration j is less than the value at iteration $j+1$, the solution cannot be improved any further.*

Proof. If $V_{j+1} > V_j$, a further left shift of the jobs does not fetch a better solution. Note that the objective function has two parts: penalty due to earliness and penalty due to tardiness. Let us consider the earliness and tardiness of the jobs after the jth iterations are E_i^j and T_i^j for $i = 1, 2, \ldots, n$. Then we have $V_j = \sum_{i=1}^{n}(\alpha_i E_i^j + \beta_i T_i^j)$ and $V^{j+1} = \sum_{i=1}^{n}(\alpha_i E_i^{j+1} + \beta_i T_i^{j+1})$. Besides, after every iteration of the *while* loop in Algorithm 1, the completion times are reduced or in other words the jobs are shifted left. This leads to an increase in the earliness and a decrease in the tardiness of the jobs. Let's say, the difference in the reduction between V^{j+1} and V^j is x. Then we have $E^{j+1} = E^j + x$ and $T_{j+1} = T_j - x$. Since $V^{j+1} > V^j$, we have: $\sum_{i=1}^{n}(\alpha_i E_i^{j+1} + \beta_i T_i^{j+1}) > \sum_{i=1}^{n}(\alpha_i E_i^j + \beta_i T_i^j)$. By substituting the values of E^{j+1} and T^{j+1} we get $\sum_{i=1}^{j+1} \alpha_i x > \sum_{i=j+2}^{n} \beta_i x$. Hence, at the $(j+1)^{th}$ iteration the total penalty due to earliness exceeds the total penalty due to tardiness. This proves that for any further reduction there cannot be an improvement in the solution because a decrease in the tardiness penalty will always be less than the increase in the earliness penalty. ■

8 Algorithm Run-Time Complexity

In this section we study and prove the run-time complexity of the Algorithms 1 and 2. We calculate the complexities of all the algorithms separately considering the worst cases for all. Let T_1 and T_2 be the run-time complexities of the algorithms, respectively.

Lemma 3. *The run-time complexities of both Algorithms 1 and 2 are* $O(n^2)$, *where* n *is the total number of jobs.*

Proof. As for Algorithm 1, the calculations involved in the initialization step and evaluation of PL, DT, ES, Sol are all of $O(n)$ complexity and their evaluation is irrespective of the any conditions unlike inside the *while* loop. The *while* loop again evaluates and updates these parameters at every step of its iteration and returns the output once there is no improvement possible. The worst case will occur when the *while* loop is iterated over all the values of j, $j = 2, 3, \ldots, n$. Hence the complexity of Algorithm 1 is $O(n^2)$ with n being the number of jobs processed by the machine. Hence, $T_1 = O(n^2)$.

Let m be the number of machines, then in the Algorithm 2, the complexity for the first two *for* loops is $O(m + (n - m)m)$ where $O(m)$ corresponds to the first *for* loop and $O((n - m)m)$ corresponds to the second *for* loop involving the calculation of λ. For the last *for* loop, we need to consider all the cases of the number of jobs processed by each machine.

Let $x_1, x_2, x_3, \ldots, x_m$ be the number of jobs processed by the machines, respectively. Then, $\sum_{i=1}^{m} x_i = n$. We make a reasonable assumption that the number of machines is less than the number of jobs, which is usually the case. In such a case the complexity of Algorithm 2 (T_2) is equal to $O(m + nm - m^2) + \sum_{i=1}^{m} O(x_i^2)$. Since $\sum_{i=1}^{m} x_i = n$, we have $\sum_{i=1}^{m} O(x_i^2) = O(n^2)$. Thus the complexity of Algorithm 2 is $O(m + nm - m^2 + n^2)$. Since we assume $m < n$ we have $T_2 = O(n^2)$. ■

9 Exponential Search: An Efficient Implementation of Algorithm 1

Algorithm 1 shifts the jobs to the left by reducing the completion times of all the jobs by $\min\{ES, DT_j\}$ on every iteration of the *while* loop. The run-time complexity of the algorithm can be improved from $O(n^2)$ to $O(n \log n)$ by implementing an exponential search instead of a step by step reduction, as in Algorithm 1. To explain this we first need to understand the slope of the objective function values for each iteration. In the proof of optimality of Algorithm 1, we proved that there is only one minimum present in $V^j \forall j$. Besides, the value of DT_j increases for every j as it depends on the completion times. Also note that the reduction in the completion times is made by $\min\{ES, DT_j\}$. Hence, if for any j, $ES \leq DT_j$ then every iteration after j will fetch the same objective function value, V^j. Hence, the solution values after each iteration will have a trend as shown below in Fig. 8.

With such a slope of the solution we can use the exponential search as opposed to a step by step search, which will in turn improve the run-time complexity of Algorithm 1. This can be achieved by increasing or decreasing the step size of the *while* loop by orders of 2 (i.e. $2, 2^2, 2^3, \ldots, n$) while keeping track of the slope of the solution. The index of the next iteration should be increased if the slope is negative and decreased if the slope is non-negative. At each step we need to keep track of the previous two indices and once the difference between the indices is less

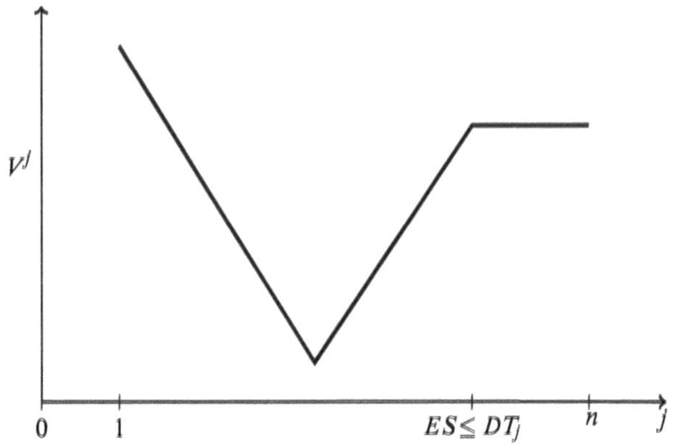

Fig. 8 The trend of the solution value against each iteration of Algorithm 1, for a job sequence. The value of the solution does not improve any further after a certain number of reductions

than the minimum of the two, then we need to perform binary search on the same lines. The optimum will be reached if both the adjacent solutions are greater than the current value. In this methodology we do not need to search for all values of j but in steps of 2^j. Hence the run-time complexity with exponential search will be $O(n \log n)$ for both the single machine and parallel machine cases.

10 A Dynamic Case of CDD

In this section we discuss about a dynamic case of the common due-date problem for the single machine case at the planning stage. Consider the case when an optimal schedule has been calculated for a certain number of jobs, and then an unknown number of jobs with unknown processing times arrive later. We assume that the original schedule is not disturbed and the new sequence of jobs can be processed after the first set of jobs. We show that in such a case the optimal schedule for the new extended job sequence can be achieved only by further reducing the completion times of all the jobs. We would like to emphasize here that we are considering the dynamic case at the planning stage when none of the jobs of the original known job sequence has gone to the processing stage.

Let us assume that at any given point of time there are a certain number of jobs (n) in a sequence J, for which the optimal schedule against a common due-date D on a machine has been already calculated using Algorithm 1. In such a case, if there are some additional jobs n' in a sequence J' to be processed against the same due-date and by the same machine without disturbing the sequence J, the optimum solution for the new sequence of $n + n'$ jobs in the extended sequence $J + J'$[1] can be found by further reducing the completion times of jobs in J and the same reduction in the completion times of jobs in J' using Algorithm 1. We prove it using Lemma 4.

Lemma 4. *Let C_i $(i = 1, 2, \ldots, n)$ be the optimal completion times of jobs in sequence J and C'_j $(j = 1, 2, \ldots, n, n+1, \ldots, n+n'-1, n+n')$ be the optimal completion times of jobs in the extended job sequence $J + J'$ with $n + n'$ jobs. Then,*

(i) $\exists\, \gamma \geq 0$ s.t. $C_i - C'_i = \gamma$ for $i = 1, 2, \ldots, n$
(ii) $C'_k = C_n - \gamma + \sum_{\tau=n+1}^{k} P_\tau$, $(k = n+1, n+2, \ldots, n+n')$.

Proof. Let Sol_J denote the optimal solution for the job sequence J. This optimal value for sequence J is calculated using Algorithm 1 which is optimal according to Theorem 1. In the optimal solution let the individual penalties for earliness and tardiness be E_i and T_i, respectively, hence

$$Sol_J = \sum_{i=1}^{n} (\alpha_i E_i + \beta_i T_i) . \tag{5}$$

Clearly, the value of Sol_J cannot be improved by either reducing the completion times any further as explained in Theorem 1. Now, processing an additional job sequence J' starting from C_n (the completion time of the last job in J) means that for the new extended sequence $J + J'$ the tardiness penalty increases further by some value, say $PT_{J'}$. Besides, the due date remains the same, the sequence J is not disturbed and all the jobs in the sequence J' are tardy. Hence the new solution value (say $V_{J+J'}$) for the new sequence $J + J'$ will be

$$V_{J+J'} = Sol_J + PT_{J'} . \tag{6}$$

For this new sequence we do not need to increase the completion times since it will only increase the tardiness penalty. This can be proved by contradiction. Let x be the increase in the completion times of all the jobs in $J + J'$ with $x > 0$. The earliness and tardiness for the jobs in J become $E_i - x$ and $T_i + x$, respectively, and the new total penalty (V_J) for the job sequence J becomes

$$\begin{aligned} V_J &= \sum_{i=1}^{n} (\alpha_i \cdot (E_i - x) + \beta_i \cdot (T_i + x)) \\ &= \sum_{i=1}^{n} (\alpha_i \cdot E_i + \beta_i \cdot T_i) + \sum_{i=1}^{n} (\beta_i - \alpha_i) \cdot x . \end{aligned} \tag{7}$$

[1] J and J' are two disjoint sets of jobs, hence $J + J'$ is the union of two sets maintaining the job sequences in each set.

Equation (5) yields

$$V_J = Sol_J + \sum_{i=1}^{n} (\beta_i - \alpha_i) \cdot x \,. \tag{8}$$

Since Sol_J is optimal $Sol_J \leq V_J$, we have

$$\sum_{i=1}^{n} (\beta_i - \alpha_i) \cdot x \geq 0 \,. \tag{9}$$

Besides, the total tardiness penalty for the sequence J' will further increase by the same quantity, say $\delta, \delta \geq 0$. With this shift, the new overall solution value $V'_{J+J'}$ will be

$$V'_{J+J'} = V_J + PT_{J'} + \delta \,. \tag{10}$$

Substituting V_J from Eq. (8) we have

$$V'_{J+J'} = Sol_J + \sum_{i=1}^{n} (\beta_i - \alpha_i) \cdot x + PT_{J'} + \delta \,. \tag{11}$$

Using Eq. (6) gives

$$V'_{J+J'} = V_{J+J'} + \sum_{i=1}^{n} (\beta_i - \alpha_i) \cdot x + \delta \,. \tag{12}$$

Using Eq. (9) and $\delta \geq 0$ we have

$$V'_{J+J'} \geq V_{J+J'} \,. \tag{13}$$

This shows that only a reduction in the completion times of all the jobs can improve the solution. Thus, there exists a $\gamma, \gamma \geq 0$ by which the completion times are reduced to achieve the optimal solution for the new job sequence $J + J'$. Clearly, $C_i - C'_i = \gamma$ for $i = 1, 2, \ldots, n$ and $C'_k = C_n - \gamma + \sum_{\tau=n+1}^{k} P_\tau$, $(k = n+1, n+2, \ldots, n+n')$ since all the jobs are processed one after another without any idle time. ■

11 Results

In this section we present our results for the single and parallel machine cases. We used our exact algorithms with simulated annealing for finding the best job sequence. All the algorithms were implemented on MATLAB® and run on a machine with a 1.73 GHz processor and 2 GB RAM. We present our results for the benchmark instances provided by Biskup and Feldmann in [4] for both the single and parallel machine cases. For brevity, we call our approach as APSA and the benchmark results as BR.

We use a modified Simulated Annealing algorithm to generate job sequences and Algorithm 1 to optimize each sequence to its minimum penalty. Our experiments show that an ensemble size of $4 + n/10$ and the maximum number of iterations as $500 \cdot n$, where n is the number of jobs, work best for the provided benchmark instances. The run-time for all the results is the time after which the solutions mentioned in Tables 2 and 3 are obtained. The initial temperature is kept as twice the standard deviation of the energy at infinite temperature: $\sigma_{E_{T=\infty}} = \sqrt{\langle E^2\rangle_{T=\infty} - \langle E\rangle^2_{T=\infty}}$. We estimate this quantity by randomly sampling the configuration space [20]. An exponential schedule for cooling is adopted with a cooling rate of 0.999. One of the modifications from the standard SA is in the acceptance criterion. We implement two acceptance criteria: the Metropolis acceptance probability, $\min\{1, \exp((-\Delta E)/T)\}$ [20] and a constant acceptance probability of 0.07. A solution is accepted with this constant probability if it is rejected by the Metropolis criterion. This concept of a constant probability is useful when the SA is run for many iterations and the metropolis acceptance probability is almost zero, since the temperature would become infinitesimally small. Apart from this, we also incorporate elitism in our modified SA. Elitism has been successfully adopted in evolutionary algorithms for several complex optimization problems [8, 14]. We observed that this concept works well for the CDD problem. Lässig and Sudholt made theoretical studies analysing speed-ups in parallel evolutionary algorithms with elitism applied to combinatorial optimization problems [15]. In [16] it is shown that for a large class of quality measures, the best possible probability distribution is a rectangular distribution over certain individuals sorted by their objective values, which can be seen as a mild form of elitism. As for the perturbation rule, we first randomly select a certain number of jobs in any job sequence and permute them randomly to create a new sequence. The number of jobs selected for this permutation is taken as $2 + \lfloor\sqrt{n/10}\rfloor$, where n is the number of jobs. For large instances the size of this permutation is quite small but we have observed that it works well with our modified simulated annealing algorithm.

In Tables 2 and 3 we present our results (APSA) for the single machine case. The results provided by Biskup and Feldmann can be found in [7]. The first 40 instances with ten jobs each have been already solved optimally by Biskup and Feldmann and we reach the optimality for all these instances within an average run-time of 0.457 s.

Among the next 160 instances we achieve equal results for 13 instances, better results for 133 instances and for the remaining 14 instances with 50, 100, and 200 jobs, our results are within a gap of 0.803 %, 0.1955 %, and 0.1958 %, respectively. Feldmann and Biskup [7] solved these instances using three metaheuristic approaches, namely: simulated annealing, evolutionary strategies, and threshold accepting; and presented the average run-time for the instances on a Pentium/90 PC.

In Table 4 we show our average run-times for the instances and compare them with the heuristic approach considered in [7]. Apparently our approach is faster

Table 2 Results obtained for the single machine case of the common due date problem and comparison with benchmark results provided in the OR Library [3]. For any given number of jobs there are ten different instances provided and each instance is designated a number k. The gray boxes indicate the instances for which our algorithm could not achieve the known solution values given in [3]

Jobs	h=0.2		h=0.4		h=0.6		h=0.8	
n=10	APSA	BR	APSA	BR	APSA	BR	APSA	BR
k=1	1,936	1,936	1,025	1,025	841	841	818	818
k=2	1,042	1,042	615	615	615	615	615	615
k=3	1,586	1,586	917	917	793	793	793	793
k=4	2,139	2,139	1,230	1,230	815	815	803	803
k=5	1,187	1,187	630	630	521	521	521	521
k=6	1,521	1,521	908	908	755	755	755	755
k=7	2,170	2,170	1,374	1,374	1,101	1,101	1,083	1,083
k=8	1,720	1,720	1,020	1,020	610	610	540	540
k=9	1,574	1,574	876	876	582	582	554	554
k=10	1,869	1,869	1,136	1,136	710	710	671	671
n=20	APSA	BR	APSA	BR	APSA	BR	APSA	BR
k=1	4,394	4,431	3,066	3,066	2,986	2,986	2,986	2,986
k=2	8,430	8,567	4,847	4,897	3,206	3,260	2,980	2,980
k=3	6,210	6,331	3,838	3,883	3,583	3,600	3,583	3,600
k=4	9,188	9,478	5,118	5,122	3,317	3,336	3,040	3,040
k=5	4,215	4,340	2,495	2,571	2,173	2,206	2,173	2,206
k=6	6,527	6,766	3,582	3,601	3,010	3,016	3,010	3,016
k=7	10,455	11,101	6,279	6,357	4,126	4,175	3,878	3,900
k=8	3,920	4,203	2,145	2,151	1,638	1,638	1,638	1,638
k=9	3,465	3,530	2,096	2,097	1,965	1,992	1,965	1,992
k=10	4,979	5,545	3,012	3,192	2,110	2,116	1,995	1,995
n=50	APSA	BR	APSA	BR	APSA	BR	APSA	BR
k=1	40,936	42,363	24,146	24,868	17,970	17,990	17,982	17,990
k=2	31,174	33,637	18,451	19,279	14,217	14,231	14,067	14,132
k=3	35,552	37,641	20,996	21,353	16,497	16,497	16,517	16,497
k=4	28,037	30,166	17,137	17,495	14,088	14,105	14,101	14,105
k=5	32,347	32,604	18,049	18,441	14,615	14,650	14,615	14,650
k=6	35,628	36,920	20,790	21,497	14,328	14,251	14,075	14,075
k=7	43,203	44,277	23,076	23,883	17,715	17,715	17,699	17,715
k=8	43,961	46,065	25,111	25,402	21,345	21,367	21,351	21,367
k=9	34,600	36,397	20,302	21,929	14,202	14,298	14,064	13,952
k=10	33,643	35,797	19,564	20,048	14,367	14,377	14,374	14,377
n=100	APSA	BR	APSA	BR	APSA	BR	APSA	BR
k=1	148,316	156,103	89,537	89,588	72,017	72,019	72,017	72,019
k=2	129,379	132,605	73,828	74,854	59,350	59,351	59,348	59,351
k=3	136,385	137,463	83,963	85,363	68,671	68,537	68,670	68,537
k=4	134,338	137,265	87,255	87,730	69,192	69,231	69,039	69,231
k=5	129,057	136,761	74,626	76,424	55,291	55,291	55,275	55,277

(continued)

Table 2 (continued)

Jobs	h=0.2		h=0.4		h=0.6		h=0.8	
n=10	APSA	BR	APSA	BR	APSA	BR	APSA	BR
k=6	145,927	151,938	81,182	86,724	62,507	62,519	62,410	62,519
k=7	138,574	141,613	79,482	79,854	62,302	62,213	62,208	62,213
k=8	164,281	168,086	95,197	95,361	80,722	80,844	80,841	80,844
k=9	121,189	125,153	72,817	73,605	58,769	58,771	58,771	58,771
k=10	121,425	124,446	72,741	72,399	61,416	61,419	61,416	61,419
k=3	136,385	137,463	83,963	85,363	68,671	68,537	68,670	68,537
k=4	134,338	137,265	87,255	87,730	69,192	69,231	69,039	69,231
k=5	129,057	136,761	74,626	76,424	55,291	55,291	55,275	55,277
k=6	145,927	151,938	81,182	86,724	62,507	62,519	62,410	62,519
k=7	138,574	141,613	79,482	79,854	62,302	62,213	62,208	62,213
k=8	164,281	168,086	95,197	95,361	80,722	80,844	80,841	80,844
k=9	121,189	125,153	72,817	73,605	58,769	58,771	58,771	58,771
k=10	121,425	124,446	72,741	72,399	61,416	61,419	61,416	61,419

Table 3 Results obtained for the single machine case of the common due date problem and comparison with benchmark results provided in the OR Library [3]. There are ten different instances provided and each instance is designated a number k.The gray boxes indicate the instances for which our algorithm could not achieve the known solution values given in [3]

Jobs	$h = 0.2$		$h = 0.4$		$h = 0.6$		$h = 0.8$	
$n = 200$	APSA	BR	APSA	BR	APSA	BR	APSA	BR
$k = 1$	523,042	526,666	300,079	301,449	254,268	254,268	254,362	254,268
$k = 2$	557,884	566,643	333,930	335,714	266,105	266,028	266,549	266,028
$k = 3$	510,959	529,919	303,924	308,278	254,647	254,647	254,572	254,647
$k = 4$	596,719	603,709	359,966	360,852	297,305	297,269	297,729	297,269
$k = 5$	543,709	547,953	317,707	322,268	260,703	260,455	260,423	260,455
$k = 6$	500,354	502,276	287,916	292,453	235,947	236,160	236,013	236,160
$k = 7$	477,734	479,651	279,487	279,576	246,910	247,555	247,521	247,555
$k = 8$	522,470	530,896	287,932	288,746	225,519	225,572	225,897	225,572
$k = 9$	561,956	575,353	324,475	331,107	254,953	255,029	254,956	255,029
$k = 10$	560,632	572,866	328,964	332,808	269,172	269,236	269,208	269,236

and achieves better results. However, there is a difference in the machines used for the implementation of the algorithms. In Table 5 we present results for the same problem but with parallel machines for the Biskup benchmark instances. The computation has been carried out for $k = 1$ up to 200 jobs and a different number of machines with restrictive factor h. We make a change in the due date as the number of machines increases and assume that the due date D is $D = \lfloor h \cdot \sum_{i=1}^{n} P_i/m \rfloor$. This assumption makes sense as an increase in the number of machines means that the jobs can be completed much faster and reducing the due-date will test the whole setup for more competitive scenarios. We implemented

Table 4 Average run-times in seconds for the single machine cases for the obtained solutions. The average run-time for any job is the average of all the 40 instances

Jobs	10	20	50	100	200
BR	0.9	47.8	87.3	284.9	955.2
APSA	0.46	1.12	22.17	55.22	132.32

Table 5 Results obtained for parallel machines for the benchmark instances for $k = 1$ with 2, 3, and 4 machines up to 200 jobs

No. of jobs	Machines	h value	Results obtained	Run-time (s)
10	2	0.4	612	0.0473
		0.8	398	0.0352
	3	0.4	507	0.0239
		0.8	256	0.0252
	4	0.4	364	0.0098
		0.8	197	0.0157
20	2	0.4	1,527	0.4061
		0.8	1,469	0.6082
	3	0.4	1,085	3.4794
		0.8	957	7.8108
	4	0.4	848	8.5814
		0.8	686	8.4581
50	2	0.4	12,911	7.780
		0.8	9,020	55.3845
	3	0.4	8,913	59.992
		0.8	6,010	125.867
	4	0.4	7,097	153.566
		0.8	4,551	22.347
100	2	0.4	45,451	101.475
		0.8	37,195	147.832
	3	0.4	31,133	159.872
		0.8	25,097	186.762
	4	0.4	23,904	236.132
		0.8	19,001	392.967
200	2	0.4	154,094	165.436
		0.8	133,848	231.768
	3	0.4	103,450	226.140
		0.8	96,649	365.982
	4	0.4	81,437	438.272
		0.8	71,263	500.00

Algorithm 2 with six different combinations of the number of machines and the restrictive factor. Since these instances have not been solved for the parallel machines, we are presenting the upper bounds achieved for these instances using Algorithm 2 and the modified simulated annealing.

12 Conclusion and Future Direction

In this paper we present two novel exact polynomial algorithms for the common due-date problem to optimize any given job sequence. We prove the optimality for the single machine case and the run-time complexity of the algorithms. We implemented our algorithms over the benchmark instances provided by Biskup and Feldmann [4] and the results obtained by using our algorithms are superior to the benchmark results in quality. We discuss how our approach can be used for non-identical parallel machines and present results for the parallel machine case for the same instances. Furthermore, we also discuss the efficiency of our algorithm for a special dynamic case of CDD at the planning stage.

Acknowledgements The research project was promoted and funded by the European Union and the Free State of Saxony, Germany. The authors take the responsibility for the content of this chapter.

References

1. Awasthi, A., Lässig, J., Kramer, O.: Common due-date problem: Exact polynomial algorithms for a given job sequence. In: 15th International Symposium on Symbolic and Numeric Algorithms for Scientific Computing (SYNASC 2013), pp. 260–266 (2013)
2. Banisadr, A.H., Zandieh, M., Mahdavi, I.: A hybrid imperialist competitive algorithm for single-machine scheduling problem with linear earliness and quadratic tardiness penalties. Int. J. Adv. Manuf. Technol. **65**(5–8), 981–989 (2013)
3. Beasley, J.E.: OR-library: Distributing test problems by electronic mail. J. Oper. Res. Soc. **41**(11), 1069–1072 (1990)
4. Biskup, D., Feldmann, M.: Benchmarks for scheduling on a single machine against restrictive and unrestrictive common due dates. Comput. Oper. Res. **28**(8), 787–801 (2001)
5. Cheng, T.C.E.: Optimal due-date assignment and sequencing in a single machine shop. Appl. Math. Lett. **2**(1), 21–24 (1989)
6. Cheng, T.C.E., Kahlbacher, H.G.: A proof for the longest-job-first policy in one-machine scheduling. Naval Res. Logist. (NRL) **38**(5), 715–720 (1991)
7. Feldmann, M., Biskup, D.: Single-machine scheduling for minimizing earliness and tardiness penalties by meta-heuristic approaches. Comput. Ind. Eng. **44**(2), 307–323 (2003)
8. Gen, M., Tsujimura, Y., Kubota, E.: Solving job-shop scheduling problems by genetic algorithm. In: IEEE International Conference on Systems, Man, and Cybernetics, 1994. Humans, Information and Technology, vol. 2, pp. 1577–1582 (1994)
9. Graham, R.L., Lawler, E.L., Lenstra, J.K., Rinnooy Kan, A.H.G.: Optimization and approximation in deterministic sequencing and scheduling: a survey. In: Discrete Optimization II Proceedings of the Advanced Research Institute on Discrete Optimization and Systems

Applications of the Systems Science Panel of NATO and of the Discrete Optimization Symposium co-sponsored by IBM Canada and SIAM Banff, Aha. and Vancouver, vol. **5**, pp. 287–326. (1979)
10. Hall, N.G., Kubiak, W., Sethi, S.P.: Earliness–tardiness scheduling problems, ii: deviation of completion times about a restrictive common due date. Oper. Res. **39**(5), 847–856 (1991)
11. Hoogeveen, J.A., Van de Velde, S.L.: Scheduling around a small common due date. Eur. J. Oper. Res. **55**(2), 237–242 (1991)
12. James, R.J.W.: Using tabu search to solve the common due date early/tardy machine scheduling problem. Comput. Oper. Res. **24**(3), 199–208 (1997)
13. Kanet, J.J.: Minimizing the average deviation of job completion times about a common due date. Naval Res. Logist. Q. **28**(4), 643–651 (1981)
14. Kim, J.L.: Genetic algorithm stopping criteria for optimization of construction resource scheduling problems. Constr. Manag. Econ. **31**(1), 3–19 (2013)
15. Lässig, J., Sudholt, D.: General Upper Bounds on the Runtime of Parallel Evolutionary Algorithms, MIT Press, Cambridge 1–33 (2013)
16. Lässig, J., Hoffmann, K.H.: Threshold-selecting strategy for best possible ground state detection with genetic algorithms, Phys. Rev. E. American Physical Society **79**(4), 046702 (2009)
17. Panwalkar, S.S., Smith, M.L., Seidmann, A.: Common due date assignment to minimize total penalty for the one machine scheduling problem. Oper. Res. **30**(2), 391–399 (1982)
18. Rebai, M., Kacem, I., Adjallah, K.H.: Earliness-tardiness minimization on a single machine to schedule preventive maintenance tasks: metaheuristic and exact methods. J. Intel. Manuf. **23**(4), 1207–1224 (2012)
19. Ronconi, D.P., Kawamura, M.S.: The single machine earliness and tardiness scheduling problem: lower bounds and a branch-and-bound algorithm. Comput. Appl. Math. **29**, 107–124 (2010)
20. Salamon, P., Sibani, P., Frost, R.: Facts, Conjectures, and Improvements for Simulated Annealing. Society for Industrial and Applied Mathematics, Philadelphia (2002). DOI10.1137/1.9780898718300
21. Seidmann, A., Panwalkar, S.S., Smith, M.L.: Optimal assignment of due-dates for a single processor scheduling problem. Int. J. Prod. Res. **19**(4), 393–399 (1981)
22. Toksari, M.D., Guner, E.: The common due-date early/tardy scheduling problem on a parallel machine under the effects of time-dependent learning and linear and nonlinear deterioration. Expert Syst. Appl. **37**(1), 92–112 (2010)
23. Yang, S.J., Lee, H.T., Guo, J.Y.: Multiple common due dates assignment and scheduling problems with resource allocation and general position-dependent deterioration effect. Int. J. Adv. Manuf. Technol. **67**(1–4), 181–188 (2013)

On Gaussian Process NARX Models and Their Higher-Order Frequency Response Functions

Keith Worden, Graeme Manson, and Elizabeth J. Cross

Abstract One of the most versatile and powerful approaches to the identification of nonlinear dynamical systems is the NARMAX (Nonlinear Auto_regressive Moving Average with eXogenous inputs) method. The model represents the current output of a system by a nonlinear regression on past inputs and outputs and can also incorporate a nonlinear noise model in the most general case. Although the NARMAX model is most often given a polynomial form, this is not a restriction of the method and other formulations have been proposed based on nonparametric machine learning paradigms, for example. All of these forms of the NARMAX model allow the computation of Higher-order Frequency Response Functions (HFRFs) which encode the model in the frequency domain and allow a direct interpretation of how frequencies interact in the nonlinear system under study. Recently, a NARX (no noise model) formulation based on Gaussian Process (GP) regression has been developed. One advantage of the GP NARX form is that confidence intervals are a natural part of the prediction process. The objective of the current paper is to discuss the GP formulation and show how to compute the HFRFS corresponding to GP NARX. Examples will be given based on simulated data.

Keywords Nonlinear system identification • NARMAX models • Higher-order Frequency Response Functions (HFRFs) • Gaussian processes

1 Introduction

The presence of a chapter here on system identification is motivated by the fact that almost all identification problems can be cast as optimisation problems. In the simplest sense, one wishes to find a mathematical model which is "closest" in some sense to the physical system of interest. In almost all cases, this is accomplished by measuring data from the system and finding the model that can reproduce that data

K. Worden (✉) • G. Manson • E.J. Cross
Dynamics Research Group, Department of Mechanical Engineering, University of Sheffield, Mappin Street, Sheffield S1 3JD, UK
e-mail: k.worden@sheffield.ac.uk; graeme.manson@sheffield.ac.uk; e.j.cross@sheffield.ac.uk

S. Koziel et al. (eds.), *Solving Computationally Expensive Engineering Problems*, Springer Proceedings in Mathematics & Statistics 97,
DOI 10.1007/978-3-319-08985-0_14

with the minimum error. Finding the optimal model is a matter of first establishing a model class or structure and then optimising the free parameters of the structure to give the best agreement with the data. It is clear at this point that one should always choose (where possible) a model class with the highest possible generality and therefore explanatory power.

Over the last 30 years one of the most versatile and enduring time series models used for nonlinear system identification has been the NARMAX (Nonlinear Auto-Regressive Moving Average with eXogenous inputs) model. The NARMAX model was introduced in 1985 [1, 2] and has been the subject of constant interest and development since. A comprehensive monograph on the theory and applications of the model recently appeared in [3]. In its full generality the model form accommodates nonlinear discrete-time process *and* noise models. However, if one can assume that the noise process is white Gaussian, one can adopt the simpler NARX form that will be discussed in this chapter. The basic principle of the NARX model is that one predicts the current value of system output using a nonlinear function F of previous inputs and outputs, i.e.

$$y_i = F(y_{i-1}, \dots, y_{i-n_y}; x_{i-1}, \dots, x_{i-n_x}) \tag{1}$$

The earliest and still most common form of the NARX model adopts a multinomial expansion basis for the function F and learns the expansion parameters by linear (but advanced) least-squares methods; however, this is by no means the only possible form. Any expansion basis which satisfies a universal approximation property can be used, and this has led to nonparametric NARX model forms based on machine learning including Multi-Layer Perceptron (MLP) and Radial Basis Function (RBF) neural networks [4, 5]. The nonparametric forms of the NARX model have at least one attractive feature in that they bypass (or rather, usually ignore) the *structure detection* problem. One can think of the problem of establishing a "traditional" NARX (or NARMAX) model in terms of two steps. The first step is *structure detection*, i.e. determining which multinomial terms should be included in the model; the second is establishing the expansion parameters for the included terms, i.e. *parameter estimation*. The nonparametric forms of the NARX models simply include all expansion terms consistent with certain *hyperparameters* of the form, e.g. number of nodes per layer in an MLP neural network. One then need only concern oneself with issues of including too many terms—leading to overfitting of models—and these issues can usually be addressed in a principled manner in a machine learning context [6].

One of the interesting features of the NARX model is that, through a connection with the Volterra series [7], it allows the construction of Higher-order Frequency Response Functions (HFRFs) that allow one to visualise how different frequencies in the input to a nonlinear system interact in forming the output [8]. In fact, the HFRFs are important, if not vital, if one wishes to extract a meaningful physical interpretation from a NARX model. Because almost all NARX models (even the multinomial ones) are nonparametric in the sense that their expansion coefficients have no physical meaning, one has to move to the frequency domain to make contact with the physics of the processes they express. In the case of the polynomial form of the NARX model, the method for determining the HFRFs—the *harmonic probing* algorithm—proved to be a simple extension of the long-held algorithm for differential equations [9, 10]. In the case of the neural network forms of the NARX model, the harmonic probing algorithm could also provide closed form expressions for the HFRFS at the expense of a little more complicated algebra [11].

A comparatively recent addition to the literature of the NARX model was the discussion of the Gaussian Process (GP) NARX model in [12]. This model form allows a number of potential advantages over the previously mentioned common forms of the NARX model, including a Bayesian framework encompassing the generation of natural confidence intervals for model predictions. The GP form of the model also suffers from a number of disadvantages relating to its tolerance of noise on training data and its computational expenses; these matters will be discussed in a little more detail later. The objective of the current paper is to discuss and illustrate the GP NARX model and to provide expressions for its HFRFs. No attempt is made here to give a comprehensive survey of the literature relating to dynamic GP models, for such a survey the reader could consult the comparatively recent [13].

The layout of the paper is as follows: Sect. 2 will provide a short summary of the relevant Gaussian process theory and how one can use it to define a NARX model. Section 3 introduces a case study and shows how the GP NARX model is applied in the context of a nonlinear Single-Degree-of-Freedom (SDOF) system. Section 4 discusses the basic principles of the Volterra series and how it leads to the definition of HFRFs. Section 5 presents the derivation of the HFRFs for the GP NARX model which is then computed for the case study system in Sect. 6. The chapter ends with a short discussion and conclusions.

2 Gaussian Process NARX Models

2.1 *Gaussian Processes*

The Gaussian Process (GP) has its roots in the geostatistics community where it was developed as a tool for interpolating the profile of landscapes considered as random fields. Although much of the main theory was developed later, it rests on early work carried out in the Masters thesis of Krige, which dates back to 1951 [14]. For this reason, the technique has long been known as *Kriging* in the geostatistics field. In more recent times, GPs were brought to the attention of the machine learning community by Neal [15] and Mackay [16], and consolidated in the recent book by Rasmussen and Williams [17]. The basic premise of the method is to perform inference over *functions* directly, as opposed to inference over *parameters* of functions.

For simplicity, the discussion here will assume that the system of interest has a single output variable. Following the notation of [17], let $X = [\underline{x}_1, \underline{x}_2 \ldots \underline{x}_N]^T$ denote a matrix of multivariate training inputs, and $\underline{y}$ denote the corresponding vector of training outputs. The input vector for a testing point will be denoted by the column vector $\underline{x}^*$ and the corresponding (unknown) output by y^*.

A Gaussian process prior is formed by assuming a (Gaussian) distribution over *functions*,

$$f(\underline{x}) \sim \mathcal{GP}(m(\underline{x}), k(\underline{x}, \underline{x})) \tag{2}$$

where $m(\underline{x})$ is the *mean function* and $k(\underline{x}, \underline{x}')$ is a positive-definite *covariance function*.

One of the defining properties of the GP is that the density of a finite number of outputs from the process is multivariate normal. Together with the known marginalisation properties of the Gaussian density, it is therefore possible to consider the value of this function only at the points of interest: training points and predictions. Allowing $\underline{f}$ to denote the function values at the training points X, and f^* to denote the predicted function value at a new point $\underline{x}^*$, one has

$$\begin{pmatrix} \underline{f} \\ f^* \end{pmatrix} \sim \mathcal{N}\left(\underline{0}, \begin{bmatrix} K(X,X) & K(X,\underline{x}^*) \\ K(\underline{x}^*,X) & K(\underline{x}^*,\underline{x}^*) \end{bmatrix}\right) \tag{3}$$

where a zero-mean prior has been used for simplicity (see [17] for a discussion), and $K(X, X)$ is a matrix whose i, j^{th} element is equal to $k(\underline{x}_i, \underline{x}_j)$. Similarly, $K(X, \underline{x}^*)$ is a column vector whose i^{th} element is equal to $k(\underline{x}_i, \underline{x}^*)$, and $K(\underline{x}^*, X)$ is the transpose of the same.

In order to relate the observed target data $\underline{y}$ to the function values $\underline{f}$, a simple Gaussian noise model can be assumed,

$$\underline{y} \sim \mathcal{N}(\underline{f}, \sigma_n^2 I) \tag{4}$$

where I is the identity matrix and σ_n^2 constitutes a hyperparameter which can easily be identified by optimisation. Since one is not interested in the variable $\underline{f}$, it can be marginalised (integrated out) from Eq. (3) [17], as the relevant integral

$$p(\underline{y}) = \int p(\underline{y}|\underline{f})p(\underline{f})d\,\underline{f} \tag{5}$$

is over a multivariate Gaussian and is therefore analytically tractable. The result is the joint distribution for the training and testing target values,

$$\begin{pmatrix} \underline{y} \\ y^* \end{pmatrix} \sim \mathcal{N}\left(\underline{0}, \begin{bmatrix} K(X,X) + \sigma_n^2 I & K(X,\underline{x}^*) \\ K(\underline{x}^*,X) & K(\underline{x}^*,\underline{x}^*) + \sigma_n^2 \end{bmatrix}\right) \tag{6}$$

In order to make use of the above, it is necessary to re-arrange the joint distribution $p(\underline{y}, y^*)$ into a conditional distribution $p(y^*|\underline{y})$. Using standard results for the conditional properties of a Gaussian reveals [17]

$$y^* \sim \mathcal{N}(m^*(\underline{x}^*), k^*(\underline{x}^*, \underline{x}^*)) \tag{7}$$

where

$$m^*(\underline{x}^*) = k(\underline{x}^*, X)[K(X, X) + \sigma_n^2 I]^{-1}\underline{y} \tag{8}$$

is the *posterior mean* of the GP and,

$$k^*(\underline{x}^*, \underline{x}') = k(\underline{x}^*, \underline{x}') - K(\underline{x}^*, X)[K(X, X) + \sigma_n^2 I]^{-1} K(X, \underline{x}') \tag{9}$$

is the posterior variance.

Thus the GP model provides a posterior distribution for the unknown quantity y^*. The mean from Eq. (7) can then be used as a "best estimate" for a regression problem, and the variance can also be used to define confidence intervals.

There does remain the question of the choice of covariance function $k(\underline{x}, \underline{x}')$. In practice, it is often useful to take a squared-exponential function of the form

$$k(\underline{x}, \underline{x}') = \sigma_f^2 \exp\left(-\frac{1}{2l^2}||\underline{x} - \underline{x}'||^2\right) \tag{10}$$

although various other forms are possible (see [17]). Equation (10) is the form adopted here. The covariance function involves the specification of two *hyper-parameters* σ_f^2 and l. The hyperparameters can be optimised using an evidence framework, along with the noise parameter σ_n^2 [17]. Denoting the complete set of these parameters as $\underline{t}$, they can be found by maximising a function,

$$f(\underline{t}) = -\frac{1}{2}\underline{y}^T[K(X, X) + \sigma_n^2 I]\underline{y} - \frac{1}{2}\log|K(X, X) + \sigma_n^2 I| \tag{11}$$

which is equal to the log of the evidence, up to some constant. Since the number of hyperparameters in this case is small, the optimisation can be carried out simply by gradient descent.

2.2 GP NARX Models

The GP models discussed so far are essentially *static* maps, learning the relationship between point inputs and point outputs. The question now arises as to how such models can be used to learn dynamical system behaviour. The answer adopted here will be to use a NARX framework. The functional form in Eq. (1) is used with the function F represented by a GP. A slight variant of the NARX form which also uses the current input for prediction will be used here.

Once the GP NARX model has been learned, there are various tests one can apply to assess the goodness of fit. The most basic is to compute *one step ahead* (OSA) predictions. In this case, using the training data, one computes the predictions for a given time using observed inputs up to that time, i.e.

$$y_i^* = F(y_{i-1}, \dots, y_{i-n_y}; x_{i-1}, \dots, x_{i-n_x}) \tag{12}$$

and compare the predicted and observed outputs. It is useful to have an objective measure of comparison, the one used here will be the *Normalised Mean-Square Error* (NMSE) defined by

$$\text{NMSE}(\hat{y}) = \frac{100}{N\sigma_y^2} \sum_{i=1}^{N} (y_i - \hat{y}_i)^2 \tag{13}$$

Experience shows that an NMSE of less than 5.0 indicates good agreement while one of less than 1.0 reflects an excellent fit.

Clearly, the OSA predictions are not a particularly stringent test of the model. A more demanding test is to compute the *Model Predicted Output* (MPO) defined by

$$y_i^* = F(y_{i-1}^*, \dots, y_{i-n_y}^*; x_{i-1}, \dots, x_{i-n_x}) \tag{14}$$

and this test can be conducted on testing data as well as training data, which is an important consideration in the more general context of machine learning.

As discussed in the introduction to this chapter, the use of the GP form in order to create a NARX model has advantages and disadvantages. Two of the main issues will be discussed briefly here with directions to the literature as to their means of solution.

The first problem is that the GP algorithm depends on the inversion of the covariance matrix K; this is an operation which costs $O(N^3)$ multiplications, where N is the number of training points. Slightly less costly is the prediction of new outputs with $O(N)$ multiplications needed for the predictive mean and $O(N^2)$ for the predictive variance. In fact, system identification with NARX models has traditionally been carried out with small training sets with a low number of thousands of data points, and this size of problem is typically feasible using a standard GP algorithm. However, if one wishes to move to larger training sets, the costs of computation can become prohibitive. This problem has led to the idea of *sparse Gaussian processes* which, as the name suggests, can establish models on reduced training sets [18]. One of the most effective methods is the so-called Fully Independent Training Conditional (FITC) model [19]. The FITC approach approximates the full GP by establishing M *pseudo-inputs* which are not restricted to be actual data points, but can be considered as hyperparameters which can be learned. In the FITC approach, the computational complexity of establishing the GP is reduced to $O(M^2N)$; the cost of computing the predictive mean is reduced to $O(M)$ and that of the predictive variance is reduced to $O(M^2)$.

The second problem with the GP NARX formulation relates to noise on the training data. The standard formulation of the GP algorithm assumes that the training inputs are noise-free and that the noise on the outputs is Gaussian with constant variance. One immediately sees an issue if one is attempting multi-step ahead predictions with a GP NARX model; because of the feeding back of the output predictions, the outputs *become* inputs and carry their predictive uncertainty with them. The most principled approach to this problem is to adopt the Bayesian approach of marginalising or integrating over the input noise distributions; however, even if these were known with complete accuracy, the computation would be intractable. One of the first comprehensive studies of this problem appears to have been the work leading to the thesis [20]. The thesis of Giraud is largely concerned with time-series predictions and the models are named GP AR there. This name makes complete sense in terms of the fact that the models are auto-regressive Gaussian processes; however, it misses the fact that the terms AR or ARX in the time series literature usually refer to linear models. The term GP NARX is preferred here as it indicates that the GP models are typically nonlinear. Of course, by appropriate choice of the covariance function one could fit linear GP AR models and the algorithm would then essentially be Bayesian linear regression [17].

Because of the highly simplified case study presented in this work, the issues referred to above have been ignored without damage to the results; however, in real engineering problems they will likely need to be addressed.

3 Case Study: An Asymmetric Duffing Oscillator

In order to illustrate the use of the GP NARX formulation, data simulated from a Duffing oscillator data system will be used. In the asymmetric case when a quadratic stiffness is present, the relevant equation of motion is

$$m\ddot{y} + c\dot{y} + ky + k_2 y^2 + k_3 y^3 = x(t) \tag{15}$$

Data were simulated by integrating the equation of motion using a fourth-order fixed step Runge-Kutta algorithm [21]. The parameters adopted were $m = 1$, $c = 20$, $k = 10^4$, $k_2 = 10^7$, and $k_3 = 5 \times 10^9$. The excitation used was a zero-mean Gaussian random sequence with a standard deviation of 2.0. The time step used was $\Delta t = 0.001$ seconds corresponding to a sampling frequency of 1 kHz. Noise was added to the data to introduce an element of reality to matters; initially in this case, Gaussian noise of 1 % RMS of the signal was added to both the excitation and response time data.

As discussed in Sect. 2.1, there are three hyperparameters for the simple GP formulation used here. Although these parameters can (and will later) be determined by optimisation, in the current case good estimates of two are directly available as σ_f^2 is essentially the response RMS and σ_n^2 is the noise RMS and both of these are known here. The parameter l is a scale for the covariance function and was established here to be 11.0 through a coarse line search. To improve the conditioning of the estimation process, all data were standardised before the

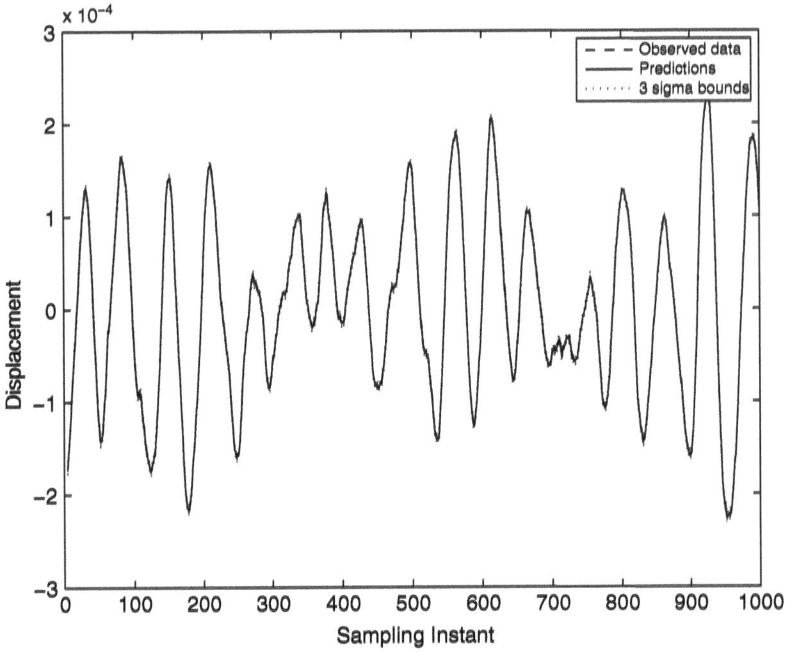

Fig. 1 OSA predictions for GP NARX model of Duffing oscillator data

computation (and this fixes $\sigma_f^2 = 1.0$), the scales for the data were reintroduced after predictions were made. Somewhat arbitrarily, the number of lags for the model used here was $n_y = 3$ and $n_x = 3$. The training data consisted of 1,000 samples of input and output data. As a more severe test of the prediction capability of the model, the results presented here are for an independent test set of data, also comprising 1,000 samples of data from the system at the same level of excitation as the training data.

The first set of results presented are for the OSA predictions on the test data set from the trained GP as shown in Fig. 1. The NMSE value for the predictions was 0.05, indicating an excellent result. The confidence intervals ($\pm$ 3 standard deviations) on the predictions in this case are so small that they are indistinguishable in the figure.

As discussed above, the MPO predictions provide a more stringent test and these are shown in Fig. 2. The corresponding NMSE in this case was 3.65, which still indicates a good fit.

As discussed in [22], the confidence intervals are still very small and do not accommodate the observed prediction errors. This is simply because not all of the uncertainty has been accounted for. In the predictions so far, the predicted outputs have been fed back into the model in order to form the MPO. This means that the only uncertainty accounted for in the predictions is the *parameter* uncertainty. In order to take a proper Bayesian viewpoint, one should allow for the fact that

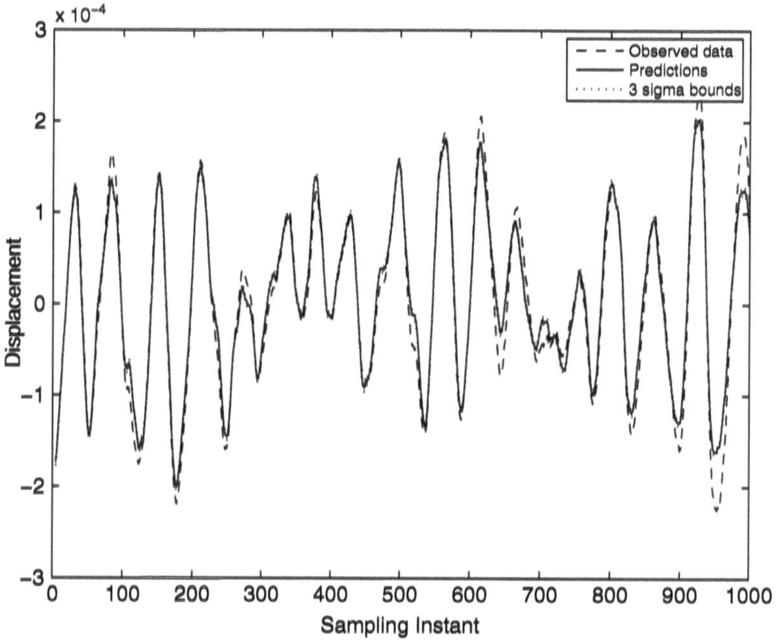

Fig. 2 MPO predictions for GP NARX model of Duffing oscillator data

each prediction is actually a sample from a distribution; this distribution being determined by the parameter distribution. To account for this, during a prediction run, at each instant i the prediction y_i^* was sampled from the distribution specified by the mean and covariance from the GP for that instant. One such run generates a single realisation of the prediction process, in order to accumulate information about the distribution of predictions with state estimation taken into account, a Monte Carlo approach was adopted here with 50 different runs conducted. Figure 3 shows 50 realisations of the predictions.

There is clearly a great deal of more uncertainty associated with the predictions now. From the MC realisations, one can compute a mean prediction and determine $\pm 3\sigma$ confidence bounds, and the result of the analysis for the case here is shown in Fig. 4. The confidence intervals are now a more appropriate assessment of the predictive capability of the model. This exercise shows clearly that the dominant contribution to uncertainty in the predictions is not the direct component from the parameter uncertainty, but the indirect component due to state estimation from the uncertain parameters. In fact, for a full treatment of uncertainty, the confidence limits should be augmented by adding in the variance identified by the σ_n^2 hyperparameter [17].

As discussed above, the hyperparameters for this example have been chosen partly by appealing to prior knowledge of the problem. When the hyperparameters are optimised through the evidence procedure, values of $l = 1265.4$, $\sigma_f^2 = 657.8$

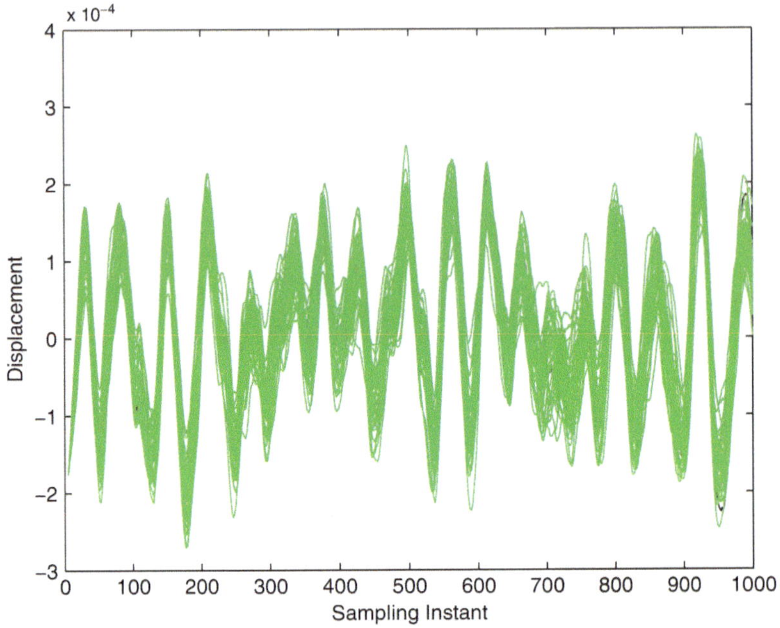

Fig. 3 MC realisations of predictions for GP NARX model of Duffing oscillator data

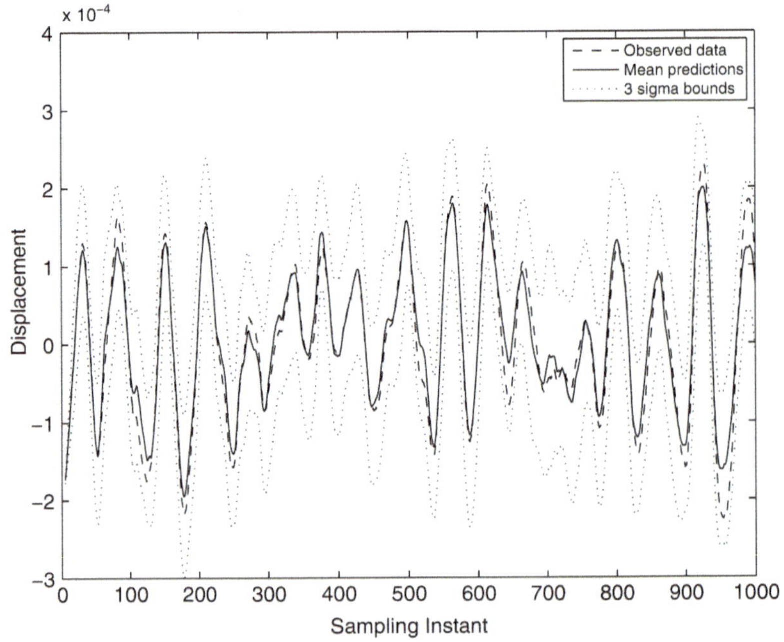

Fig. 4 MC predictions for GP NARX model of Duffing oscillator data

and $\sigma_n^2 = 0.0184$ are obtained. While the value of σ_n^2 is reasonably close to the true noise variance, the other two values are rather counter-intuitive, but do lead to a lower MSE of 3.5 on the test set. For reasons discussed later, the option of setting the hyperparameters partly "by hand" is pursued in the sequel.

4 The Volterra Series and Higher-Order FRFs

In the time-domain analysis of linear dynamical systems, the *impulse response function* $h(\tau)$ is known to characterise the system completely. For such a system, excited by an input signal $x(t)$, the response $y(t)$ is given by the convolution integral,

$$y(t) = \int_{-\infty}^{\infty} d\tau \; h(\tau)x(t-\tau) \tag{16}$$

This relationship is manifestly linear and will not hold for nonlinear systems; however, the theory was extended by Volterra [23] in the early part of the last century to cover the more general case. The output of a nonlinear system is composed of additional higher-order contributions. Volterra showed that the total response, $y(t)$, is given by

$$y(t) = y_0 + y_1(t) + y_2(t) + y_3(t) + \ldots + y_n(t) \tag{17}$$

where y_0 is a constant and,

$$y_1(t) = \int_{-\infty}^{\infty} d\tau \; h_1(\tau_1)x(t-\tau_1) \tag{18}$$

$$y_2(t) = \int_{-\infty}^{\infty}\int_{-\infty}^{\infty} d\tau_1 d\tau_2 \; h_2(\tau_1,\tau_2)x(t-\tau_1)x(t-\tau_2) \tag{19}$$

and the general term is

$$y_n(t) = \int_{-\infty}^{\infty} \cdots \int_{-\infty}^{\infty} d\tau_1 d\tau_2 \ldots d\tau_n \; h_n(\tau_1,\tau_2 \ldots \tau_n)x(t-\tau_1)x(t-\tau_2)\ldots x(t-\tau_n) \tag{20}$$

This is essentially a generalisation of the standard Taylor series to the case of *functionals*, i.e. mappings between functions. The generalised coefficients of the series h_n are the nth-order *Volterra kernels*, and these can be thought of as multi-dimensional, or higher-order, impulse response functions [7]. The series provides a representation of a given functional or system $y(t) = S[x(t)]$, which is insensitive to the input $x(t)$, provided that the system is time-invariant and contains only analytic nonlinearities [24].

The Volterra series is thus a time-domain representation for nonlinear systems. As in the case of linear systems, a dual frequency-domain representation exists which can give a clearer perspective of system behaviour in some respects. For a linear system, Eq. (16) shows how to compute the response $y(t)$ for any input $x(t)$, given the system impulse response function $h(t)$. The corresponding frequency-domain expression is simply obtained by taking the Fourier transform of both sides, noting that the RHS is a convolution. The result is

$$Y(\omega) = H(\omega)X(\omega) \tag{21}$$

where

$$H(\omega) = \int_{-\infty}^{\infty} d\omega \; e^{-i\omega t} h(t) \tag{22}$$

is the system *Frequency Response Function*, and $Y(\omega)$ and $X(\omega)$ have similar definitions. By direct extension of the linear case, the higher-order FRFs (HFRFs) $H_n(\omega_1, \ldots, \omega_n)$ can be defined as the multi-dimensional Fourier transforms of the kernels,

$$H_n(\omega_1, \ldots, \omega_n) = \int_{-\infty}^{+\infty} \cdots \int_{-\infty}^{+\infty} d\tau_1 \ldots d\tau_n h_n(\tau_1, \ldots, \tau_n) e^{-i(\omega_1\tau_1 + \ldots + \omega_n\tau_n)} \tag{23}$$

with inverse

$$h_n(\tau_1, \ldots, \tau_n) = \frac{1}{(2\pi)^n} \int_{-\infty}^{+\infty} \cdots \int_{-\infty}^{+\infty} d\omega_1 \ldots d\omega_n H_n(\omega_1, \ldots, \omega_n) e^{+i(\omega_1\tau_1 + \ldots + \omega_n\tau_n)} \tag{24}$$

It is then a straightforward matter to obtain the frequency-domain dual of expression (17),

$$Y(\omega) = Y_1(\omega) + Y_2(\omega) + Y_3(\omega) + \ldots \tag{25}$$

where

$$Y_1(\omega) = H_1(\omega)X(\omega) \tag{26}$$

$$Y_2(\omega) = \frac{1}{2\pi} \int_{-\infty}^{+\infty} d\omega_1 \, H_2(\omega_1, \omega - \omega_1) X(\omega_1) X(\omega - \omega_1) \tag{27}$$

$$Y_3(\omega) = \frac{1}{(2\pi)^2} \times$$

$$\int_{-\infty}^{+\infty} \int_{-\infty}^{+\infty} d\omega_1 d\omega_2 H_3(\omega_1, \omega_2, \omega - \omega_1 - \omega_2) X(\omega_1) X(\omega_2) X(\omega - \omega_1 - \omega_2) \tag{28}$$

The interpretation of these quantities is well established, a description can be found in [8]. If the equations of motion are known for a system, the method of harmonic probing can be used in order to compute the HFRFs [9]. Harmonic probing for the Gaussian process NARX models is discussed in the next section and the technique is illustrated using the Duffing oscillator system in Eq. (15); the exact results in this case are well known as [8],

$$H_1(\omega) = \frac{1}{-m\omega^2 + ic\omega + k} \tag{29}$$

$$H_2(\omega_1, \omega_2) = -k_2 H_1(\omega_1) H_1(\omega_2) H_1(\omega_1 + \omega_2) \tag{30}$$

and,

$$H_3(\omega_1, \omega_2, \omega_3) = -\frac{1}{6} H_1(\omega_1 + \omega_2 + \omega_3)\times$$

$$\{4k_2 \left(H_1(\omega_1) H_2(\omega_2, \omega_3) + H_1(\omega_2) H_2(\omega_3, \omega_1) + H_1(\omega_3) H_2(\omega_1, \omega_2)\right) +$$

$$6k_3 H_1(\omega_1) H_1(\omega_2) H_1(\omega_3)\} \tag{31}$$

In order to see the important structure in the HFRFs, it is often sufficient to plot only the leading diagonal, i.e. $H_2(\omega, \omega)$. This format also allows simple comparisons between the functions.

5 Harmonic Probing of the GP NARX Model

If the governing equations of motion are known, the HFRFs of a system can be obtained analytically by the use of the *harmonic probing* algorithm, introduced by Bedrosian and Rice [9]. Although this was originally designed for continuous-time systems, the algorithm was extended to the type of discrete-time systems considered here by Billings and Tsang [4].

Before proceeding, it is necessary to determine the explicit form of the GP NARX model. First of all, one observes, following [17], that the GP is essentially an expansion in terms of basis functions fixed by the covariance kernel and the training data, the predicted output y^* corresponding to a new input $\underline{x}^*$ is given by

$$y^* = \sum_{l=1}^{N} a_l k(\underline{x}^*, \underline{x}_l) \tag{32}$$

where according to Eq. (8),

$$\underline{a} = [k(X, X) + \sigma_n^2 I]^{-1} \underline{y} \tag{33}$$

and this is fixed by the training data. If one adopts the squared exponential covariance function of (10), one arrives at the GP NARX form,

$$y_i = \sigma_f^2 \sum_{j=1}^{N} a_j \exp\left\{-\frac{1}{2l^2}\left[\sum_{k=1}^{n_y}(y_{i-k} - v_{jk})^2 + \sum_{m=0}^{n_x}(x_{i-m} - u_{jm})^2\right]\right\} \tag{34}$$

where the matrix $V = \{v_{ij}\}$ is formed from the first n_y columns of the matrix X and $U = \{u_{ij}\}$ is formed from the remaining $n_x + 1$ columns of X.

Note that this expression is essentially that of the radial basis function neural network considered in [11]; this means that the HFRFs derived in that paper are applicable here. However, the analysis here presents a more direct approach in terms of homogeneous ARX and NARX model coefficients at each polynomial order; the expressions here also correct some typographical errors in [11]. The first issue which arises is that the function in (34) must be expanded as a polynomial in order to apply harmonic probing. As observed in [11], direct expansion means that the term of order n will contain powers of all orders up to n and this makes it impossible to group linear terms, etc. The solution is simple, a trivial rearrangement yields the more amenable form,

$$y_i = \sigma_f^2 \sum_{j=1}^{N-p} a_j \gamma_j \exp\left\{-\frac{1}{2l^2}\left[\sum_{k=1}^{n_y}(y_{i-k}^2 - 2v_{jk}y_{i-k}) + \sum_{m=0}^{n_x}(x_{i-m}^2 - 2u_{jm}x_{i-m})\right]\right\} \tag{35}$$

where

$$\gamma_j = \exp\left\{-\frac{1}{2l^2}\left[\sum_{k=1}^{n_y} v_{jk}^2 + \sum_{m=0}^{n_x} u_{jm}^2\right]\right\} \tag{36}$$

Now, the basis of the harmonic probing method is to examine the response of the system to certain very simple inputs. In order to identify $H_1(\omega)$, for example, the system is "probed" with the single harmonic,

$$x_i^p = e^{i\Omega t} \tag{37}$$

Substituting this expression into the Volterra series (17), the corresponding response is [8]

$$y_i^p = H_1(\Omega)e^{i\Omega t} + H_2(\Omega,\Omega)e^{2i\Omega t} + H_3(\Omega,\Omega,\Omega)e^{3i\Omega t} + \dots \tag{38}$$

Now, consider the consequences of substituting the expressions (37) and (38) into the network function (35) and expanding it as a polynomial. None of the higher-order terms in (38) can combine in any way to generate a component at the fundamental frequency of excitation Ω. As a result, if the coefficient of $e^{i\Omega t}$ is extracted from the resulting expression, the *only* HFRF which can appear is $H_1(\Omega)$;

thus, the expression can be rearranged to give an analytical expression for H_1. In fact, one need to only consider the linear terms in the expansion in order to extract H_1, so one essentially considers the ARX model,

$$y_i = \sigma_f^2 \sum_{j=1}^{N-p} \frac{a_j \gamma_j}{l^2} \left\{ \sum_{k=1}^{n_y} v_{jk} y_{i-k} + \sum_{m=0}^{n_x} u_{jm} x_{i-m} \right\} \tag{39}$$

Changing the order of summation here results in the standard ARX form,

$$y_i = \sum_{j=1}^{n_y} \alpha_j y_{i-j} + \sum_{j=0}^{n_x} \beta_j x_{i-j} \tag{40}$$

where

$$\alpha_j = \frac{\sigma_f^2}{l^2} \sum_{i=1}^{N-p} a_i \gamma_i v_{ij} \tag{41}$$

$$\beta_j = \frac{\sigma_f^2}{l^2} \sum_{i=1}^{N-p} a_i \gamma_i u_{ij} \tag{42}$$

Harmonic probing of this expression is straightforward; one substitutes the probing expressions (37) and (38) into (40) and collects together all the coefficients of $e^{i\Omega t}$. In doing this, account must be taken of the effect of time-delays on the harmonic signals, this is straightforward to compute as

$$x_{i-k} = \Delta^k x_i = \Delta^k e^{i\Omega t} = e^{-ki\Omega\Delta t} e^{i\Omega t} \tag{43}$$

$$y_{i-k} = \Delta^k y_i = \Delta^k H_1(\Omega) e^{i\Omega t} = e^{-ki\Omega\Delta t} H_1(\Omega) e^{i\Omega t} \tag{44}$$

where Δ is the backward shift operator. The result of the calculation is

$$H_1(\Omega) = \frac{\sum_{j=0}^{n_x} \beta_j e^{-ij\Delta t\Omega}}{1 - \sum_{j=1}^{n_y} \alpha_j e^{-ij\Delta t\Omega}} \tag{45}$$

with the α_j and β_j as defined in Eqs. (41) and (42).

The extraction of H_2 is a little more complicated, this requires probing with two independent harmonics, so,

$$x_i^p = e^{i\Omega_1 t} + e^{i\Omega_2 t} \tag{46}$$

The standard computation using based on (17) shows that the corresponding response is [8]

$$y_i^p = H_1(\Omega_1)e^{i\Omega_1 t} + H_1(\Omega_2)e^{i\Omega_2 t} + 2H_2(\Omega_1,\Omega_2)e^{i(\Omega_1+\Omega_2)t} + \dots \tag{47}$$

The argument proceeds as for H_1; if these expressions are substituted into the network function (35), the only HFRFs to appear in the coefficient of the sum harmonic $e^{i(\Omega_1+\Omega_2)t}$ are H_1 and H_2, where H_1 is already known from Eq. (45). As before, the coefficient can be rearranged to give an expression for H_2 in terms of the GP parameters and H_1. The only terms in the expansion of (35) which are relevant for the calculation are those at first and second-order. The calculation is straightforward but tedious and yields

$$H_2(\Omega_1,\Omega_2) = \frac{A+B+C}{D} \tag{48}$$

where

$$A = \sum_{k=1}^{n_y}\sum_{l=1}^{n_y} \alpha_{kl} H_1(\Omega_1)H_1(\Omega_2)\left(e^{-i\Omega_1 k\Delta t}.e^{-i\Omega_2 l\Delta t} + e^{-i\Omega_2 k\Delta t}.e^{-i\Omega_1 l\Delta t}\right) \tag{49}$$

$$B = \sum_{k=1}^{n_y}\sum_{l=0}^{n_x} \beta_{kl}\left(H_1(\Omega_1)e^{-i\Omega_1 k\Delta t}.e^{-i\Omega_2 l\Delta t} + H_1(\Omega_2)e^{-i\Omega_2 k\Delta t}.e^{-i\Omega_1 l\Delta t}\right) \tag{50}$$

$$C = \sum_{k=0}^{n_x}\sum_{l=0}^{n_x} \gamma_{kl}\left(e^{-i\Omega_1 k\Delta t}.e^{-i\Omega_2 l\Delta t} + e^{-i\Omega_2 k\Delta t}.e^{-i\Omega_1 l\Delta t}\right) \tag{51}$$

and,

$$D = 1 - \sum_{k=1}^{n_y} \alpha_k e^{-i(\Omega_1+\Omega_2)k\Delta t} \tag{52}$$

The coefficients in the above expressions are given by

$$\alpha_{jm} = \frac{\sigma_f^2}{4l^4}\sum_{i=1}^{N-p} a_i \gamma_i v_{ij} v_{im} - \delta_{jm}\frac{\sigma_f^2}{2l^2}\sum_{i=1}^{N-p} a_i \gamma_i \tag{53}$$

$$\beta_{jm} = \frac{\sigma_f^2}{2l^4}\sum_{i=1}^{N-p} a_i \gamma_i v_{ij} u_{im} \tag{54}$$

$$\gamma_{jm} = \frac{\sigma_f^2}{4l^4}\sum_{l=1}^{N-p} a_i \gamma_i u_{ij} u_{im} - \delta_{jm}\frac{\sigma_f^2}{2l^2}\sum_{i=1}^{N-p} a_i \gamma_i \tag{55}$$

where δ_{jm} is the standard Kronecker delta.

Derivation of H_3 is considerably more lengthy and requires probing with three harmonics, the expression is not given here for reasons of space. The following results section of this paper will present examples of these calculations for H_1 and H_2.

6 HFRF Results for Case Study System

In this section, the HFRFs for the asymmetric Duffing oscillator system of Eq. (15) are estimated from the GP NARX model fit to the simulated data. In fact, a subtlety forces a reanalysis of the data. Usually, if input and output data are known to be corrupted by noise, best practice demands that a NARMAX model with nonlinear noise model is fitted to the data in order to avoid the possibility of bias in the parameter estimates [8]. However, when the HFRFs are to be computed, the noise model is discarded, leaving a NARX model for harmonic probing. In the case of the GP model, the noise variance σ_n^2 is essentially built in to the parameter estimates as it effectively acts as a regularisation parameter in inverting the $K(X, X)$ matrix to form the parameters $\underline{a}$. In order to mimic the usual practice of discarding the noise information, the hyperparameter σ_n^2 was set by hand here to a value of 0.0001 and the training data were regenerated without added noise; this means that the GP could achieve very low training and test errors. Because of standardisation of the data, the prescription $\sigma_f^2 = 1.0$ was again used and a line search gave a value of $l = 11.0$. With these values for the hyperparameters, the parameters $\underline{a}$ were estimated and the GP gave an OSA error of 9×10^{-6} and an MPO error of 0.008.

The comparisons between predicted and measured response are not given as the curves are not distinguishable given the accuracy of the predictions. However, it is meaningful to give comparisons between the exact HFRFs, given by Eqs. (29) and (30), and those estimated from the GP. Figure 5 shows a comparison between the exact and estimated $H_1(\omega)$; it is clear that the estimate is very accurate indeed.

Figures 6 and 7 show comparisons between the exact and estimated H_2 functions in terms of magnitude and phase, respectively. Because a direct visual comparison is subjective when the surfaces are displayed, the exact and estimated diagonals $H_2(\omega, \omega)$ are shown in Fig. 8, the accuracy of the estimates is clearly excellent.

7 Conclusions

The main aim of this paper has been to present analytical expressions for the HFRFs of Gaussian process NARX models specific to the squared exponential covariance function. The expressions have been validated on simulated data from an SDOF nonlinear system. The excellent agreement between exact HFRFs and those extracted from the GP fitted to simulated data confirms that the expressions are correct. Perhaps more importantly, the results show that it is possible to obtain

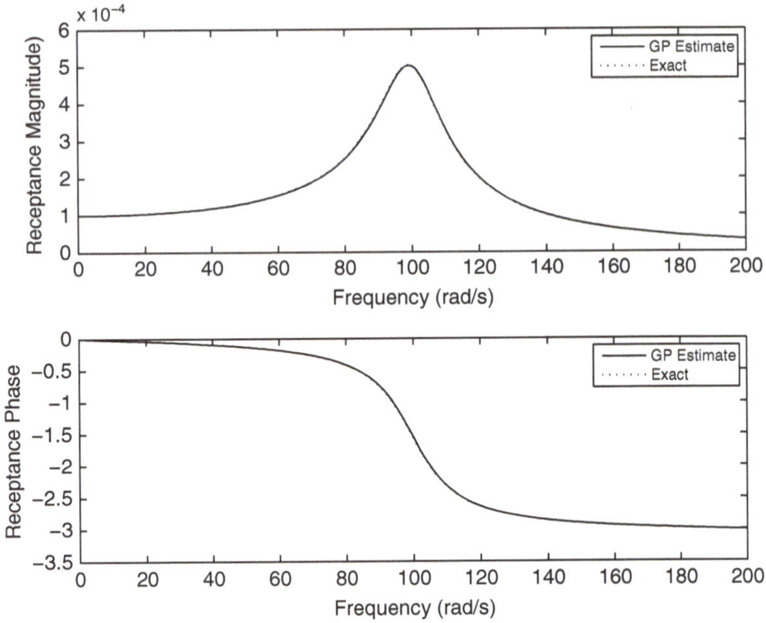

Fig. 5 Gaussian Process estimate of $H_1(\omega)$ compared to exact result.

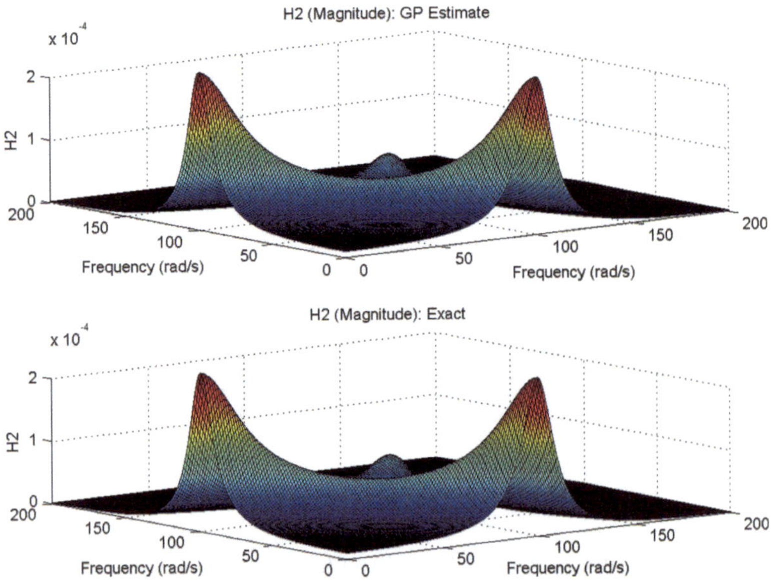

Fig. 6 Gaussian Process estimate of $H_2(\omega_1, \omega_2)$ magnitude compared to exact result

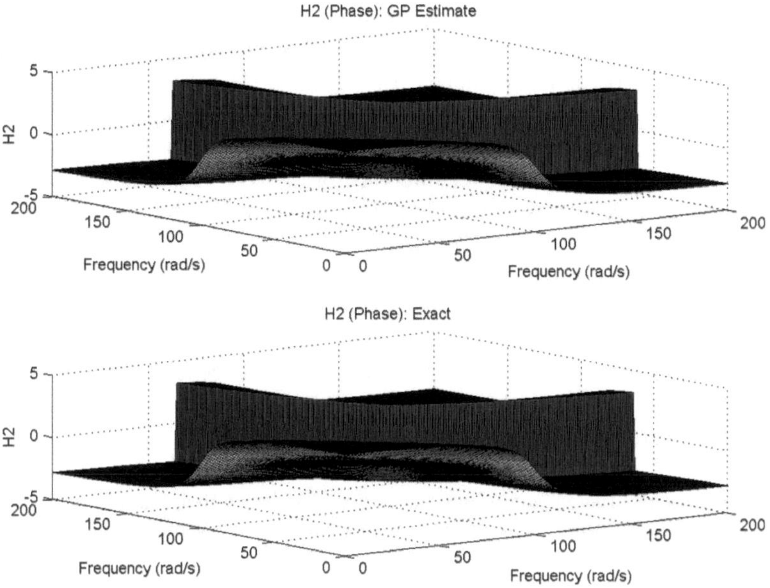

Fig. 7 Gaussian Process estimate of $H_2(\omega_1, \omega_2)$ phase compared to exact result

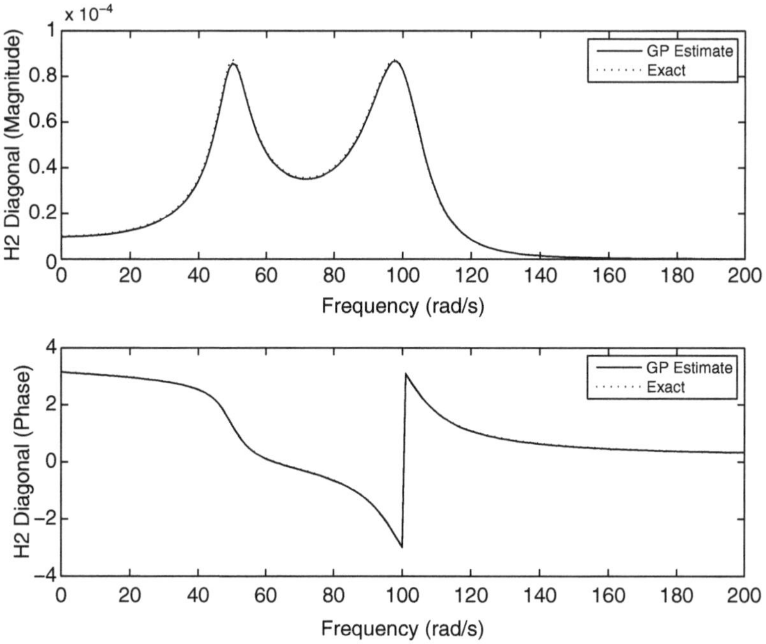

Fig. 8 Gaussian Process estimate of $H_2(\omega, \omega)$ magnitude and phase compared to exact result

accurate estimates of system HFRFs by using GP NARX models. This in itself is not a surprise as previous work had shown that the HFRFs could be extracted from neural network NARX models learnt from data. However, the GP form of the NARX model may have advantages over the previous crisp neural network models in allowing the computation of confidence intervals for predictions; this may lead to a means of establishing confidence intervals for the HFRFs and this is a possibility that is under further investigation. One of the issues raised by the analysis here is concerned with the fact that the "noise model" in the GP NARX model cannot be simply discarded as it can in the polynomial NARMAX case. Because of this issue, the results for the HFRFs presented here were given for noise-free training data and one might argue that it is therefore no surprise that accurate estimates were obtained. Further analysis is required in terms of how noise variance is accommodated in the HFRF estimation; it may be that estimation of the confidence intervals provides the means of answering this question.

Acknowledgements The authors would like to thank Dr James Hensman of the University of Sheffield Centre for Translational Neuroscience for a number of interesting and useful discussions.

References

1. Leontaritis, I.J., Billings, S.A.: Input-output parametric models for nonlinear systems, part I: deterministic nonlinear systems. Int. J. Control **41**, 303–328 (1985)
2. Leontaritis, I.J., Billings, S.A.: Input-output parametric models for nonlinear systems, part II: stochastic nonlinear systems. Int. J. Control **41**, 329–344 (1985)
3. Billings, S.A.: Nonlinear System Identification: NARMAX, Methods in the Time, Frequency, and Spatio-Temporal Domains. Wiley, Hoboken (2013)
4. Billings, S.A., Jamaluddin, H.B., Chen, S.: Properties of neural networks with applications to modelling non-linear dynamical systems. Int. J. Control **55**, 193–224 (1992)
5. Chen, S., Billings, S.A., Cowan, C.F.N., Grant, P.M.: Practical identification of NARMAX models using radial basis functions. Int. J. Control **52**, 1327–1350 (1990)
6. Bishop, C.M.: Neural Networks for Pattern Recognition. Oxford University Press, New York (1998)
7. Schetzen, M.: The Volterra and Wiener Theories of Nonlinear Systems. John Wiley Interscience Publication, New York (1980).
8. Worden, K., Tomlinson, G.R.: Nonlinearity in Structural Dynamics: Detection, Modelling and Identification. Institute of Physics Press, Bristol (2001)
9. Bedrosian, E., Rice, S.O.: The output properties of Volterra systems driven by harmonic and Gaussian inputs. Proc. IEEE **59**, 1688–1707 (1971)
10. Billings, S.A., Tsang, K.M.: Spectral analysis for nonlinear systems, part I: parametric nonlinear spectral analysis. Mech. Syst. Signal Process. **3**, 319–339 (1989)
11. Chance, J.E., Worden, K., Tomlinson, G.R.: Frequency domain analysis of NARX neural networks. J. Sound Vib. **213**, 915–941 (1997)
12. Murray-Smith, R., Johansen, T. A., Shorten, R.: On transient dynamics, off-equilibrium behaviour and identification in blended multiple model structures. In: European Control Conference, Karlsruhe, BA-14 (1999)
13. Kocijan, J.: Dynamic GP models: an overview and recent developments. In: ASM12, Proc. 6th Int. Conf. Appl. Maths Sim. Mod., pp. 38–43 (2012)

14. Krige, D.G.: A Statistical Approach to Some Mine Valuations and Allied Problems at the Witwatersrand. Master's Thesis, University of Witwatersrand (1951)
15. Neal, R.M.: Monte Carlo implementation of Gaussian process models for Bayesian regression and classification. Arxiv preprint physics/9701026, (1997)
16. MacKay, D.J.C.: Gaussian processes - a replacement for supervised neural networks. Lecture Notes for Tutorial at Int. Conf. Neural Inf. Proc. Sys. (1997)
17. Rasmussen, C.E., Williams, C.K.I.: Gaussian Processes for Machine Learning. MIT Press, Cambridge (2005)
18. Quinonero-Candelo, Q. Rasmussen, C.E.: A unifying view of sparse approximation Gaussian process regression. J. Mach. Learn. Res. **6**, 1939–1959 (2005)
19. Snelson, E. Ghahramani, Z.: Sparse Gaussian processes using pseudo-inputs. In: AAdvances in Neural Information Processing Systems. MIT Press, Cambridge (2006)
20. Giraud, A.: Approximate Methods for Propagation of Uncertainty with Gaussian Process Models. PhD Thesis, University of Glasgow (2004)
21. Press, W.H., Teukolsky, S.A., Vetterling, W.T., Flannery, B.P.: Numerical Recipes: The Art of Scientific Computing, 3rd edn. Cambridge University Press, New York (2007)
22. Worden, K., Hensman, J.J.: Parameter estimation and model selection for a class of hysteretic systems using Bayesian inference. Mech. Syst. Signal Process. **32**, 153–169 (2011)
23. Volterra, V.: Theory of Functionals and of Integral and Integro-Differential Equations. Dover, New York (1959)
24. Palm, G., Poggio, T.: The Volterra representation and the Wiener expansion: validity and pitfalls. SIAM J. Appl. Math. **33**, 195–216 (1997)

MIX
Papier aus verantwortungsvollen Quellen
Paper from responsible sources
FSC® C105338

If you have any concerns about our products,
you can contact us on
ProductSafety@springernature.com

In case Publisher is established outside the EU,
the EU authorized representative is:
Springer Nature Customer Service Center GmbH
Europaplatz 3, 69115 Heidelberg, Germany

Printed by Libri Plureos GmbH
in Hamburg, Germany